Nonlinear Diffusion of Electromagnetic Fields

WITH APPLICATIONS TO EDDY CURRENTS AND SUPERCONDUCTIVITY

This is a volume in

ELECTROMAGNETISM

Isaak Mayergoyz, Series Editor
University of Maryland
College Park, Maryland

Nonlinear Diffusion of Electromagnetic Fields

WITH APPLICATIONS TO EDDY CURRENTS AND SUPERCONDUCTIVITY

ISAAK MAYERGOYZ

University of Maryland
Department of Electrical Engineering
College Park, Maryland

ACADEMIC PRESS

San Diego London Boston
New York Sydney Tokyo Toronto

This book is printed on acid-free paper. ♾

ACADEMIC PRESS
525 B Street, Suite 1900, San Diego, CA 92101, USA
1300 Boylston Street, Chestnut Hill, MA 02167, USA
http://www.apnet.com

ACADEMIC PRESS LIMITED
24–28 Oval Road, London NW1 7DX, UK
http://www.hbuk.co.uk/ap/

Library of Congress Cataloging-in-Publication Data

Mayergoyz, I. D.
Nonlinear diffusion of electromagnetic fields : with applications to eddy currents and superconductivity / Isaak Mayergoyz
p. cm. — (Electromagnetism)
Includes bibliographical references and index.
ISBN 0-12-480870-0 (alk. paper)
1. Electromagnetic fields. 2. Eddy currents (Electric)
3. Superconductivity. I. Title. II. Series.
QC665.E4M4 1998
537—dc21 98-12650
CIP

Printed in the United States of America
98 99 00 01 02 IC 9 8 7 6 5 4 3 2 1

To the Memory of My Father

CONTENTS

PREFACE

This book covers diffusion of electromagnetic fields in magnetically nonlinear conductors and electrically nonlinear superconductors. This diffusion is described by nonlinear partial differential equations, and for this reason it is termed "nonlinear" diffusion. Nonlinear diffusion has many qualitative features that are not observed for linear diffusion, which explains why the study of nonlinear diffusion of electromagnetic fields is of significant theoretical interest. At the same time, the study of nonlinear diffusion is very important in many practical applications. Indeed, analysis of electromagnetic field diffusion in magnetically nonlinear conductors is, in a way, analysis of eddy currents in those conductors. The latter analysis is very instrumental in such diverse applications as: design of electric machines, transformers and actuators, induction heating, nondestructive testing, electromagnetic shielding, development of inductive writing heads for magnetic recording, and design of magnetic components in power electronics. On the other hand, the study of nonlinear diffusion of electromagnetic fields in superconductors is instrumental for the analysis of magnetic hysteresis in those superconductors as well as for the understanding of creep phenomena.

In spite of significant theoretical and practical interests, nonlinear diffusion of electromagnetic fields has not been extensively studied, and currently no book exists that covers this topic in depth. It is hoped that this book will bridge this gap.

The book has the following salient and novel features.

- Extensive use of **analytical** techniques for the solution of nonlinear partial differential equations, which describe electromagnetic field diffusion in nonlinear media.
- Simple analytical formulas for surface impedances of nonlinear and hysteretic media.
- Analytical analysis of nonlinear diffusion for linear, circular, and elliptical polarizations of electromagnetic fields.
- Novel and extensive analysis of eddy current losses in steel laminations for unidirectional and rotating magnetic fields.

- Preisach approach to the modeling of eddy current hysteresis and superconducting hysteresis.
- Extensive analytical study of nonlinear diffusion in superconductors with gradual resistive transitions (scalar and vectorial problems).
- Scalar potential formulations of nonlinear impedance boundary conditions and their finite element implementations.

The book contains five chapters and one appendix.

Chapter 1 deals with the analytical study of electromagnetic field diffusion in magnetically nonlinear conducting media in the case of linear polarization of magnetic fields. This diffusion is described by **scalar** nonlinear partial differential equations of the parabolic type. Discussions start with the case of abrupt magnetic transition (abrupt saturation) and proceed to the case of gradual magnetic transition (gradual saturation). For the latter case, first self-similar analytical solutions are found, which reveal that nonlinear diffusion occurs as an inward progress of almost rectangular profiles of magnetic flux density of variable height. These almost rectangular profiles of magnetic flux density represent an intrinsic feature of nonlinear diffusion in the case of sufficiently strong magnetic fields, and they occur because magnetic permeability (or differential permeability) is increased as the magnetic fields are attenuated. The analysis of the self-similar solutions suggests the idea of rectangular profile approximation of actual magnetic flux density profiles. This approximation is used to derive simple analytical expressions for the surface impedance. Chapter 1 also contains discussions of the "standing" mode of nonlinear diffusion, applications of nonlinear diffusion to circuit analysis, and the representation of eddy current hysteresis in terms of the Preisach model. The last representation reveals the remarkable fact that nonlinear (and dynamic) eddy current hysteresis can be fully characterized by its step response.

In Chapter 2, diffusion of circularly and elliptically polarized electromagnetic fields in magnetically nonlinear conducting media is discussed. This diffusion is described by **vector** (rather than scalar) nonlinear partial differential equations, which naturally raises the level of mathematical difficulties. However, it is shown that these difficulties can be completely circumvented in the case of circular polarizations and isotropic media. Simple and exact analytical solutions are obtained for the above case by using power law approximations for magnetization curves. These solutions reveal the remarkable fact that there is no generation of higher-order harmonics despite nonlinear magnetic properties of conducting media. This is because of the high degree of symmetry that exists in the case of circular polarizations and isotropic media. Elliptical polarizations and anisotropic media are then treated as perturbations of circular polarizations and isotropic media, respectively. On the basis of this treatment, the perturbation technique is developed and simple analytical solutions of perturbed problems are found.

The chapter concludes with an extensive analysis of eddy current losses in steel laminations caused by rotating magnetic fields.

Chapter 3 presents analysis of nonlinear diffusion of weak magnetic fields. In the case of weak magnetic fields, magnetic permeability (or differential permeability) is decreased as the magnetic fields are attenuated. As a result, physical features of this nonlinear diffusion are quite different from those in the case of strong magnetic fields. However, the same mathematical machinery that has been developed in the first two chapters can be used for the analysis of nonlinear diffusion of weak magnetic fields. As a result, many formal arguments and derivations presented in Chapter 3 are in essence slightly modified repetitions of what has been already discussed in the first and second chapters. These arguments and derivations are presented (albeit in concise form) for the sake of completeness of exposition.

Chapter 4 deals with nonlinear diffusion of electromagnetic fields in type-II superconductors. Phenomenologically, type-II superconductors can be treated as conductors with strongly nonlinear constitutive relations $E(J)$. These relations are usually approximated by sharp (ideal) resistive transitions or by "power" laws (gradual resistive transitions). Discussions start with the case of ideal resistive transitions and the critical state model for superconducting hysteresis. It is shown that this model is a very particular case of the Preisach model of hysteresis and, on this basis, it is strongly advocated to use the Preisach model for the description of superconducting hysteresis. For the case of gradual resistive transitions described by the power laws, analysis of nonlinear diffusion in superconductors has many mathematical features in common with the analysis of nonlinear diffusion in magnetically nonlinear conductors. For this reason, the analytical techniques that have been developed in the first two chapters are extensively applied to the analysis of nonlinear diffusion in superconductors. Thus, our discussion of this diffusion inevitably contains some repetitions; however, it is deliberately more concise and it stresses the points that are distinct to superconductors.

In Chapter 5, nonlinear impedance boundary conditions are introduced and extensively used for the solution of nonlinear eddy current problems. These boundary conditions are based on the expressions for nonlinear surface impedances derived in the previous chapters. The main emphasis in this chapter is on scalar potential formulations of impedance boundary conditions and their finite element implementations. However, the discussion presented in the chapter is much broader than this. It encompasses such related and important topics as: a general mathematical structure of 3-D eddy current problems, calculation of source fields, analysis of eddy currents in thin nonmagnetic conducting shells, derivations of easily computable estimates for eddy current losses, and analysis of thin magnetic shells subject to static magnetic fields.

Finally, Appendix A covers the basic facts related to the Preisach model of hysteresis. This model is treated as a general mathematical tool that can be used for the description of hysteresis of various physical origins. In this way, the physical universality of the Preisach model is clearly revealed and strongly emphasized.

In the book, no attempt is made to refer to all relevant publications. For this reason, the reference lists given at the end of each chapter are not exhaustive but rather suggestive. The presentation of the material in the book is largely based on the author's publications that have appeared over the last thirty years.

In writing this book, I have been assisted by Mrs. Patricia Keehn who patiently, diligently and professionally typed several versions of the manuscript. In preparation of the manuscript, I have also been assisted by my students Chung Tse and Michael Neely. I am very grateful to these individuals for their invaluable help in my work on this book. The main part of the book was written during my sabbatical leave at the Laboratory for Physical Sciences at College Park, Maryland, and I am very thankful to Dr. Thomas Beahn for the given opportunity. My work on this book was strongly encouraged and supported by Dr. Oscar Manley and Dr. Robert Price from the U.S. Department of Energy. I gratefully acknowledge their encouragements as well as the financial support for my research on nonlinear diffusion from the U.S. Department of Energy, Engineering Research Program.

Nonlinear Diffusion of Electromagnetic Fields

WITH APPLICATIONS TO EDDY CURRENTS AND SUPERCONDUCTIVITY

CHAPTER 1

Diffusion of Electromagnetic Fields in Magnetically Nonlinear Conducting Media (Linear Polarization)

1.1 STATEMENT OF THE PROBLEM

This chapter, as well as the next two chapters, will be concerned with the penetration of electromagnetic fields in magnetically nonlinear conducting media. This penetration process is described by the following Maxwell equations:

$$\text{curl } \mathbf{H} = \sigma \mathbf{E}, \tag{1.1}$$

$$\text{curl } \mathbf{E} = -\frac{\partial \mathbf{B}(\mathbf{H})}{\partial t}. \tag{1.2}$$

Here $\mathbf{H}$ and $\mathbf{E}$ are magnetic and electric fields, respectively; σ is the conductivity of media; and $\mathbf{B}(\mathbf{H})$ stands for a nonlinear (and possibly hysteretic) constitutive relationship between magnetic flux density $\mathbf{B}$ and magnetic field $\mathbf{H}$.

In Eq. (1.1), displacement currents were disregarded. This is because these currents are usually small in comparison with conduction currents $\sigma\mathbf{E}$.

The above two equations can be reduced to one equation with respect to the magnetic field:

$$\text{curl curl } \mathbf{H} = -\sigma \frac{\partial \mathbf{B}(\mathbf{H})}{\partial t}. \tag{1.3}$$

By using the well-known expression for the curl curl-operator, the last equation can be written as follows:

$$-\nabla^2 \mathbf{H} + \text{ grad div } \mathbf{H} = -\sigma \frac{\partial \mathbf{B}(\mathbf{H})}{\partial t}. \tag{1.4}$$

Expression (1.4) is a nonlinear vector partial differential equation. In general, its solution is affected (and complicated) by a particular geometric shape of magnetic conductor as well as by its nonlinear magnetic properties. To make the problem more or less analytically tractable, we consider the case of normal penetration of a plane electromagnetic wave into a semi-infinite magnetically nonlinear conducting half-space shown in Fig. 1.1. Naturally, this is the simplest problem that can be posed for Eq. (1.4). Nevertheless, the solution to this problem is of strong interest for the following two reasons. First, the solution to this problem will not depend on a particular shape of magnetic conductor and, in this sense, it will reveal in pure terms the effects of nonlinear properties of magnetic conductors on the penetration process. Second and more important, the results obtained for the plane wave normally penetrating in magnetically nonlinear conducting half-space can be used for the derivation of nonlinear impedance boundary conditions. These boundary conditions can then be applied to the analysis of the penetration process in magnetic conductors of complex shapes provided that the penetration ("skin") depth is small.

In the case of normal penetration of a plane electromagnetic wave, the magnetic field can be represented in the form:

$$\mathbf{H}(z,t) = \mathbf{a}_x H_x(z,t) + \mathbf{a}_y H_y(z,t), \tag{1.5}$$

where $\mathbf{a}_x$ and $\mathbf{a}_y$ are unit vectors directed along x and y Cartesian axes, respectively.

It is apparent from (1.5) that

$$\nabla^2 \mathbf{H} = \frac{\partial^2 \mathbf{H}}{\partial z^2}, \tag{1.6}$$

$$\text{div } \mathbf{H} = 0. \tag{1.7}$$

By using expressions (1.6) and (1.7) in Eq. (1.4), the latter equation can be appreciably simplified as follows:

$$\frac{\partial^2 \mathbf{H}}{\partial z^2} = \sigma \frac{\partial \mathbf{B}(\mathbf{H})}{\partial t}. \tag{1.8}$$

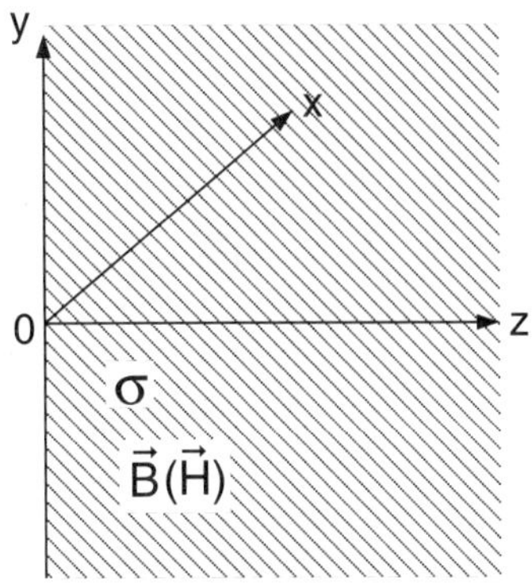

Fig. 1.1

The last equation is a nonlinear vector diffusion equation. For this reason, the penetration of electromagnetic fields in conducting media is often termed as diffusion of electromagnetic fields. The penetration process has indeed many physical features of diffusion. This is especially true in the case of linear conducting media. It will be shown later in this chapter that in the case of nonlinear media the diffusion (penetration) process may deviate from its conventional properties and exhibit some features of a wave propagation process.

There are also essential differences between linear and nonlinear diffusion of electromagnetic fields that can be directly ascertained from the very form of Eq. (1.8). To do this, we consider magnetically linear conducting media described by the constitutive equation:

$$\mathbf{B} = \mu \mathbf{H}, \tag{1.9}$$

where μ is the magnetic permeability of media.

By substituting expression (1.9) into (1.8), we end up with the linear vector diffusion equation:

$$\frac{\partial^2 \mathbf{H}}{\partial z^2} = \mu\sigma \frac{\partial \mathbf{H}}{\partial t}. \tag{1.10}$$

This vector equation can be written as two scalar diffusion equations:

$$\frac{\partial^2 H_x}{\partial z^2} = \mu\sigma \frac{\partial H_x}{\partial t}, \tag{1.11}$$

$$\frac{\partial^2 H_y}{\partial z^2} = \mu\sigma \frac{\partial H_y}{\partial t}, \tag{1.12}$$

which are completely decoupled (independent from one another). For this reason, these two equations can be solved separately. As a result, the penetration process of an arbitrarily polarized plane electromagnetic wave can be viewed as a superposition of two diffusion processes for linearly polarized electromagnetic waves. This is not the case for nonlinear media. Indeed, the nonlinear vector diffusion Eq. (1.8) can be written as the following two scalar nonlinear equations:

$$\frac{\partial^2 H_x}{\partial z^2} = \sigma \frac{\partial B_x(H_x, H_y)}{\partial t}, \tag{1.13}$$

$$\frac{\partial^2 H_y}{\partial z^2} = \sigma \frac{\partial B_y(H_x, H_y)}{\partial t}. \tag{1.14}$$

It is clear that the above scalar equations are coupled through nonlinear constitutive relations $B_x(H_x, H_y)$ and $B_y(H_x, H_y)$. For this reason, these equations cannot be solved separately. As a result, the case of arbitrary polarization of electromagnetic waves is not reducible to the superposition of two linear polarizations.

The previous discussion clearly reveals the main mathematical difficulties encountered in the analysis of nonlinear diffusion of plane electromagnetic waves. These difficulties are related to the nonlinear nature of partial differential Eqs. (1.13)-(1.14) and their mathematical coupling. There is, however, an additional difficulty of proper description of nonlinear and hysteretic magnetic properties of media. This is a challenging (and not completely solved) problem of appropriate specification of constitutive relations $B_x(H_x, H_y)$ and $B_y(H_x, H_y)$. This difficulty is especially pronounced in the case of hysteretic media.

In the view of the difficulties just described, we shall first consider the simplest case when the plane wave is linearly polarized. In this case, the magnetic field is constrained to vary in time along one direction, which is designated as the direction of y-axis. Thus, we have:

$$\mathbf{H}(z, t) = \mathbf{a}_y H(z, t). \tag{1.15}$$

It will also be assumed that the magnetic flux density has the same direction as $\mathbf{H}$:

$$\mathbf{B} = \mathbf{a}_y B(H), \tag{1.16}$$

where $B(H)$ is a scalar nonlinear (and hysteretic) relation.

By using expressions (1.15) and (1.16), the nonlinear vector diffusion Eq. (1.8) can be reduced to the following scalar nonlinear diffusion equation:

$$\frac{\partial^2 H}{\partial z^2} = \sigma \frac{\partial B(H)}{\partial t}. \tag{1.17}$$

Analytical techniques for the solution of the above scalar equations will be the main topic of our discussion in this chapter. In the next chapter, our focus will be on the analytical solution of nonlinear vector diffusion Eq. (1.8) or coupled Eqs. (1.13)–(1.14). This will require the development of different mathematical machinery than that used in this chapter.

1.2 NONLINEAR DIFFUSION IN THE CASE OF ABRUPT (SHARP) MAGNETIC TRANSITIONS

It has already been pointed out that the analytical solution of nonlinear diffusion Eq. (1.17) encounters formidable mathematical difficulties. In the past, these difficulties were fully circumvented only for the case of very simple magnetic nonlinearities describing abrupt (sharp) magnetic transitions. Such a transition for nonhysteretic media is shown in Fig. 1.2. It can be mathematically represented by the following expression:

$$B(H) = B_m \operatorname{sign}(H), \tag{1.18}$$

where, as usual, sign(H) is defined by:

$$\operatorname{sign}(H) = \begin{cases} 1, & \text{if } H > 0, \\ -1, & \text{if } H < 0. \end{cases} \tag{1.19}$$

The development of the analytical technique for the solution of nonlinear diffusion problems with constitutive relation (1.18) can be traced back to the landmark paper of W. Wolman and H. Kaden [21] published more than sixty years ago. This technique was afterwards independently rediscovered and further extended by V. Arkad'ev [2] in Russia and by W. MacLean [10], H.M. McConnell [17], and P. Agarwal [1] in the United States. This technique is traditionally derived by using integral forms of Maxwell's equations (such as Ampére's Law and Faraday's Law of electromagnetic induction) rather than by directly solving the nonlinear diffusion Eq. (1.17). Below, we deviate from this tradition and give a simple derivation of this technique based upon the solution of Eq. (1.17). To this end, we shall first modify this equation by introducing shifted magnetic flux density of $b(H)$ defined as follows:

$$b(H) = B(H) + B_m = 2B_m s(H), \tag{1.20}$$

where $s(H)$ is the unit step function

$$s(H) = \begin{cases} 1, & \text{if } H > 0, \\ 0, & \text{if } H < 0. \end{cases} \tag{1.21}$$

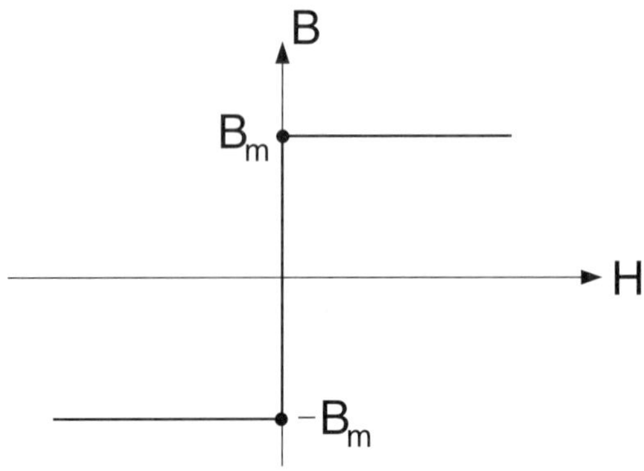

Fig. 1.2

In terms of $b(H)$, the nonlinear diffusion Eq. (1.17) takes the form:

$$\frac{\partial^2 H}{\partial z^2} = \sigma \frac{\partial b(H)}{\partial t}. \tag{1.22}$$

We consider the solution of this equation for the following initial and boundary conditions:

$$H(z, 0) = 0, \tag{1.23}$$

$$B(z, 0) = -B_m \qquad \text{or} \qquad b(z, 0) = 0, \tag{1.24}$$

$$H(0, t) = H_0(t) > 0. \tag{1.25}$$

It is clear that magnetic flux density B and shifted flux density b have spatial distributions as shown in Figs. 1.3 a and 1.3 b, respectively. Indeed, as the magnetic field $H_0(t)$ is increased at the boundary $z = 0$, this increase

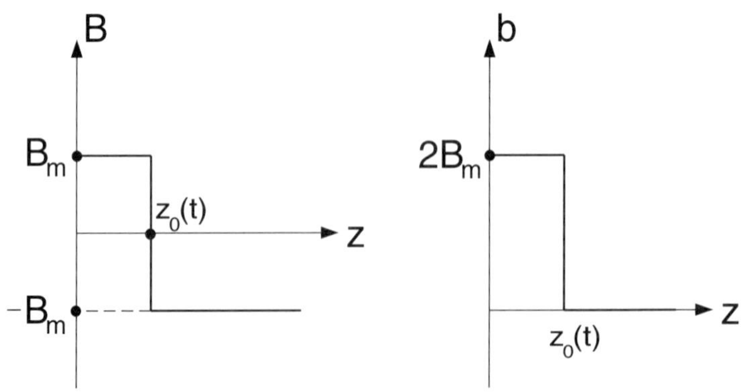

Fig. 1.3

extends inside the media causing B and b to switch from $-B_m$ to B_m and from 0 to $2B_m$, respectively. The distributions of $B(z)$ and $b(z)$ will be fully described if we find the expression for the front $z_0(t)$ in terms of $H_0(t)$, B_m, and σ. Indeed, if $z_0(t)$ is known, then

$$B(z,t) = \begin{cases} B_m, & \text{if } z < z_0(t), \\ -B_m, & \text{if } z > z_0(t), \end{cases} \tag{1.26}$$

and

$$b(z,t) = \begin{cases} 2B_m, & \text{if } z < z_0(t), \\ 0, & \text{if } z > z_0(t). \end{cases} \tag{1.27}$$

To find $z_0(t)$, we shall represent the nonlinear diffusion Eq. (1.22) as two coupled first-order partial differential equations:

$$\frac{\partial w}{\partial z} = -\sigma b(H), \tag{1.28}$$

$$\frac{\partial H}{\partial z} = -\frac{\partial w}{\partial t}. \tag{1.29}$$

It is easy to see that partial differential Eqs. (1.28) and (1.29) are formally equivalent to Eq. (1.22). Indeed, by formally differentiating Eq. (1.28) with respect to t and Eq. (1.29) with respect to z and then subtracting the results, we arrive at Eq. (1.22). However, Eqs. (1.28) and (1.29) have some mathematical advantages over Eq. (1.22). First, Eq. (1.22) contains the time derivative of the discontinuous function $b(H)$ and, for this reason, this equation is not rigorously defined (in a classical sense) for abrupt magnetic transitions. Equations (1.28) and (1.29) do not contain the derivative of discontinuous functions and retain mathematical sense for abrupt magnetic transitions. Actually, a solution to nonlinear diffusion Eq. (1.22) can be defined as a solution to coupled Eqs. (1.28) and (1.29). Second and more important, coupled Eqs. (1.28) and (1.29) are easy to solve. Indeed, from the definition of $b(H)$, we have:

$$\frac{\partial w}{\partial z} = \begin{cases} -2\sigma B_m, & \text{if } z < z_0(t). \\ 0, & \text{if } z > z_0(t). \end{cases} \tag{1.30}$$

Because function $w(z,t)$ is defined by Eqs. (1.28) and (1.29) up to a constant, from expression (1.30) we find that $w(z,t)$ is linear with respect to z when $0 \leq z \leq z_0(t)$ and it can be assumed to be equal to zero when $z \geq z_0(t)$:

$$w(z,t) = \begin{cases} w(0,t)[1 - \frac{z}{z_0(t)}], & \text{if } z \leq z_0(t), \\ 0, & \text{if } z \geq z_0(t). \end{cases} \tag{1.31}$$

It is clear from (1.31) that the slope of $w(z,t)$ is equal to $-\frac{w(0,t)}{z_0(t)}$ for $0 \leq z \leq z_0(t)$. According to Eq. (1.30), the same slope is equal to $-2\sigma B_m$. Thus:

$$\frac{w(0,t)}{z_0(t)} = 2\sigma B_m, \tag{1.32}$$

and

$$w(0,t) = 2\sigma B_m z_0(t). \tag{1.33}$$

By using expression (1.32) in formula (1.31), we find:

$$w(z,t) = \begin{cases} w(0,t) - 2\sigma B_m z, & \text{if } z \leq z_0(t), \\ 0, & \text{if } z \geq z_0(t). \end{cases} \tag{1.34}$$

From the last relation, we obtain:

$$\frac{\partial w(z,t)}{\partial t} = \begin{cases} \frac{dw(0,t)}{dt}, & \text{if } z \leq z_0(t), \\ 0, & \text{if } z \geq z_0(t). \end{cases} \tag{1.35}$$

By substituting expression (1.35) into Eq. (1.29), we arrive at:

$$\frac{\partial H}{\partial z}(z,t) = \begin{cases} -\frac{dw(0,t)}{dt}, & \text{if } z \leq z_0(t), \\ 0, & \text{if } z \geq z_0(t). \end{cases} \tag{1.36}$$

This means that at every instant of time $H(z,t)$ has a constant negative slope with respect to z for $0 \leq z \leq z_0(t)$ and the zero slope for $z \geq z_0(t)$. The latter is consistent with the fact that $H(z,t) = 0$ for $z \geq z_0(t)$. Thus:

$$H(z,t) = \begin{cases} H_0(t)[1 - \frac{z}{z_0(t)}], & \text{if } z \leq z_0(t), \\ 0, & \text{if } z \geq z_0(t). \end{cases} \tag{1.37}$$

By comparing expressions (1.36) and (1.37), we find:

$$\frac{H_0(t)}{z_0(t)} = \frac{dw(0,t)}{dt}. \tag{1.38}$$

According to formula (1.33), we have:

$$\frac{dw(0,t)}{dt} = 2\sigma B_m \frac{dz_0(t)}{dt}. \tag{1.39}$$

By substituting the last relation into (1.38), we obtain:

$$H_0(t) = 2\sigma B_m z_0(t) \frac{dz_0(t)}{dt}, \tag{1.40}$$

or

$$H_0(t) = \sigma B_m \frac{dz_0^2(t)}{dt}. \tag{1.41}$$

By integrating Eq. (1.41) and taking into account that $z_0(0) = 0$, we finally arrive at:

$$z_0(t) = \left(\frac{\int_0^t H_0(\tau)d\tau}{\sigma B_m} \right)^{1/2}. \tag{1.42}$$

Expression (1.42) together with (1.37) fully describe the solution of nonlinear diffusion Eq. (1.17) in the case of abrupt magnetic transitions. By using this solution as well as the expressions

$$j = -\frac{\partial H}{\partial z}, \qquad j = \sigma E, \tag{1.43}$$

we can derive the following formulas for the induced (eddy) current density j and electric field E:

$$j(z,t) = \begin{cases} \frac{H_0(t)}{z_0(t)}, & \text{if } z \leq z_0(t), \\ 0, & \text{if } z \geq z_0(t), \end{cases} \tag{1.44}$$

$$E(z,t) = \begin{cases} \frac{H_0(t)}{\sigma z_0(t)}, & \text{if } z \leq z_0(t), \\ 0, & \text{if } z \geq z_0(t). \end{cases} \tag{1.45}$$

At first, it may seem that formula (1.45) is in contradiction with the continuity condition for tangential components of electric fields. However, this continuity is valid only for stationary boundaries. In the case of moving boundaries, the above continuity condition is replaced by (see J.A. Kong [8]):

$$\vec{\nu} \times (\mathbf{E}^+ - \mathbf{E}^-) = (\vec{\nu} \cdot \mathbf{v})(\mathbf{B}^+ - \mathbf{B}^-),$$

where: $\vec{\nu}$ is a unit normal to a moving boundary, $\mathbf{v}$ is its local velocity, while $\mathbf{E}^+, \mathbf{E}^-, \mathbf{B}^+$, and $\mathbf{B}^-$ are the vector values of electric field and magnetic flux density on two sides of the moving boundary, respectively.

For our problem, the last boundary condition yields

$$E(z_0(t)) = 2B_m \frac{dz_0(t)}{dt},$$

which, according to formula (1.40), leads to

$$E(z_0(t)) = \frac{H_0(t)}{\sigma z_0(t)}.$$

The last formula is consistent with Eq. (1.45).

Spatial distributions of $H(z,t)$ and $j(z,t)$ are shown in Figs. 1.4 a and 1.4 b, respectively. It is clear that positive rectangular fronts of B, j, E and linear front of H move inside the medium as long as $H_0(t)$ remains positive. As soon as $H_0(t)$ reaches zero value and then becomes negative, the above motion is terminated and rectangular B- and j-fronts and linear H-fronts of opposite polarity are formed and continuously move inside the conducting medium. By using literally the same line of reasoning as before, it can be shown that the same expression (1.42) is valid for a new zero front, $z_0(t)$, with only one correction: a minus sign appears in front of the integral.

Now, we can consider the important case when the magnetic field at the boundary is sinusoidal:

$$H_0(t) = H_m \sin \omega t. \tag{1.46}$$

It is clear that, during the positive half-cycle, the positive rectangular fronts of B propagate inside the medium (see Fig. 1.5 a). This inward motion of $z_0^+(t)$ is terminated at $t = \frac{T}{2}$. During the negative half-cycle, the negative rectangular front of B is formed and it moves inside the medium (see Fig. 1.5 b). At $t = T$, this inward motion of $z_0^-(t)$ completely wipes out the positive rectangular wave of B. During subsequent cycles, the situation repeats itself.

Next, we want to find the relation between electric and magnetic fields at the boundary $z = 0$. We consider only the positive half-cycle; for the negative half-cycle this relation remains the same. By combining formulas (1.45) and (1.42), we obtain:

$$E_0(t) = E(0,t) = \frac{H_0(t)}{\sigma z_0(t)} = H_0(t) \left(\frac{B_m}{\sigma \int_0^t H_0(\tau) d\tau} \right)^{1/2}. \tag{1.47}$$

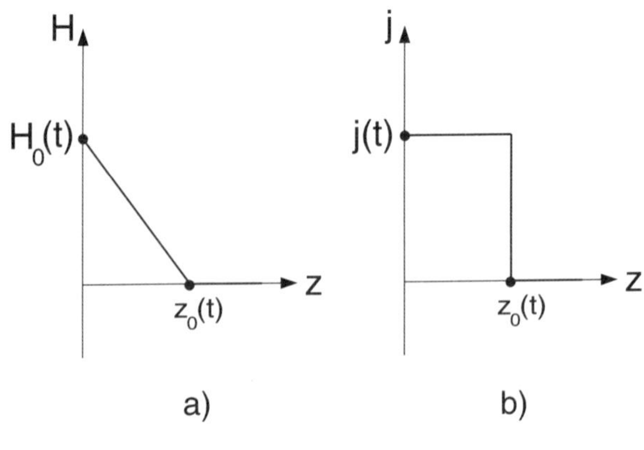

Fig. 1.4

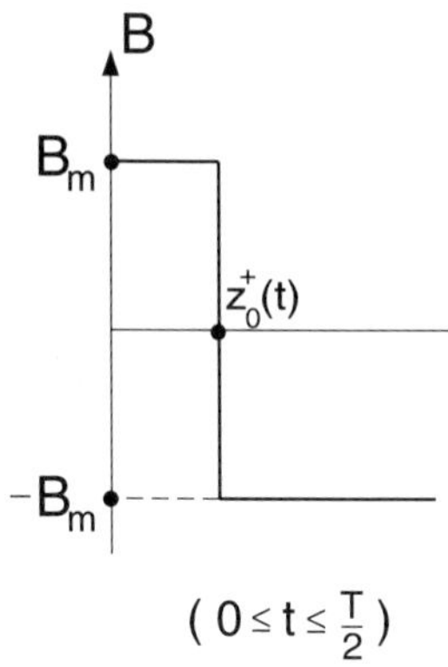

Fig. 1.5a

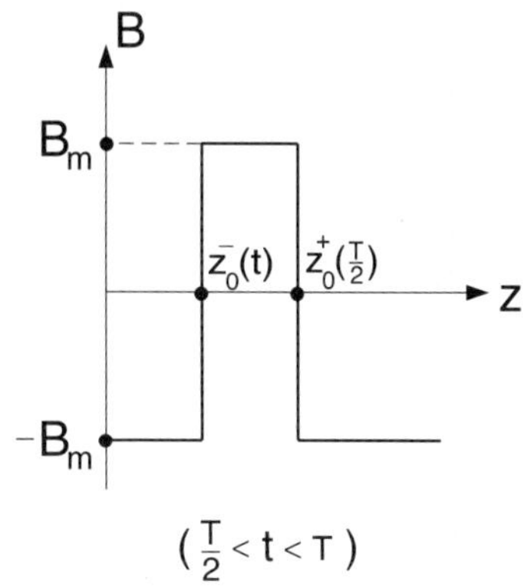

Fig. 1.5b

By substituting (1.46) into (1.47), performing integration, and introducing the notation

$$\mu_m = \frac{B_m}{H_m}, \tag{1.48}$$

we arrive at

$$E_0(t) = H_m\sqrt{\frac{\omega\mu_m}{\sigma}}\frac{\sin\omega t}{\sqrt{1-\cos\omega t}}. \tag{1.49}$$

Thus, we can see that the electric field, $E_0(t)$, at the boundary is not purely sinusoidal and contains higher-order harmonics. This generation of higher-order harmonics can be attributed to the nonlinear magnetic properties of the conducting media. It is interesting to point out that this is not always the case, and it will be shown in the next chapter that for circular polarization of electromagnetic fields there is no generation of higher-order harmonics despite the nonlinearity of media.

By using expression (1.49), we can find the first harmonic $E_0^{(1)}(t)$ of the electric field at the boundary

$$E_0^{(1)}(t) = H_m\sqrt{\frac{\omega\mu_m}{\sigma}}(a\cos\omega t + b\sin\omega t), \tag{1.50}$$

where coefficients a and b are given by the following integrals:

$$a = \frac{4}{T}\int_0^{\frac{T}{2}}\frac{\sin\omega t\cos\omega t}{\sqrt{1-\cos\omega t}}dt = \frac{2}{\pi}\int_0^{\pi}\frac{\sin\zeta\cos\zeta}{\sqrt{1-\cos\zeta}}d\zeta, \tag{1.51}$$

$$b = \frac{4}{T}\int_0^{\frac{T}{2}}\frac{\sin^2\omega t}{\sqrt{1-\cos\omega t}}dt = \frac{2}{\pi}\int_0^{\pi}\frac{\sin^2\zeta}{\sqrt{1-\cos\zeta}}d\zeta, \tag{1.52}$$

By performing integration in (1.51) and (1.52), we arrive at

$$a = \frac{4\sqrt{2}}{3\pi}, \qquad b = \frac{8\sqrt{2}}{3\pi}. \tag{1.53}$$

Expression (1.50) can also be written in the following form:

$$E_0^{(1)}(t) = \sqrt{a^2 + b^2} H_m \sqrt{\frac{\omega \mu_m}{\sigma}} \sin(\omega t + \varphi). \tag{1.54}$$

According to (1.53), we find

$$\sqrt{a^2 + b^2} = 1.34, \qquad \tan\varphi = \frac{a}{b} = 0.5, \tag{1.55}$$

which leads to the expression

$$E_0^{(1)}(t) = (1.34)\sqrt{\frac{\omega \mu_m}{\sigma}} H_m \sin(\omega t + 26^\circ). \tag{1.56}$$

The last formula can be rewritten in the phasor form

$$\hat{E}_0^{(1)} = \left(1.34\sqrt{\frac{\omega \mu_m}{\sigma}} e^{j\frac{\pi}{6.77}}\right) \hat{H}_0, \tag{1.57}$$

where the symbol " ^ " is used for the notation of phasors, while $j = \sqrt{-1}$.

The last expression can be represented in terms of surface impedance η:

$$\hat{E}_0^{(1)} = \eta \hat{H}_0, \tag{1.58}$$

where

$$\eta = 1.34\sqrt{\frac{\omega \mu_m}{\sigma}} e^{\frac{j\pi}{6.77}}. \tag{1.59}$$

It is instructive to recall that in the case of linear conducting media the surface impedance is given by

$$\eta^{(\ell)} = \sqrt{\frac{\omega \mu}{\sigma}} e^{j\frac{\pi}{4}}. \tag{1.60}$$

By using expressions (1.42), (1.46), and (1.48), the penetration depth $z_0\left(\frac{T}{2}\right)$ of an electromagnetic field in magnetically nonlinear conducting media can be found:

$$z_0\left(\frac{T}{2}\right) = \sqrt{\frac{2}{\omega \sigma \mu_m}} \tag{1.61}$$

The last expression has the same "appearance" as the classical formula for the penetration depth, δ, in linear conducting media:

$$\delta = \sqrt{\frac{2}{\omega\sigma\mu}}. \tag{1.62}$$

However, in spite of formal similarities, there are two essential differences between formulas (1.61) and (1.62). First, formula (1.61) gives a complete penetration depth; there is no time-varying electromagnetic field beyond $z_0\left(\frac{T}{2}\right)$, that is, for $z > z_0\left(\frac{T}{2}\right)$. On the other hand, formula (1.62) gives a distance at which the electromagnetic field is attenuated only to e^{-1} times its value at the boundary. Second, in formula (1.62) μ is constant and the penetration depth is field independent, while in expression (1.61) μ_m is inversely proportional to H_m (see (1.48)), which makes the penetration depth field dependent. The last remark is also valid as far as comparison of expressions (1.59) and (1.60) for surface impedances is concerned. In the case of linear conducting media, the surface impedance (1.60) is field independent, while for magnetically nonlinear conducting media the surface impedance (1.59) is a function of H_m.

It is also important to stress that the surface impedance for nonlinear conducting media is defined as the ratio of first harmonic phasors of electric and magnetic fields. For this reason, the value of the surface impedance may depend on the boundary conditions for H. To illustrate this fact as well as to appreciate the range of possible variations of η, consider the case when the magnetic field at the boundary varies with time as follows:

$$H_0(t) = \begin{cases} \frac{H_m}{1.3}\left(\sin\omega t - \frac{1}{2}\sin 2\omega t\right), & \text{if } 0 \le t \le \frac{T}{2}, \\ \frac{H_m}{1.3}\left(\sin\omega t + \frac{1}{2}\sin 2\omega t\right), & \text{if } \frac{T}{2} \le t \le T. \end{cases} \tag{1.63}$$

Here, 1.3 is the maximum value of $\sin\omega t \pm \frac{1}{2}\sin 2\omega t$; consequently, H_m has the meaning of the peak value of $H_0(t)$.

This boundary condition is chosen because it leads to the sinusoidal electric field $E_0(t)$ at the boundary. To demonstrate this, we substitute (1.63) into (1.42) and after integration we obtain:

$$z_0^{\pm}(t) = \sqrt{\frac{H_m}{2.6\omega\sigma B_m}}\left(1 \mp \cos\omega t\right), \tag{1.64}$$

where the superscripts "+" and "–" correspond to positive and negative half-cycles, respectively.

Now, by using expressions (1.63) and (1.64) in formula (1.45) and taking into account the definition (1.48) of μ_m, we end up with:

$$E_0(t) = 1.24\sqrt{\frac{\omega\mu_m}{\sigma}}H_m \sin\omega t. \tag{1.65}$$

Next, in order to find the surface impedance η that corresponds to the boundary condition (1.63), we determine the first harmonic $H_0^{(1)}(t)$ of $H_0(t)$:

$$H_0^{(1)}(t) = \frac{H_m}{1.3}(a \cos \omega t + b \sin \omega t), \tag{1.66}$$

where coefficients a and b are given by the following integrals:

$$a = \frac{2}{T} \int_0^T 1.3 \frac{H_0(t)}{H_m} \cos \omega t dt, \tag{1.67}$$

$$b = \frac{2}{T} \int_0^T 1.3 \frac{H_0(t)}{H_m} \sin \omega t dt. \tag{1.68}$$

By substituting expression (1.63) into formulas (1.67) and (1.68) and by performing integration, we obtain:

$$a = -\frac{4}{3\pi}, \qquad b = 1. \tag{1.69}$$

By using these values of a and b, we transform expression (1.66) as follows:

$$H_0^{(1)}(t) = \frac{H_m}{1.19} \sin(\omega t - \varphi), \tag{1.70}$$

where

$$\tan \varphi = \frac{4}{3\pi} = 0.424, \qquad \varphi = 23^\circ. \tag{1.71}$$

Now, by transforming expressions (1.65) and (1.70) into phasor forms, we compute the surface impedance:

$$\eta = \frac{\hat{E}_0}{\hat{H}_0^{(1)}} = 1.47 \sqrt{\frac{\omega \mu_m}{\sigma}} e^{j \frac{\pi}{7.83}}. \tag{1.72}$$

By employing expression (1.64), we can find the penetration depth, $z_0\left(\frac{T}{2}\right)$, in the case of sinusoidal variation of the electric field on the boundary:

$$z_0(T/2) = \sqrt{\frac{1.57}{\omega \sigma \mu_m}}. \tag{1.73}$$

Comparison of expressions (1.59) and (1.61) with expressions (1.72) and (1.73) is suggestive of to what extent the surface impedance and the penetration depth may depend on a particular time variation of the magnetic field on the boundary.

The results of the previous analysis can be extended to the practically important case of magnetically nonlinear conducting laminations (plates). Such laminations are used in many applications. For instance, steel laminations are stacked together to form magnetic cores of transformers, electric machines, and various actuators. Laminated permalloy heads as well as thin film heads are widely used in magnetic recording. In all these designs, magnetic laminations are employed for flux-guiding purposes. For this reason, it is desirable that cross-sections of magnetic laminations are utilized effectively. To check this, distributions of magnetic flux density over lamination cross-sections can be computed by using the previously derived expressions. Indeed, during an initial stage of positive half-cycle, magnetic fields penetrate from both sides of the laminations in the same way as in the case of semi-infinite half-space (see Fig. 1.6 a). The motion of the positive front $z_0^+(t)$ can be determined by using formula (1.42) if the magnetic field $H_0(t)$ on the boundary of the lamination is known. This is usually the case when the current through the coil, which creates the magnetic flux, is known. When the voltage applied to the coil is known, then the boundary value $E_0(t)$ of the electric field can be determined. By using $E_0(t)$, the motion of the zero front can be found according to the formula:

$$E_0(t) = 2B_m \frac{dz_0^+(t)}{dt}, \tag{1.74}$$

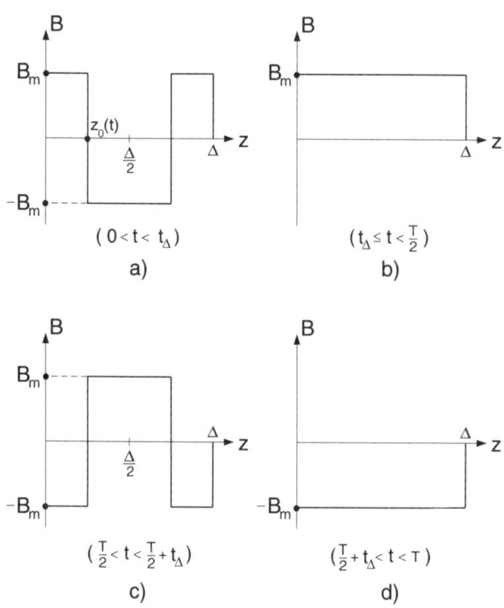

Fig. 1.6

which leads to

$$z_0^+(t) = \frac{1}{2B_m} \int_0^t E_0(\tau)d\tau. \qquad (1.75)$$

We note that formula (1.74) is easily derived from expressions (1.36), (1.39), and (1.43).

At the instant of time t_Δ when

$$z_0^+(t_\Delta) = \frac{\Delta}{2}, \qquad (1.76)$$

(where Δ is the lamination thickness), the two positive fronts are merged together (see Fig. 1.6 b) and the distribution of magnetic flux density over a lamination cross-section is uniform. It remains this way during the rest of the positive half-cycle. With the commencement of the negative half-cycle, negative fronts of magnetic flux density are formed and they penetrate from both sides of the lamination (see Fig. 1.6 c). At the instant of time $\frac{T}{2} + t_\Delta$ these negative fronts are merged together (see Fig. 1.6 d) and the distribution of magnetic flux density remains uniform during the rest of the negative half-cycle. At subsequent cycles, the situation repeats itself. It is clear from the above discussion that the lamination cross-section will be effectively utilized if t_Δ is substantially smaller than $T/2$. The validity of this fact can be evaluated for every particular case by using formula (1.76) along with expression (1.42) or (1.75).

The analytical technique just presented can be generalized to the case when abrupt magnetic transitions are described by a rectangular hysteresis loop as shown in Fig. 1.7. Again, we begin with the case when the initial value of the magnetic field throughout conducting media is equal to zero, while the initial value of magnetic flux density is equal to $-B_m$. Suppose the magnetic field $H_0(t)$ at the boundary is increased. Until this field reaches the coercive value, H_c, nothing happens. As soon as $H_0(t)$ exceeds the coercive value H_c, the rectangular front of magnetic flux density is formed and it moves inside the medium. To compute the zero front $z_0(t)$, we introduce the shifted magnetic field $h(z,t)$:

$$h(z,t) = H(z,t) - H_c, \qquad (1.77)$$

and rewrite the nonlinear diffusion Eq. (1.22) as follows:

$$\frac{\partial^2 h}{\partial z^2} = \sigma \frac{\partial b(h)}{\partial t}. \qquad (1.78)$$

Now, by literally repeating the same line of reasoning as before, we can derive the following expression for $z_0(t)$:

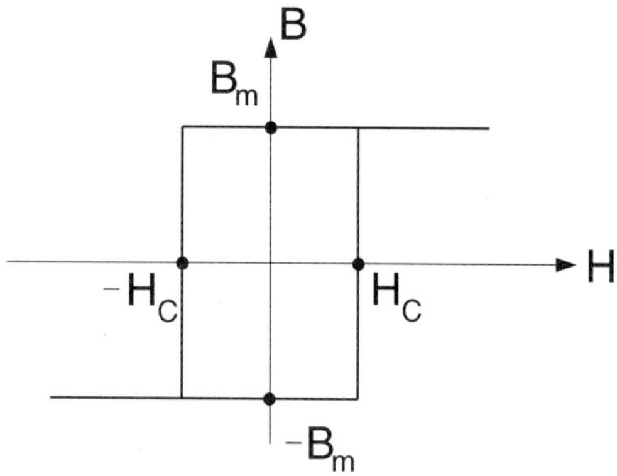

Fig. 1.7

$$z_0(t) = \left(\frac{\int_{t_c}^{t} h_0(\tau) d\tau}{\sigma B_m} \right)^{1/2}. \tag{1.79}$$

Here t_c is the time when

$$H_0(t_c) = H_c, \tag{1.80}$$

while $h_0(t)$ is the boundary value of the magnetic field, $h(0,t)$. As far as the distributions of $h(z,t), j(z,t)$, and $E(z,t)$ are concerned, the same formulas (1.37), (1.44), and (1.45) are valid, however, $H_0(t)$ in these formulas must be replaced by $h_0(t)$.

The propagation of the positive rectangular front of magnetic flux density will continue until the magnetic field at the boundary is reduced back to its coercive value H_c (or $h_0(t)$ is reduced to zero). As the magnetic field at the boundary is reduced from H_c to $-H_c$, nothing happens and induced eddy currents and electric fields are equal to zero. As soon as the magnetic field on the boundary is reduced below $-H_c$, the negative front of magnetic flux density is formed and it moves inside the medium. The motion of this negative front can be determined by using the same formula (1.79) with the following corrections: (a) the minus sign appears in front of the integral in (1.79), (b) $h_0(\tau)$ is defined as $H_0(t) + H_c$, and (c) the time t_c is determined from the equation $H_0(t_c) = -H_c$.

Next, consider the example when the magnetic field at the boundary is sinusoidal and given by expression (1.46). We want to find the surface impedance η. To this end, we shall first find the electric field $E_0(t)$ at the boundary. It is clear from the previous discussion that

$$E_0(t) = 0, \qquad \text{if } 0 \le t \le t_c, \tag{1.81}$$

where

$$t_c = \frac{1}{\omega} \arcsin \left(\frac{H_c}{H_m} \right), \tag{1.82}$$

$$E_0(t) = h_0(t) \left(\frac{B_m}{\sigma \int_{t_c}^{t} h_0(\tau) d\tau} \right)^{1/2}, \quad \text{if } t_c \leq t \leq \frac{T}{2} - t_c, \tag{1.83}$$

and

$$E_0(t) = 0, \quad \text{if } \frac{T}{2} - t_c \leq t \leq \frac{T}{2}. \tag{1.84}$$

Similar expressions can be written for the negative half-cycle.

By substituting formula (1.46) into (1.83), by performing integration and taking into account the definition of $h(t), \mu_m$, and t_c, we derive

$$E_0(t) = H_m \sqrt{\frac{\omega \mu_m}{\sigma}} \frac{\sin \omega t - \sin \omega t_c}{\sqrt{\cos \omega t_c - \cos \omega t - (\omega t - \omega t_c) \sin \omega t_c}}. \tag{1.85}$$

The first harmonic of $E_0(t)$ can be written in the form (1.50), where coefficients a and b are determined by the following integrals:

$$a = \frac{2}{\pi} \int_{\zeta_c}^{\pi - \zeta_c} \frac{(\sin \zeta - \sin \zeta_c) \cos \zeta d\zeta}{\sqrt{\cos \zeta_c - \cos \zeta - (\zeta - \zeta_c) \sin \zeta_c}}, \tag{1.86}$$

$$b = \frac{2}{\pi} \int_{\zeta_c}^{\pi - \zeta_c} \frac{(\sin \zeta - \sin \zeta_c) \sin \zeta d\zeta}{\sqrt{\cos \zeta_c - \cos \zeta - (\zeta - \zeta_c) \sin \zeta_c}}, \tag{1.87}$$

and

$$\zeta_c = \arcsin \left(\frac{H_c}{H_m} \right). \tag{1.88}$$

In terms of a and b, the surface impedance η is given by

$$\eta = \sqrt{a^2 + b^2} \sqrt{\frac{\omega \mu_m}{\sigma}} e^{j\varphi}, \tag{1.89}$$

where

$$\tan \varphi = \frac{a}{b}. \tag{1.90}$$

By using formulas (1.86) and (1.87), the values of $\sqrt{a^2 + b^2}$, $\tan \varphi$, and φ have been computed as functions of ζ_c. The results of the computations are shown in Figs. 1.8 a, 1.8 b, and 1.8 c. By using these figures, the dependence of surface impedance η on the coercivity H_c can be evaluated.

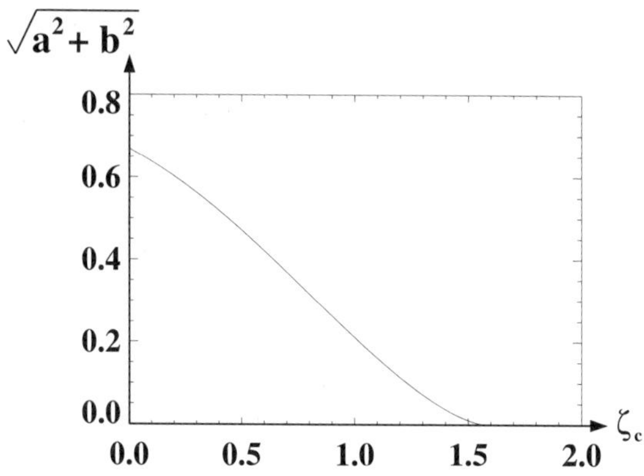

Fig. 1.8 a

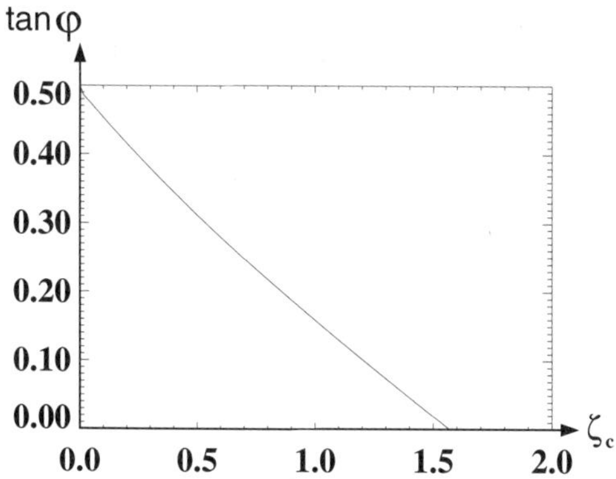

Fig. 1.8 b

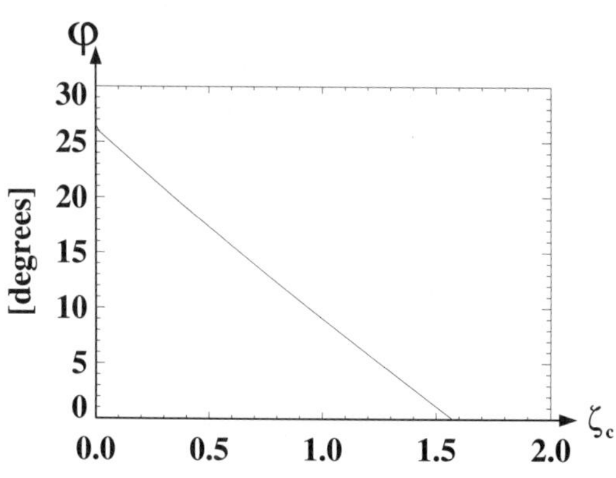

Fig. 1.8 c

1.3 MODEL PROBLEM FOR GRADUAL MAGNETIC TRANSITIONS

In Section 1.2, nonlinear diffusion of electromagnetic fields in conducting media with abrupt magnetic transitions was discussed and simple analytical solutions were derived. However, these solutions do not allow one to understand how actual gradual magnetic transitions (or actual shapes of hysteresis loops) may affect the diffusion process. For this reason, the analytical study of nonlinear diffusion of electromagnetic fields in conducting media with gradual (and more or less realistic) magnetic transitions is an important problem. Next, an attempt will be made to solve this problem for the case of hysteresis loops that are exemplified by Fig. 1.9. These hysteresis loops are characterized by the property that their ascending (and descending) branches can be subdivided into two distinct parts: part I of slow increase of magnetic flux density B from $-B_m$ to $-B_c$ and part II of steep increase of B from $-B_c$ to B_m. Such hysteresis loops are typical for most ferromagnetic materials in the case of sufficiently large values of H_m and they are encountered in many applications.

To attempt the analytical solution of nonlinear diffusion Eq. (1.17), we adopt a "flat-power" approximation of a hysteresis loop shown in Fig. 1.9. This approximation is illustrated in Fig. 1.10 and it is analytically described by the following equations:

$$B = -B_m, \quad \text{if } -H_m < H \leq H_c, \tag{1.91}$$

$$B + B_m = [k(H - H_c)]^{\frac{1}{n}}, \quad \text{if } H_c \leq H \leq H_m \tag{1.92}$$

in the case of the ascending branch, and

$$B = B_m, \quad \text{if } -H_c \leq H \leq H_m, \tag{1.93}$$

$$B_m - B = [-k(H + H_c)]^{\frac{1}{n}}, \quad \text{if } -H_m \leq H \leq -H_c \tag{1.94}$$

in the case of the descending branch. In other words, part I of the ascending

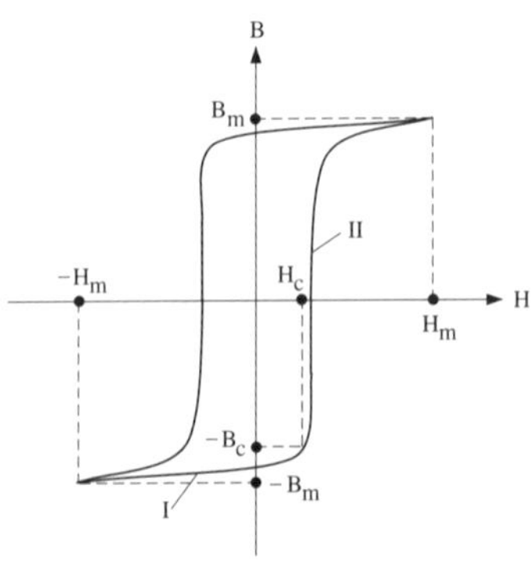

Fig. 1.9

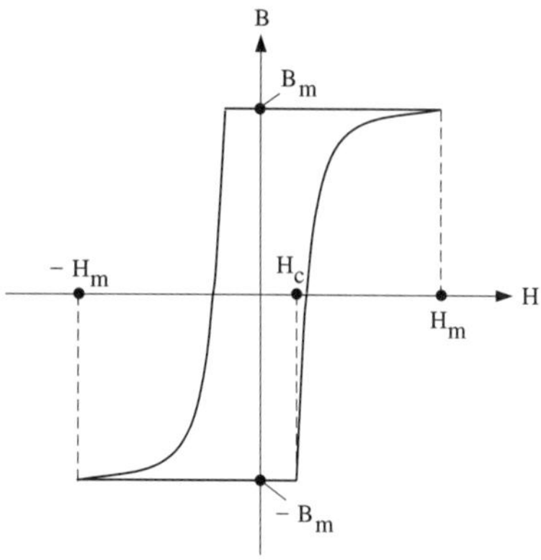

Fig. 1.10

branch is approximated by a "flat" straight line parallel to the H-axis, while part II is approximated by the "power" expression (1.92).

In the above formulas, coefficient k coordinates the dimensions of both sides of expressions (1.92) and (1.94), while the exponent n is a measure of the sharpness of magnetic transition. It is important to note that in applications the exponent n is usually larger than 7 ($n \geq 7$). This fact is essential and it will be used in our subsequent discussions in order to simplify relevant analytical expressions and to achieve some universality in the final form of the solution to the nonlinear diffusion equation. By introducing the "shifted" magnetic field h and magnetic flux density b

$$h = H - H_c, \qquad b = B + B_m, \tag{1.95}$$

expression (1.92) can be rewritten as follows:

$$h = \frac{b^n}{k}. \tag{1.96}$$

Next, we shall consider the following "model" problem. It will be assumed that at time $t = 0$ the magnetic flux density B is equal to $-B_m$ throughout the conducting half-space:

$$B(z, 0) = -B_m. \tag{1.97}$$

It will also be assumed that the magnetic flux density at the boundary of the conducting half-space is monotonically increased with time as follows:

$$B(0, t) = -B_m + ct^p, \qquad (p > 0) \tag{1.98}$$

By using the nonlinear diffusion Eq. (1.17) as well as expressions (1.95) and (1.96), the stated model problem can be reduced to the following initial boundary value problem: find the solution of the nonlinear diffusion equation

$$\frac{\partial^2 b^n}{\partial z^2} = k\sigma \frac{\partial b}{\partial t}, \tag{1.99}$$

subject to the following initial and boundary conditions:

$$b(z, 0) = 0, \tag{1.100}$$

$$b(0, t) = ct^p. \tag{1.101}$$

It is worthwhile to mention that these boundary conditions are chosen for the following two reasons. First, it will be demonstrated that it is possible to find simple analytical solutions for these boundary conditions. Second, these boundary conditions describe a broad class of monotonically increasing functions as p varies from 0 to ∞ (see Fig. 1.11). It will be shown in the sequel that for all these monotonically increasing boundary conditions the distribution (profile) of the magnetic flux density as a function of z remains practically the same. This observation will suggest using the same profile of magnetic flux density for arbitrary monotonically increasing (between $-B_m$ and B_m) boundary conditions. This, in turn, will lead to very general and simple analytical solutions, which can then be extended to periodic in time boundary conditions.

Next, we shall show that the initial boundary value problem (1.99)–(1.101) has an exact and very simple analytical solution in the case when $p = \frac{1}{n-1}$. This solution describes a wave of magnetic flux density that moves inside the conducting medium with constant velocity v. To find this solution, we look for $b(z,t)$ in the form:

$$b(z, t) = \varphi(z - vt). \tag{1.102}$$

By introducing the variable

$$\zeta = z - vt, \tag{1.103}$$

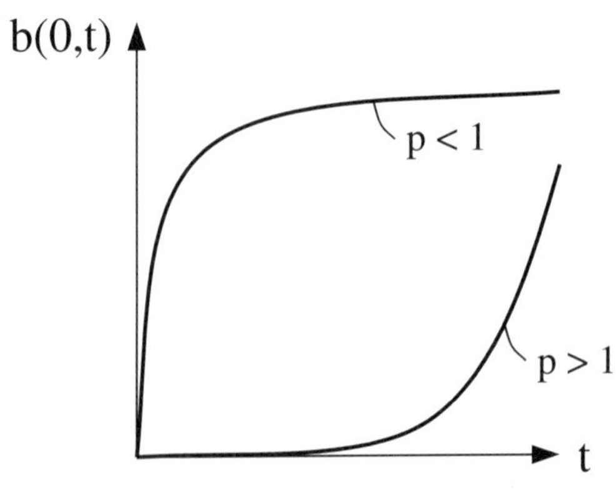

Fig. 1.11

we have:

$$b(z,t) = \varphi(\zeta). \tag{1.104}$$

Now, we reduce the nonlinear partial differential Eq. (1.99) to the ordinary differential equation with respect to $\varphi(\zeta)$. To this end, we note that

$$\frac{\partial^2 b^n}{\partial z^2} = \frac{d^2 \varphi^n}{d\zeta^2}, \tag{1.105}$$

$$\frac{\partial b}{\partial t} = -v\frac{d\varphi}{d\zeta}. \tag{1.106}$$

By substituting expressions (1.105) and (1.106) into nonlinear diffusion Eq. (1.99), we arrive at:

$$\frac{d^2\varphi^n}{d\zeta^2} + k\sigma v\frac{d\varphi}{d\zeta} = 0. \tag{1.107}$$

By integrating Eq. (1.107), we obtain:

$$\frac{d\varphi^n}{d\zeta} + k\sigma v\varphi = C_1, \tag{1.108}$$

where C_1 is an integration constant. From formula (1.102) and initial condition (1.100), for $t = 0$ we have:

$$b(z,0) = \varphi(\zeta) = 0, \quad (\zeta > 0), \tag{1.109}$$

$$\frac{d\varphi^n}{d\zeta} = \frac{d\varphi^n(z)}{dz} = 0. \tag{1.110}$$

Thus, φ and $\frac{d\varphi^n}{d\zeta}$ must be equal to zero simultaneously. This is only possible if

$$C_1 = 0, \tag{1.111}$$

which leads to

$$\frac{d\varphi^n}{d\zeta} + k\sigma v\varphi = 0. \tag{1.112}$$

The last equation can be rewritten as follows:

$$n\varphi^{n-1}\frac{d\varphi}{d\zeta} = -k\sigma v\varphi. \tag{1.113}$$

By separating the variables, we obtain:

$$-\frac{n}{k\sigma v}\varphi^{n-2}d\varphi = d\zeta. \tag{1.114}$$

By integrating the last expression, we arrive at

$$-\frac{n}{n-1}\frac{\varphi^{n-1}}{k\sigma v} = \zeta + C_2, \tag{1.115}$$

where C_2 is an integration constant.

From formulas (1.100), (1.101), (1.103), and (1.104) we find that at $z = 0$ and $t = 0$ we have:

$$\zeta = 0, \qquad \varphi(0) = 0. \tag{1.116}$$

According to equalities (1.116), we conclude that the integration constant C_2 in Eq. (1.115) is equal to zero:

$$C_2 = 0. \tag{1.117}$$

Consequently,

$$\varphi^{n-1} = \frac{n-1}{n} k\sigma v(-\zeta), \tag{1.118}$$

which leads to

$$b(z,t) = \left(\frac{n-1}{n} k\sigma v\right)^{\frac{1}{n-1}} (vt - z)^{\frac{1}{n-1}}. \tag{1.119}$$

The last expression can be further transformed as follows:

$$b(z,t) = ct^{\frac{1}{n-1}} \left(1 - \frac{z}{vt}\right)^{\frac{1}{n-1}}, \tag{1.120}$$

where

$$c = \left(\frac{n-1}{n} k\sigma v^2\right)^{\frac{1}{n-1}}. \tag{1.121}$$

It is apparent from the solution (1.120) that this solution corresponds to the following boundary condition:

$$b_0(t) = b(0,t) = ct^{\frac{1}{n-1}}. \tag{1.122}$$

Now, we shall use the physical reasoning to demonstrate that formula (1.120) gives the right expression for $b(z,t)$ only for $z \leq vt$, whereas for $z \geq vt$ the magnetic flux density should be equal to zero. Indeed, since the magnetic flux density $b_0(t)$ on the boundary of conducting media is nonnegative and $b(z,t)$ is assumed to be equal to zero at $t = 0$ throughout the media, the magnetic flux density should remain nonnegative at all instants of time everywhere within the media. In addition, since the electromagnetic field is attenuated in the conducting media, $b(z,t)$ should be a monotonically decreasing function of z at any t. This means that if $b(z,t)$ reaches zero at some point z_0 it should remain equal to zero for $z \geq z_0$. From the last fact and formula (1.120), we conclude that the correct expression for the magnetic flux density is

$$b(z,t) = \begin{cases} ct^{\frac{1}{n-1}} \left(1 - \frac{z}{vt}\right)^{\frac{1}{n-1}}, & \text{if } 0 \leq z \leq vt, \\ 0, & \text{if } z \geq vt. \end{cases} \tag{1.123}$$

It is also worthwhile to note that the second line on the right-hand side of expression (1.123) guarantees the zero initial condition for $b(z,t)$. Without it, the solution given by expression (1.120) or (1.119) does not satisfy the zero initial condition. This is another (mathematical) reason behind the formula (1.123).

Solution (1.123) is illustrated in Fig. 1.12 where profiles of magnetic flux density $b(z,t)$ are shown for different instants of time. It is clear from this figure as well as from expression (1.123) that there exists a zero front $z_0(t)$ of electromagnetic field that moves with constant velocity v:

$$z_0(t) = vt. \tag{1.124}$$

The value of this velocity depends on electric and magnetic properties of the conducting media as well as on the value of coefficient c in the boundary condition (1.122). Indeed, from expression (1.121) we derive

$$v = \sqrt{\frac{n}{(n-1)k\sigma}} c^{\frac{n-1}{2}}. \tag{1.125}$$

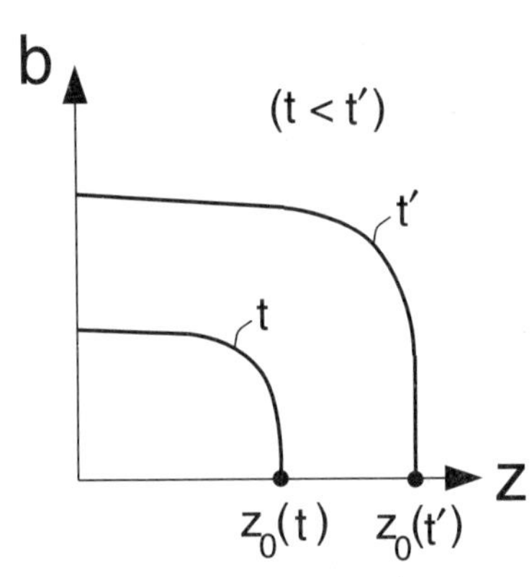

Fig. 1.12

The existence of zero fronts of electromagnetic fields is typical for wave propagation problems, however, it is unusual for diffusion problems. This is because displacement currents are neglected in derivations of diffusion equations. For this reason, we should expect the infinite velocity of propagation of zero front. To illustrate and highlight the last point, consider a stepwise change of magnetic flux density at the boundary of conducting half-space with linear magnetic and electric properties. It can be shown (or found in the literature) that in this case the diffusion of the magnetic field in the conducting media is described by the formula:

$$B(z,t) = \frac{2B_0}{\sqrt{\pi}} \int_u^\infty e^{-\alpha^2} d\alpha, \tag{1.126}$$

where

$$u = \frac{\sqrt{\mu\sigma}z}{2\sqrt{t}}, \tag{1.127}$$

while B_0 is a "step" value of magnetic flux density at the boundary.

It is apparent from formulas (1.126) and (1.127) that the magnetic flux density differs from zero at any (however small) instant of time t and at any (however remote) point z. This suggests that the zero front velocity is infinite. This fact is illustrated in Fig. 1.13 where distributions (profiles) of $B(z,t)$ are shown at various instants of time.

Now, the question can be posed why in the case of magnetically nonlinear media the zero front velocity is finite. The answer is that the finite velocity of the zero front appears as a result of idealization introduced by the "power" approximation (1.92) or (1.96). This approximation leads to the infinite growth of differential magnetic permeability $\mu_d = \frac{dB}{dH}$ as h approaches zero. This infinite growth of μ_d causes the complete attenuation of

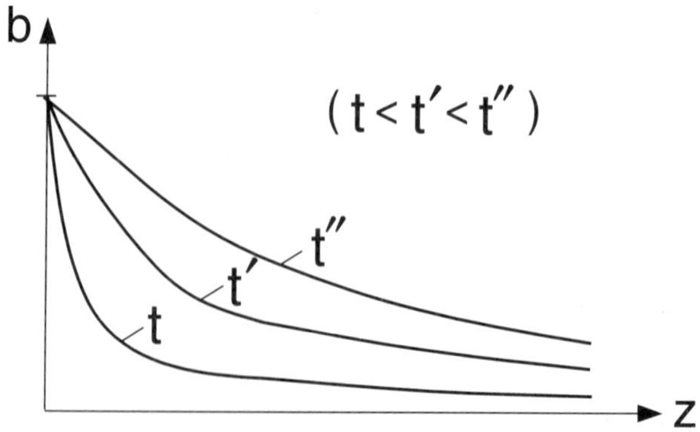

Fig. 1.13

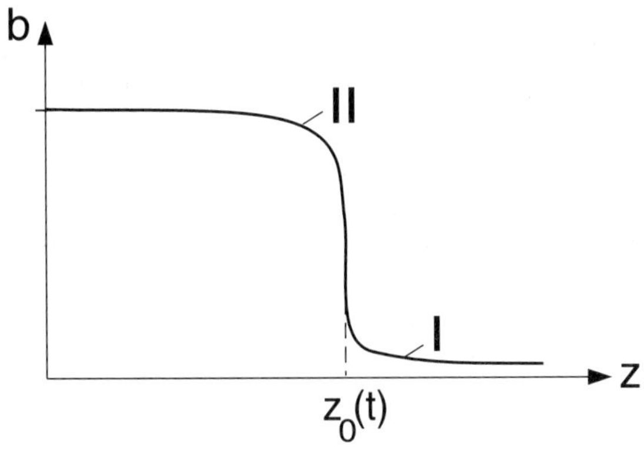

Fig. 1.14

the magnetic field at some finite distance $z_0(t)$. The actual profile of $b(z,t)$, schematically shown in Fig. 1.14, has a "tail" I of small values of magnetic flux density $b(z,t)$, which asymptotically approaches zero. This tail appears because in actual materials the above-mentioned steep growth of μ_d is moderated for small values of h. This tail is usually of no practical significance and can be neglected. As a result, the zero front velocity attains the physical meaning of the velocity of the inward progress of the bulk part (part II) of the magnetic flux density profile.

The exact analytical solution (1.123) of the initial boundary value problem (1.99)–(1.101) has been found for a very specific boundary condition (1.122). However, it is clear from the previous discussion that the phenomenon of finite velocity of zero front is not related to a specific boundary condition but rather it is caused by the idealization introduced by the "power" approximation (1.92). Thus, it can be concluded that the finite velocity of zero front exists for any boundary condition (1.101), that is, for any p. The specific nature of the boundary condition (1.123) reveals itself in the constant velocity of the zero front. For other boundary conditions of the type (1.101), this velocity will vary with time.

1.4 SOLUTION OF THE MODEL PROBLEM (SELF-SIMILAR SOLUTIONS)

In the previous section, we found the exact analytical solution (1.123) of the nonlinear diffusion Eq. (1.99) by reducing this partial differential equation to the ordinary differential Eq. (1.107). This reduction was performed for the particular value of p, namely $p = \frac{1}{n-1}$. It turns out that the model problem (1.99)–(1.101) can be reduced to the boundary value problem for a certain ordinary differential equation for any value of p in (1.101). This can be achieved through dimensional analysis.

The notation [a] will be used for the dimension of a physical quantity a. From formulas (1.99) and (1.101), we find the following dimensional relations:

$$\frac{[b]^n}{[z]^2} = [k][\sigma]\frac{[b]}{[t]}, \tag{1.128}$$

$$[b] = [c][t]^p. \tag{1.129}$$

By substituting relation (1.129) into formula (1.128), we obtain

$$\frac{[c]^n[t]^{np}}{[z]^2} = [k][\sigma]\frac{[c][t]^p}{[t]}. \tag{1.130}$$

From the last expression we conclude that the ratio

$$\frac{[z]^2[k][\sigma]}{[c]^{n-1}[t]^{p(n-1)+1}} \tag{1.131}$$

is dimensionless. By using this observation, we introduce the following dimensionless variable:

$$\xi = \frac{z}{\sqrt{k^{-1}\sigma^{-1}c^{n-1}t^m}}, \tag{1.132}$$

where

$$m = \frac{p(n-1)+1}{2}. \tag{1.133}$$

By using this variable ξ, we shall look for the solution of the initial boundary value problem (1.99)–(1.101) in the following form:

$$b(z,t) = ct^p f(\xi), \tag{1.134}$$

where $f(\xi)$ is some dimensionless function of variable ξ.

The main idea that we shall pursue here is to reduce the initial boundary value problem (1.99)–(1.101) to the boundary value problem for some ordinary differential equation with respect to function $f(\xi)$. To this end, we shall first evaluate the derivatives involved in the partial differential Eq. (1.99).

From (1.134) and (1.132), we derive

$$\frac{\partial b}{\partial t} = pct^{p-1}f(\xi) + ct^p\frac{df}{d\xi}\frac{d\xi}{dt}, \tag{1.135}$$

$$\frac{\partial b}{\partial t} = pct^{p-1}f(\xi) - ct^p\frac{mz}{\sqrt{k^{-1}\sigma^{-1}c^{n-1}}t^{m+1}}\frac{df}{d\xi}, \tag{1.136}$$

which leads to

$$\frac{\partial b}{\partial t} = ct^{p-1}\left(pf(\xi) - m\xi\frac{df}{d\xi}\right). \tag{1.137}$$

Next, we obtain

$$b^n(z,t) = c^n t^{pn} f^n(\xi), \tag{1.138}$$

which yields

$$\frac{\partial^2 b^n}{\partial z^2} = c^n t^{pn}\frac{d^2 f^n}{d\xi^2}\left(\frac{d\xi}{dz}\right)^2. \tag{1.139}$$

By invoking the formula (1.132), we find

$$\left(\frac{d\xi}{dz}\right)^2 = \frac{k\sigma}{c^{n-1}t^{p(n-1)+1}},$$

which leads to

$$\frac{\partial^2 b^n}{\partial z^2} = k\sigma c t^{p-1}\frac{d^2 f^n}{d\xi^2}. \tag{1.140}$$

By substituting formulas (1.137) and (1.140) in the Eq. (1.99), we arrive at the following differential equation for $f(\xi)$

$$\frac{d^2 f^n}{d\xi^2} + m\xi\frac{df}{d\xi} - pf = 0. \tag{1.141}$$

By using expressions (1.132) and (1.134), we can easily conclude that $b(z,t)$ given by (1.134) will satisfy the initial and boundary conditions (1.100) and (1.101), respectively, if the function $f(\xi)$ satisfies the boundary conditions:

$$f(0) = 1, \tag{1.142}$$

$$f(\infty) = 0. \tag{1.143}$$

Thus, the initial boundary value problem (1.99)–(1.101) is reduced to the boundary value problem (1.141)–(1.143) for the nonlinear ordinary differential Eq. (1.141). This nonlinear equation has some interesting properties. For instance, we shall prove that if $f(\xi)$ is a solution to Eq. (1.141), then the function

$$F(\xi) = \lambda^{-\frac{2}{n-1}} f(\lambda\xi) \tag{1.144}$$

is also a solution to the same equation for any constant λ.

To establish that, we introduce a new variable:

$$\zeta = \lambda\xi, \qquad \xi = \frac{\zeta}{\lambda}. \tag{1.145}$$

Then, we find

$$\frac{dF}{d\xi} = \lambda^{-\frac{2}{n-1}} \frac{df}{d\zeta} \cdot \frac{d\zeta}{d\xi} = \lambda^{\frac{n-3}{n-1}} \frac{df}{d\zeta}. \tag{1.146}$$

Similarly,

$$\frac{d^2 F^n}{d\xi^2} = \lambda^{-\frac{2n}{n-1}} \frac{d^2 f^n}{d\zeta^2} \cdot \left(\frac{d\zeta}{d\xi}\right)^2 = \lambda^{-\frac{2}{n-1}} \frac{d^2 f^n}{d\zeta^2}. \tag{1.147}$$

By using the last two formulas, we derive

$$\begin{aligned} &\frac{d^2 F^n}{d\xi^2} + m\xi \frac{dF}{d\xi} - pF = \\ &\lambda^{-\frac{2}{n-1}} \frac{d^2 f^n}{d\zeta^2} + \lambda^{\frac{n-3}{n-1}} m \frac{\zeta}{\lambda} \frac{df}{d\zeta} - \lambda^{-\frac{2}{n-1}} p f(\zeta) = \\ &\lambda^{-\frac{2}{n-1}} \left(\frac{d^2 f^n}{d\zeta^2} + m\zeta \frac{df}{d\zeta} - pf \right) = 0. \end{aligned} \tag{1.148}$$

Thus, we proved that the function $F(\xi)$ given by expression (1.144) is indeed a solution to Eq. (1.141). This fact can be utilized as follows. Suppose we can find some solution $f(\xi)$ to Eq. (1.141) that satisfies the boundary condition (1.143), but does not satisfy the boundary condition (1.142):

$$f(0) = q \neq 1 \tag{1.149}$$

Then, by using

$$\lambda = q^{\frac{n-1}{2}}, \tag{1.150}$$

we observe that the function

$$F(\xi) = \frac{1}{q} f(q^{\frac{n-1}{2}} \xi) \tag{1.151}$$

will be the solution to Eq. (1.141), that satisfies the boundary condition (1.143) and, in addition,

$$F(0) = \frac{1}{q} f(0) = 1. \tag{1.152}$$

This demonstrates that we can first find a solution to Eq. (1.141) satisfying the boundary condition (1.143), and then, by using the transformation (1.144), we can always map this solution into the solution that satisfies the boundary condition (1.142) as well. We shall illustrate this fact by deriving the solution (1.123) from the previous section by using formulas (1.132), (1.134), and Eq. (1.141). In the case of the solution (1.123), we have

$$p = \frac{1}{n-1} \quad \text{and} \quad m = 1, \tag{1.153}$$

which leads to the following form of Eq. (1.141):

$$\frac{d^2 f^n}{d\xi^2} + \xi \frac{df}{d\xi} - \frac{1}{n-1} f = 0. \tag{1.154}$$

The last equation can be rewritten as follows:

$$\frac{d^2 f^n}{d\xi^2} + \frac{df}{d\xi} = (1-\xi)\frac{df}{d\xi} + \frac{1}{n-1} f. \tag{1.155}$$

We shall look for the solution of the above equation in the form:

$$f(\xi) = \begin{cases} a(1-\xi)^\alpha, & \text{if } 0 \le \xi \le 1, \\ 0, & \text{if } \xi \ge 1, \end{cases} \tag{1.156}$$

where a and α are some unknown constant and exponent, while the second line of the right-hand side of formula (1.156) underlines the fact that a solution in the form (1.134) has a finite velocity of zero front.

By substituting (1.156) into (1.155), we obtain

$$\begin{aligned} &\alpha n(\alpha n - 1)a^n(1-\xi)^{\alpha n - 2} - \alpha a(1-\xi)^{\alpha-1} = \\ &-\alpha a(1-\xi)^\alpha + \frac{a}{n-1}(1-\xi)^\alpha. \end{aligned} \tag{1.157}$$

It can be easily observed that the right-hand side and the left-hand side of expression (1.157) will be equal to zero if α and a are given by the following formulas:

$$\alpha = \frac{1}{n-1}, \tag{1.158}$$

$$a = \left(\frac{n-1}{n}\right)^{\frac{1}{n-1}}. \tag{1.159}$$

Thus, we have found a solution to Eq. (1.141), which satisfies the boundary condition (1.143). However, this solution does not satisfy the boundary condition (1.142). This can be corrected by using the transformation (1.144) with $\lambda = a^{\frac{n-1}{2}}$. This leads to the following solution:

$$f(\xi) = \begin{cases} \left(1 - \sqrt{\frac{n-1}{n}}\xi\right)^{\frac{1}{n-1}}, & \text{if } 0 \le \xi \le \sqrt{\frac{n}{n-1}}, \\ 0, & \text{if } \xi \ge \sqrt{\frac{n}{n-1}}, \end{cases} \tag{1.160}$$

which satisfies both boundary conditions (1.142) and (1.143).

By substituting expression (1.160) into formula (1.134) and by taking into account (1.132) and (1.153), we end up with

$$b(z,t) = \begin{cases} ct^{\frac{1}{n-1}} \left(1 - \frac{z}{vt}\right)^{\frac{1}{n-1}}, & \text{if } 0 \le z \le vt, \\ 0, & \text{if } z \ge vt, \end{cases} \tag{1.161}$$

where

$$v = \sqrt{\frac{n}{(n-1)k\sigma}} c^{\frac{n-1}{2}}. \tag{1.162}$$

The last two expressions are identical to the previously found solution (1.123) and (1.125).

Next, we proceed to the analytical solution of the boundary value problem (1.141)–(1.142) with respect to the function $f(\xi)$ for arbitrary positive value of p (that is for arbitrary value of m). We shall look for this solution in the form:

$$f(\xi) = \begin{cases} a(1-\xi)^{\alpha}[1 + a_1(1-\xi) + a_2(1-\xi)^2 + \ldots], & \text{if } 0 \le \xi \le 1, \\ 0, & \text{if } \xi \ge 1. \end{cases} \tag{1.163}$$

Here, as before, the second line of the right-hand side guarantees that a solution in the form (1.134) has a finite velocity of zero front as it must.

The factor $a(1-\xi)^{\alpha}$ in (1.163) describes the asymptotic behavior of $f(\xi)$ near its zero value. It can be conjectured on the physical grounds that this asymptotic behavior should be the same as in the case of the exact solution (1.160), that is, $\alpha = \frac{1}{n-1}$. This is because the asymptotic behavior of $b(z,t)$ near its zero value should not depend on a particular boundary condition (a particular value of p), but rather it should be determined by the properties of media, that is, by the rate of increase of differential magnetic permeability μ_d as $b(z,t)$ approached zero. This fact can be supported by the following mathematical arguments. Suppose that

$$f(\xi) \sim a(1-\xi)^{\alpha}, \tag{1.164}$$

where

$$0 < \alpha < 1, \tag{1.165}$$

and sign "$\sim$" means asymptotic equality, that is the equality up to terms of higher order of smallness with respect to $(1-\xi)$.

We transform Eq. (1.141) to the form:

$$\frac{d^2 f^n}{d\xi^2} + m\frac{df}{d\xi} - m(1-\xi)\frac{df}{d\xi} - pf = 0. \tag{1.166}$$

By substituting expression (1.164) into the last equation, we obtain

$$\alpha n(\alpha n-1)a^n(1-\xi)^{\alpha n-2} - ma\alpha(1-\xi)^{\alpha-1} + \\ ma\alpha(1-\xi)^{\alpha} - pa(1-\xi)^{\alpha} \sim 0. \tag{1.167}$$

According to formula (1.165), the second term in the last expression goes to infinity as ζ approaches 1. Thus, the asymptotic equality can only be valid if the first two terms are cancelled out. This leads to

$$\alpha n - 2 = \alpha - 1, \tag{1.168}$$

which yields

$$\alpha = \frac{1}{n-1}, \tag{1.169}$$

and

$$\alpha n(\alpha n-1)a^n = ma\alpha. \tag{1.170}$$

By substituting expression (1.169) into (1.170), we find:

$$a = [\frac{m(n-1)}{n}]^{\frac{1}{n-1}}. \tag{1.171}$$

By using the established formula (1.169), we can rewrite expression (1.163) as follows:

$$f(\xi) = \\ \begin{cases} a(1-\xi)^{\frac{1}{n-1}}[1+a_1(1-\xi)+a_2(1-\xi)^2+\ldots], & \text{if } 0 \le \xi \le 1, \\ 0, & \text{if } \xi \ge 1. \end{cases} \tag{1.172}$$

To find the unknown coefficients $a_1, a_2, \ldots,$ we shall first evaluate the derivatives of f for $0 \le \xi \le 1$:

$$\frac{df}{d\xi} = \frac{-a}{n-1}(1-\xi)^{\frac{2-n}{n-1}}[1+a_1(1-\xi)+a_2(1-\xi)^2+\ldots] \\ + a(1-\xi)^{\frac{1}{n-1}}[-a_1-2a_2(1-\xi)-\ldots], \tag{1.173}$$

$$\frac{d^2 f^n}{d\xi^2} = \frac{na^n}{(n-1)^2}(1-\xi)^{\frac{2-n}{n-1}}[1+a_1(1-\xi)+a_2(1-\xi)^2+\ldots] \\ -2\frac{n^2}{n-1}a^n(1-\xi)^{\frac{1}{n-1}}[1+a_1(1-\xi)+a_2(1-\xi)^2+\ldots]^{n-1} \\ \times[-a_1-2a_2(1-\xi)-\ldots]+ \\ n(n-1)a^n(1-\xi)^{\frac{n}{n-1}}[1+a_1(1-\xi)+a_2(1-\xi)^2+\ldots]^{n-2} \\ \times[-a_1-2a_2(1-\xi)-\ldots]^2+ \\ na^n(1-\xi)^{\frac{n}{n-1}}[1+a_1(1-\xi)+a_2(1-\xi)^2+\ldots]^{n-1} \\ \times[2a_2+\ldots]. \tag{1.174}$$

By substituting formulas (1.172), (1.173), and (1.174) into Eq. (1.141) written in the form (1.166), we find:

$$
\begin{aligned}
&\frac{na^n}{(n-1)^2}(1-\xi)^{\frac{2-n}{n-1}}[1+a_1(1-\xi)+a_2(1-\xi)^2+\ldots]^n- \\
&\frac{2n^2a^n}{n-1}(1-\xi)^{\frac{1}{n-1}}[1+a_1(1-\xi)+a_2(1-\xi)^2+\ldots]^{n-1} \\
&\times[-a_1-2a_2(1-\xi)-\ldots]+ \\
&n(n-1)a^n(1-\xi)^{\frac{n}{n-1}}[1+a_1(1-\xi)+a_2(1-\xi)^2+\ldots]^{n-2} \\
&\times[-a_1-2a_2(1-\xi)-\ldots]^2+ \\
&na^n(1-\xi)^{\frac{n}{n-1}}[1+a_1(1-\xi)+a_2(1-\xi)^2+\ldots]^{n-1} \\
&\times[2a_2+\ldots]- \\
&\frac{ma}{n-1}(1-\xi)^{\frac{2-n}{n-1}}[1+a_1(1-\xi)+a_2(1-\xi)^2+\ldots]+ \\
&ma(1-\xi)^{\frac{1}{n-1}}[-a_1-2a_2(1-\xi)-\ldots]+ \\
&\frac{ma}{n-1}(1-\xi)^{\frac{1}{n-1}}[1+a_1(1-\xi)+a_2(1-\xi)^2+\ldots]- \\
&ma(1-\xi)^{\frac{n}{n-1}}[-a_1-2a_2(1-\xi)-\ldots]- \\
&pa(1-\xi)^{\frac{1}{n-1}}[1+a_1(1-\xi)+a_2(1-\xi)^2+\ldots]=0.
\end{aligned}
\tag{1.175}
$$

By collecting in formula (1.175) all terms with $(1-\xi)^{\frac{2-n}{n-1}}$ and by equating them to zero, we find:

$$\frac{na^n}{(n-1)^2}-\frac{ma}{n-1}=0, \tag{1.176}$$

which yields:

$$a=\left[\frac{m(n-1)}{n}\right]^{\frac{1}{n-1}}. \tag{1.177}$$

This is the same expression as (1.171), which has been previously derived from asymptotic analysis. Next, by collecting in formula (1.175) all terms with $(1-\xi)^{\frac{1}{n-1}}$ and by equating them to zero, we obtain:

$$\frac{n^2a^n}{(n-1)^2}a_1+\frac{2n^2a^n}{n-1}a_1-\frac{ma}{n-1}a_1-maa_1+\frac{ma}{n-1}-pa=0. \tag{1.178}$$

By dividing the last expression by a, we find:

$$\frac{n^2a^{n-1}}{(n-1)^2}a_1+\frac{2n^2a^{n-1}}{n-1}a_1-\frac{ma_1}{n-1}-ma_1+\frac{m}{n-1}-p=0. \tag{1.179}$$

By using formula (1.177) in (1.179), we arrive at:

$$a_1\left(\frac{mn}{n-1}+2nm-\frac{m}{n-1}-m\right)=\frac{p(n-1)-m}{n-1}, \tag{1.180}$$

which leads to the following final expression:

$$a_1=\frac{p(n-1)-m}{2mn(n-1)}. \tag{1.181}$$

By collecting in formula (1.175) all terms with $(1-\xi)^{\frac{n}{n-1}}$ and by equating them to zero, after somewhat lengthy and tedious calculations (which are omitted here) we derive:

$$a_2=-a_1\frac{1+0.5a_1[(2n-1)(3n-2)-4n]}{3(2n-1)}. \tag{1.182}$$

The above calculations can be continued and higher-order terms in (1.172) can be determined.

In the particular case of $p=\frac{1}{n-1}$, from formulas (1.133), (1.177), (1.181), and (1.182) we find:

$$a=\left(\frac{n-1}{n}\right)^{\frac{1}{n-1}}, \quad a_1=0, \quad a_2=0, \quad \ldots, \tag{1.183}$$

which coincides with the previously derived analytical solution (1.155), (1.158), and (1.159).

Solution (1.172) satisfies the boundary condition (1.143), however, it does not satisfy the boundary condition (1.142). This can be corrected by using transformation (1.144) with

$$\lambda=[a(1+a_1+a_2+\ldots)]^{\frac{n-1}{2}}. \tag{1.184}$$

This leads to the following solution of the boundary value problem (1.141)–(1.143):

$$f(\xi)=\begin{cases}(1-\lambda\xi)^{\frac{1}{n-1}}\frac{1+a_1(1-\lambda\xi)+a_2(1-\lambda\xi)^2+\ldots}{1+a_1+a_2+\ldots}, & \text{if } 0\le\lambda\xi\le 1,\\ 0, & \text{if } \lambda\xi\ge 1.\end{cases} \tag{1.185}$$

It is clear from (1.181) and (1.182) that a_1 and a_2 depend on n and p. However, it is possible to derive inequalities for these coefficients expressed only in terms of n. To do this, we invoke the definition (1.133) of m and use it in the expression (1.181):

$$a_1 = \frac{p(n-1)-1}{2[p(n-1)+1]n(n-1)} = \frac{1}{2n(n-1)} - \frac{1}{[p(n-1)+1]n(n-1)}. \tag{1.186}$$

From the last formula we derive:

$$|a_1| \le \frac{1}{2n(n-1)}. \tag{1.187}$$

By using inequality (1.187) in formula (1.182), we obtain:

$$\begin{aligned} |a_2| &\le \left|\frac{a_1}{3(2n-1)}\right| + \left|\frac{a_1^2[(2n-1)(3n-2)-4n]}{6(2n-1)}\right| \le \\ &\le \frac{1}{6n(n-1)(2n-1)} + \frac{1}{8n^2(n-1)^2}\left[\frac{(6n-2)(n-1)}{6n-3}\right], \end{aligned} \tag{1.188}$$

which leads to the following inequality:

$$|a_2| \le \frac{1}{6(n-1)(2n-1)n} + \frac{1}{8n^2(n-1)}. \tag{1.189}$$

It has been stressed in the previous section that the exponent n in the "power" approximation (1.96) is usually larger than 7. By using this fact, from inequalities (1.187) and (1.189) we derive:

$$|a_1| < 0.012, \quad |a_2| < 0.00075. \tag{1.190}$$

The above estimates suggest the following simplification of formula (1.184) and solution (1.185):

$$\lambda = a^{\frac{n-1}{2}} = \sqrt{\frac{m(n-1)}{n}}, \tag{1.191}$$

$$f(\xi) = \begin{cases} \left(1 - \sqrt{\frac{m(n-1)}{n}}\,\xi\right)^{\frac{1}{n-1}}, & \text{if } 0 \le \xi \le \sqrt{\frac{n}{m(n-1)}}, \\ 0, & \text{if } \xi \ge \sqrt{\frac{n}{m(n-1}}. \end{cases} \tag{1.192}$$

By substituting the last expression into formula (1.134) and taking into account definition (1.132) of ξ, we end up with the following analytical solution for the model problem:

$$b(z,t) = \begin{cases} ct^p\left(1 - \frac{z}{dt^m}\right)^{\frac{1}{n-1}}, & \text{if } 0 \le z \le dt^m, \\ 0, & \text{if } z \ge dt^m, \end{cases} \tag{1.193}$$

where

$$d = \left(\frac{nc^{n-1}}{k\sigma m(n-1)} \right)^{\frac{1}{2}}. \tag{1.194}$$

In the conclusion of this section, it is relevant to point out that solutions of the model problem exhibit an interesting property. It is clear from formulas (1.132) and (1.134) that z-profiles of magnetic flux density at various instants of time can be obtained from one another by dilation (or contraction) along b- and z-axes. In other words, those z-profiles remain similar to one another. This explains why solutions of the type (1.134) and (1.132) are called **self-similar** solutions. The property of self-similarity is closely related to the choice of "power" approximation (1.96) and boundary conditions (1.101) that makes the problem susceptible to the dimensional analysis. The intrinsic property of the self-similar solutions is that they are dimensionally deficient. This property allowed us to reduce the nonlinear partial differential Eq. (1.99) to the ordinary differential Eq. (1.141). It is also clear that the self-similar solutions are invariant under certain scaling transformations. For this reason, they are often called group-invariant solutions.

The self-similar solutions discussed in this section have been derived by using dimensional analysis. For this reason, they are regarded as self-similar solutions of the first kind. There are, however, self-similar solutions that cannot be obtained by using dimensional analysis alone. These solutions contain additional parameters, which are called anomalous dimensions. These are self-similar solutions of the second kind, and they are physically significant because they describe intermediate asymptotics [3]. The interesting treatment of these solutions by using the machinery of the renormalization group is presented in the book [7].

The self-similar solutions for nonlinear diffusion Eq. (1.99) were first studied by Ya. Zeldovich and A. Kompaneyets [21] for the radiative heat conduction problem and by G. Barenblatt [4] for problems of gas flow in porous media. The discussion presented in this section closely parallels in some respects the work of G. Barenblatt.

1.5 GENERALIZATION OF SELF-SIMILAR SOLUTIONS

A brief examination of the obtained self-similar solutions (1.193) leads to the following observation: Profiles of magnetic flux density $b(z,t)$ as

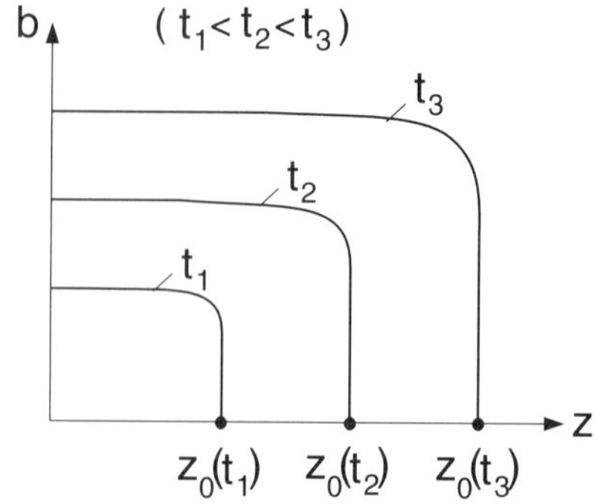

Fig. 1.15

functions of z remain approximately the same (see Fig. 1.15 as well as formula (1.193)) for wide-ranging variations of the boundary conditions (1.101) (see Fig. 1.11). For typical values of $n(n \geq 7)$, those profiles are very close to rectangular ones. That insensitivity of self-similar solutions profiles to a particular boundary condition suggests that actual profiles of magnetic flux density will be close to rectangular ones for any monotonically increasing boundary conditions $b_0(t) = b(0,t)$. Thus, we arrive at the following generalization of self-similar solutions (1.193).

The actual profiles of magnetic flux density $b(z,t)$ are approximated by rectangular ones with the height equal to the instantaneous boundary value $b_0(t)$:

$$b(z,t) = \begin{cases} b_0(t), & \text{if } 0 \leq z \leq z_0(t), \\ 0, & \text{if } z \geq z_0(t). \end{cases} \tag{1.195}$$

This generalization is illustrated by Fig. 1.16.

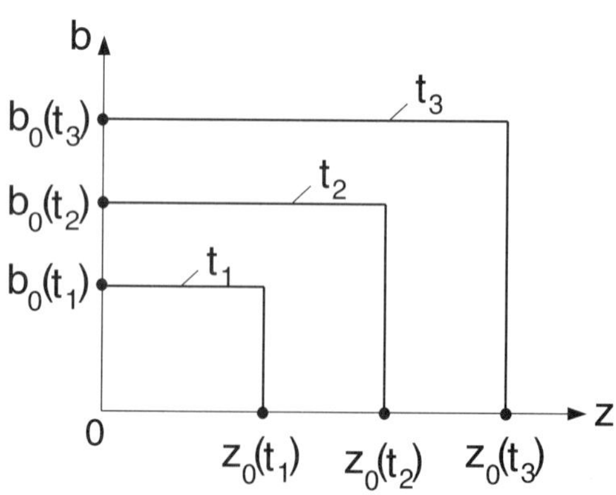

Fig. 1.16

We recall that rectangular profiles of magnetic flux density were encountered in Section 1.2 when we discussed nonlinear diffusion in media with abrupt magnetic transitions. For those transitions, rectangular profiles of magnetic flux density can be attributed to abrupt magnetic saturation. The self-similar solutions (1.193) found in the previous section show that b-profiles are close to rectangular ones even if media are not saturated. Rectangular-like shapes of b-profiles can be explained as follows. In the process of diffusion, magnetic field h is attenuated as z is increased. The attenuation of h results in the increase in magnetic permeability (defined as $\mu = \frac{b}{h} = kh^{\frac{1}{n}-1}$). This increase, at first, compensates for the decrease in h and leads to more or less "flat" values of b. When values of z are sufficiently close to z_0, the very fast attenuation of h cannot be compensated for by the increase in μ and this results in the precipitous drop in magnetic flux density b.

Next, we shall derive the expression for the zero front $z_0(t)$ of $b(z,t)$ in (1.195). To this end, we shall write the nonlinear diffusion Eq. (1.99) in the form

$$\frac{\partial^2 h}{\partial z^2} = \sigma \frac{\partial b}{\partial t}, \tag{1.196}$$

and split this equation into two first-order partial differential equations (compare with Section 1.2):

$$\frac{\partial w}{\partial z} = -\sigma b(z,t), \tag{1.197}$$

$$\frac{\partial h}{\partial z} = -\frac{\partial w}{\partial t}. \tag{1.198}$$

By using the rectangular profile approximation (1.195), Eq. (1.97) can be rewritten as follows:

$$\frac{\partial w}{\partial z} = \begin{cases} -\sigma b_0(t), & \text{if } 0 \le z \le z_0(t), \\ 0, & \text{if } z \ge z_0(t). \end{cases} \tag{1.199}$$

From the last formula we conclude that at every instant of time function $w(z,t)$ has a constant negative slope with respect to z for $0 \le z \le z_0(t)$ and the zero slope for $z \ge z_0(t)$. Thus, we have:

$$w(z,t) = \begin{cases} w(0,t)\left[1 - \frac{z}{z_0(t)}\right], & \text{if } 0 \le z \le z_0(t), \\ 0, & \text{if } z \ge z_0(t). \end{cases} \tag{1.200}$$

It is clear from relation (1.200) that the slope of $w(z,t)$ is equal to $-\frac{w(0,t)}{z_0(t)}$ for $z \le z_0(t)$. On the other hand, according to Eq. (1.199), the same slope is equal to $-\sigma b_0(t)$. Consequently,

$$\frac{w(0,t)}{z_0(t)} = \sigma b_0(t), \tag{1.201}$$

and

$$w(0,t) = \sigma b_0(t) z_0(t). \tag{1.202}$$

By substituting the last expression into formula (1.200), we obtain

$$w(z,t) = \begin{cases} \sigma b_0(t) z_0(t) - \sigma b_0(t) z, & \text{if } 0 \le z \le z_0(t) \\ 0, & \text{if } z \ge z_0(t). \end{cases} \tag{1.203}$$

By using relation (1.203) in Eq. (1.198), we end up with

$$\frac{\partial h}{\partial z} = \begin{cases} -\sigma \frac{d}{dt}[b_0(t) z_0(t)] + \sigma z \frac{db_0(t)}{dt}, & \text{if } 0 \le z \le z_0(t), \\ 0, & \text{if } z \ge z_0(t). \end{cases} \tag{1.204}$$

According to the second line of Eq. (1.204), we have:

$$h_0(t) = h(0,t) = -\int_0^{z_0(t)} \frac{\partial h}{\partial z} dz. \tag{1.205}$$

According to the first line of Eq. (1.204) and formula (1.205), we find:

$$h_0(t) = \sigma z_0(t) \frac{d}{dt}[b_0(t) z_0(t)] - \sigma \frac{z_0^2(t)}{2} \frac{db_0(t)}{dt}. \tag{1.206}$$

By using simple calculus, we can transform the last formula as follows:

$$h_0(t) = \sigma b_0(t) z_0(t) \frac{dz_0(t)}{dt} + \frac{\sigma z_0^2(t)}{2} \frac{db_0(t)}{dt}, \tag{1.207}$$

which, in turn, can be rewritten as follows:

$$h_0(t) = \sigma \frac{d}{dt} \left[\frac{b_0(t) z_0^2(t)}{2} \right]. \tag{1.208}$$

By integrating the last equation and by taking into account that $z_0(0) = 0$, we finally arrive at the following important formula:

$$z_0(t) = \left[\frac{2 \int_0^t h_0(\tau) d\tau}{\sigma b_0(t)} \right]^{\frac{1}{2}}. \tag{1.209}$$

The last formula can also be derived by using the first moment relation for nonlinear diffusion Eq. (1.196). Indeed, let us multiply Eq. (1.196) by z and integrate from 0 to $z_0(t)$:

$$\int_0^{z_0(t)} z \frac{\partial^2 h}{\partial z^2} dz = \sigma \int_0^{z_0(t)} z \frac{\partial b}{\partial t} dz. \tag{1.210}$$

By integrating twice by parts in the left-hand side of Eq. (1.210) and by taking into account that

$$\frac{\partial h}{\partial z}(z_0(t), t) = 0, \qquad h(z_0(t), t) = 0, \tag{1.211}$$

we obtain

$$\int_0^{z_0(t)} z\frac{\partial^2 h}{\partial z^2}dz = h(0, t) = h_0(t). \tag{1.212}$$

We remark here that the first equality in (1.211) comes from the fact that the electric field is equal to zero at $z = z_0(t)$ and $\frac{\partial h}{\partial z} = \frac{\partial H}{\partial z} = -\sigma E$.

By using the formula of differentiation of integral dependent on parameter, we obtain

$$\frac{d}{dt}\int_0^{z_0(t)} zb(z, t)dz = \int_0^{z_0(t)} z\frac{\partial b}{\partial t}dz + z_0(t)b(z_0(t), t)\frac{dz_0(t)}{dt}. \tag{1.213}$$

Since

$$b(z_0(t), t) = 0, \tag{1.214}$$

from formula (1.213), we derive

$$\int_0^{z_0(t)} z\frac{\partial b}{\partial t}dz = \frac{d}{dt}\int_0^{z_0(t)} zb(z, t)dz. \tag{1.215}$$

By substituting expressions (1.212) and (1.215) into formula (1.210), we arrive at the following first moment equation:

$$\int_0^t h_0(\tau)d\tau = \sigma\int_0^{z_0(t)} zb(z, t)dz. \tag{1.216}$$

By using the rectangular profile approximation (1.195) in formula (1.216), we obtain

$$\int_0^t h_0(\tau)d\tau = \sigma b_0(t)\frac{z_0^2(t)}{2}, \tag{1.217}$$

which again leads to formula (1.209).

The "rectangular profile" approximation just discussed is very suitable for the derivation of time periodic (steady state) solutions of nonlinear diffusion problems. Consider periodic time variations of magnetic field $H_0(t)$ at the boundary of magnetically nonlinear conducting half-space. Suppose that at time t_0 initial condition $B(z, t_0) = -B_m$ is in effect. Furthermore, suppose that magnetic field $H_0(t)$ is increased from H_c to H_m during the time interval $t_0 \leq t \leq t_m$, then it is decreased from H_m to H_c during the time interval $t_m \le t \le \tilde{t}_0$, and finally it is decreased from H_c to $-H_c$ during the time interval $\tilde{t}_0 \le t \le \frac{T}{2} + t_0$, where T is a period of $H_0(t)$. As the magnetic field $H_0(t)$ is increased from H_c to H_m, the rectangular profile of magnetic flux density is formed and it moves inside the conducting media. The front $z_0(t)$ of this profile can be found by using formula (1.209), which can be rewritten in terms of $H_0(t)$ and $B_0(t)$ as follows (see expression (1.95)):

$$z_0(t) = \left[\frac{2\int_{t_0}^{t} [H_0(\tau) - H_c]d\tau}{\sigma(B_0(t) + B_m)} \right]^{\frac{1}{2}}. \tag{1.218}$$

When the magnetic field at the boundary reaches the value of H_m, the height of the rectangular profile becomes equal to B_m. As the magnetic field at the boundary is decreased from H_m to H_c, the inward progress of the rectangular profile is continued and its height remains the same and equal to B_m. The latter is in accordance with the "flat-power" approximation of hysteresis loops (see Fig. 1.9). The front, $z_0(t)$, of the rectangular profile can now be found by replacing $B_0(t)$ in formula (1.218) by B_m, which leads to

$$z_0(t) = \left[\frac{\int_{t_0}^{t} [H_0(\tau) - H_c]d\tau}{\sigma B_m} \right]^{\frac{1}{2}}. \tag{1.219}$$

As the magnetic field at the boundary is further reduced from H_c to $-H_c$, nothing happens. This means that the rectangular profile of magnetic flux density remains still because induced eddy currents and electric fields are equal to zero.

As the magnetic field at the boundary is reduced from $-H_c$ to $-H_m$ and then increased from $-H_m$ to H_c during the time interval $t_0 + \frac{T}{2} \le t \le t_0 + T$, the rectangular profile of "negative" polarity is formed and it moves inside the conducting media. Its inward progress is fully analogous to the progress of the rectangular profile of "positive" polarity described above for the time interval $t_0 \le t \le t_0 + \frac{T}{2}$. The front, $z_0(t)$, of the rectangular profile of "negative" polarity can be determined by using the formula

$$z_0(t) = \left[\frac{2\int_{t_0+\frac{T}{2}}^{t} [H_0(\tau) + H_c]d\tau}{\sigma(B_0(t) - B_m)} \right]^{\frac{1}{2}}. \tag{1.220}$$

During subsequent cycles, the situation repeats itself.

In formulas (1.218)–(1.220) for the front $z_0(t)$, magnetic field $H_0(t)$

and magnetic flux density $B_0(t)$ at the boundary are related by the "flat-power" approximation of the hysteresis loop. Thus, if $H_0(t)$ is known, then, by using formulas (1.91)-(1.94), we can find $B_0(t)$, which, in turn, can be used for calculations of $z_0(t)$.

The rectangular profile approximation can be further extended to make it directly applicable to actual hysteresis loops of the type shown in Fig. 1.9. In this extension, it is assumed that as the magnetic field at the boundary is increased from H_c to H_m, the rectangular profile of magnetic flux density is formed and moves inside the media. This assumption is supported by the derived self-similar solutions and their "rectangular profile" approximation. As the magnetic field at the boundary is decreased from H_m to $-H_c$, it is assumed that the profile of magnetic flux density retains its rectangular shape as well as its inward progress (see Fig. 1.17a). That assumption is justified by the fact that the magnetic flux density varies slightly as the magnetic field varies from H_m to $-H_c$. This prevents appreciable deformations of magnetic flux density profiles. Actually, this profile deformation may even improve the resemblance of actual magnetic flux density profiles to rectangular ones. Indeed, when the magnetic field at the boundary is increased from H_c to H_m, the boundary values of magnetic flux density are larger than those within the media and a "flat" part of magnetic flux density profile exhibits some small "downward" slope. As the magnetic field at the boundary is decreased from H_m to $-H_c$, the magnetic flux density at the boundary is reduced faster than within the conducting media, and this may result in the flattening out of the above "downward" slope and in better resemblance of actual profiles to rectangular ones.

Diffusion of rectangular profiles of magnetic flux density of opposite polarity occurs in a similar way during the next half-period. This is shown in Fig. 1.17 b.

The front, $z_0(t)$, of the rectangular profiles can be determined by using formulas (1.218) and (1.220). However, in these formulas $H_0(\tau)$ and $B_0(\tau)$ are now related through the actual shapes of hysteresis loops rather than by their "flat-power" approximations. This is justified on the grounds that the derivation of formulas (1.218) and (1.220) was based on general nonlinear diffusion Eq. (1.196) and the "rectangular" profile assumption. This derivation did not use the "flat-power" approximation of hysteresis loops. The last approximation was instrumental in the derivation of self-similar solutions and, in this way, it paved the road for the notion of rectangular profiles of magnetic flux density. Now, the "flat-power" approximation of hysteresis loops can be passed into oblivion.

The described model of nonlinear diffusion implies that at every point of conducting media the magnetic field and magnetic flux density are related by the same hysteresis loop as at the boundary. This is a natural

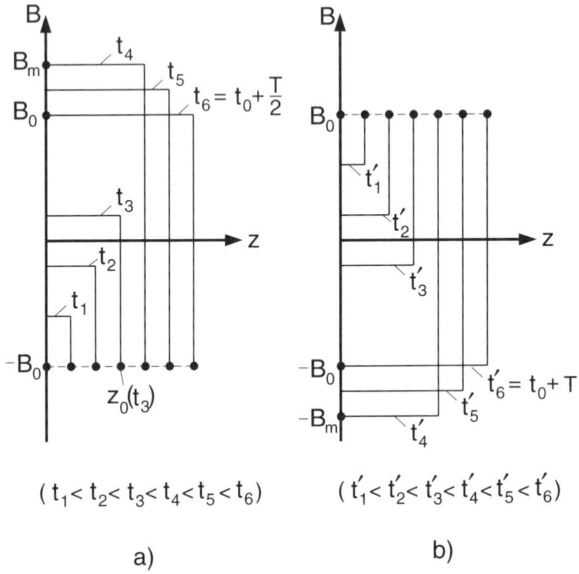

Fig. 1.17

consequence of rectangular profile approximation. In reality, at different points of conducting media the magnetic field and magnetic flux density are related by different hysteresis loops. However, because actual profiles of magnetic flux density are close to rectangular ones, these hysteresis loops are almost the same as the loop at the boundary. This is true everywhere within the conducting media except for a very narrow region where the precipitous drop in the magnetic flux density occurs.

Next, we shall derive the impedance-type relation between electric field $E_0(t)$ and magnetic field $H_0(t)$ at the boundary of conducting media. To this end, we consider the half-cycle $t_0 \le t \le t_0 + \frac{T}{2}$ and recall that

$$E_0(t) = E(0,t) = -\frac{1}{\sigma}\frac{\partial H}{\partial z}(0,t) = -\frac{1}{\sigma}\frac{\partial h}{\partial z}(0,t). \tag{1.221}$$

By using expression (1.204) in the last formula, we find:

$$E_0(t) = \frac{d}{dt}[b_0(t)z_0(t)]. \tag{1.222}$$

By taking into account that

$$b_0(t) = B_0(t) + B_c, \quad h_0(t) = H_0(t) - H_c, \tag{1.223}$$

and by using formula (1.209), from the last expression we derive

$$E_0(t) = \frac{d}{dt}\left[\frac{2}{\sigma}(B_0(t) + B_c)\int_{t_0}^{t}(H_0(\tau) - H_c)d\tau\right]^{\frac{1}{2}}. \tag{1.224}$$

Now, we introduce the following functions:

$$f_H(t) = \frac{H_0(t) - H_c}{H_m}, \tag{1.225}$$

$$f_B(t) = \frac{B_0(t) + B_c}{B_m}. \tag{1.226}$$

It is assumed that magnetic field $H_0(t)$ at the boundary is known. Then, by using the actual shape of the hysteresis loop, we can find $B_0(t)$. Next, by employing the last two formulas, the functions $f_H(t)$ and $f_B(t)$ can be figured out. Thus, it will be assumed in the subsequent discussion that functions $f_H(t)$ and $f_B(t)$ are known. By using the definitions (1.225) and (1.226) of these functions as well as the following definition of magnetic permeability μ_m:

$$\mu_m(H_m) = \frac{B_m}{H_m}, \tag{1.227}$$

the formula(1.224) can be transformed as follows:

$$E_0(t) = H_m\sqrt{\frac{\mu_m}{\sigma}}\frac{d}{dt}\left[2f_B(t)\int_{t_0}^{t}f_H(\tau)d\tau\right]^{\frac{1}{2}}. \tag{1.228}$$

Next, we shall scale the time t by using the formula

$$t' = \frac{2\pi t}{T} = \omega t, \tag{1.229}$$

where

$$\omega = \frac{2\pi}{T} \tag{1.230}$$

is the frequency of the fundamental (first) harmonic of $H_0(t)$, which is periodic (but may or may not be sinusoidal).

By using the scaling defined by (1.229), formula (1.228) can be further transformed as follows:

$$E_0(t') = H_m\sqrt{\frac{\omega\mu_m}{\sigma}}\frac{d}{dt'}\left[2\tilde{f}_B(t')\int_{t'_0}^{t'}\tilde{f}_H(\tau')d\tau'\right]^{\frac{1}{2}}, \tag{1.231}$$

where we use the following notations:

$$\tilde{f}_B(t') = f_B\left(\frac{t'}{\omega}\right), \qquad \tilde{f}_H(t') = f_H\left(\frac{t'}{\omega}\right). \tag{1.232}$$

By introducing the function

$$f_E(t') = \frac{d}{dt'}[2\tilde{f}_B(t')\int_{t'_0}^{t'} \tilde{f}_H(\tau')d\tau']^{\frac{1}{2}}, \tag{1.233}$$

we present the formula (1.231) as follows:

$$E_0(t') = H_m\sqrt{\frac{\omega\mu_m}{\sigma}}f_E(t'). \tag{1.234}$$

Formulas (1.234) and (1.233) constitute one of the most important results of this chapter. These formulas represent a nonlinear impedance-type relation between tangential components of electric and magnetic fields at the boundary of conducting media. This relation is nonlinear because μ_m is a function of H_m. Formulas (1.234) and (1.233) are very general in nature. They are valid for arbitrary periodic (not only sinusoidal) boundary condition $H_0(t)$ with only one restriction: the total cycle T can be subdivided into half-cycles of monotonic variations of $H_0(t)$. Another distinct feature of the above impedance-type relation is that it directly relates the time variations of $E_0(t')$ to the time variations of $H_0(t')$ as well as to actual shapes of hysteresis loops. The latter is accomplished through function $\tilde{f}_B(t')$.

To illustrate how formulas (1.234) and (1.233) can be used in calculations, consider a particular case when

$$H_0(t) = H_m \sin\omega t. \tag{1.235}$$

In this case, we will be interested in the first harmonic of $E_0(t)$, which can be written as follows:

$$E_0^{(1)}(t') = H_m\sqrt{\frac{\omega\mu_m}{\sigma}}(a\cos t' + b\sin t'), \tag{1.236}$$

where coefficients a and b are given by the formulas

$$a = \frac{2}{\pi}\int_{t'_0}^{\pi+t'_0} f_E(t')\cos t' dt', \tag{1.237}$$

$$b = \frac{2}{\pi}\int_{t'_0}^{\pi+t'_0} f_E(t')\sin t' dt'. \tag{1.238}$$

To simplify the calculations, we assume that $H_c \ll H_m$, which is typical in many applications. Because time t_0 is determined by the equation $H_0(t_0) = H_c$, from the last inequality and formula (1.235) we conclude that $t_0 \approx 0$. We shall also adopt the following power approximations:

$$B_0(t) + B_c = (B_m + B_c)\left(\frac{H_0(t) - H_c}{H_m - H_c}\right)^{\frac{1}{n}}, \quad (n > 1), \tag{1.239}$$

for the "steep" part of hysteresis loop traced when $H_0(t)$ is increased from H_c to H_m, and

$$B_m - B_0(t) = (B_m - B_c)\left(1 - \frac{H_0(t) + H_c}{H_m + H_c}\right)^{n_1}, \quad (n_1 > 1), \tag{1.240}$$

for the "flat" part of hysteresis loop traced when H is reduced from H_m to $-H_c$.

By using approximations (1.239) and (1.240) and the assumption $H_c \ll H_m$, from formulas (1.225), (1.226), (1.232), and (1.235), we derive:

$$\int_{t_0'}^{t'} \tilde{f}_H(\tau')d\tau' = \int_0^{t'} \sin\tau' d\tau' = 1 - \cos t' = 2(\sin\frac{t'}{2})^2, \tag{1.241}$$

$$\tilde{f}_B(t') = (1+\chi)(\sin t')^{\frac{1}{n}}, \quad \text{if } 0 \le t' \le \frac{\pi}{2}, \tag{1.242}$$

$$\tilde{f}_B(t') = (1+\chi) - (1-\chi)(1-\sin t')^{n_1}, \quad \text{if } \frac{\pi}{2} \le t' \le \pi, \tag{1.243}$$

where

$$\chi = \frac{B_c}{B_m} \tag{1.244}$$

is the "squareness" factor.

By substituting expression (1.241), (1.242), and (1.243) into formula (1.233) and then plugging the result of substitution into formulas (1.237) and (1.238), after integration by parts we derive:

$$b = -\frac{4}{\pi}\left[\int_0^{\frac{\pi}{2}} (1+\chi)^{\frac{1}{2}}(\sin t')^{\frac{1}{2n}} \sin\frac{t'}{2}\cos t' dt' + \int_{\frac{\pi}{2}}^{\pi} [(1+\chi) - (1-\chi)(1-\sin t')^{n_1}]^{\frac{1}{2}} \sin\frac{t'}{2}\cos t' dt'\right], \tag{1.245}$$

$$a = -\frac{4}{\pi}\sqrt{2\chi} + \frac{4}{\pi}\left[\int_0^{\frac{\pi}{2}} (1+\chi)^{\frac{1}{2}}(\sin t')^{\frac{1}{2n}} \sin\frac{t'}{2}\sin t' dt' + \int_{\frac{\pi}{2}}^{\pi} [(1+\chi) - (1-\chi)(1-\sin t')^{n_1}]^{\frac{1}{2}} \sin\frac{t'}{2}\sin t' dt'\right]. \tag{1.246}$$

Formula (1.236) can be represented in the phasor form

$$\hat{E}_0^{(1)} = \eta \hat{H}_0, \tag{1.247}$$

where the surface impedance η is given by

$$\eta = \sqrt{a^2 + b^2}\sqrt{\frac{\omega\mu_m}{\sigma}}e^{j\varphi}, \tag{1.248}$$

$$\tan\varphi = \frac{a}{b}. \tag{1.249}$$

By using formulas (1.245) and (1.246), we can compute $\tan\varphi$ and $\sqrt{a^2+b^2}$ for various values of χ, n, and n_1. In this way, we can evaluate to what extent the surface impedance depends on a particular shape of hysteresis loop. Computations show that the surface impedance is not very sensitive to variations of n and n_1, whereas variations of χ may appreciably affect the surface impedance, especially the value of $\tan\varphi$. The results of calculations of $\tan\varphi$ and $\sqrt{a^2+b^2}$ as functions of χ are shown in Fig. 1.18 a and 1.18 b, respectively. These calculations have been performed for $n = 10$ and $n_1 = 4$. It is apparent from Fig. 1.18 a that $\tan\varphi$ varies from 0.5 to 0.71. There is an extensive body of experimental data published in Russian literature [18], which suggests that $\tan\varphi$ varies between 0.5 and 0.69. Thus, our computational results are consistent with these experimental data.

The curves shown in Figs. 1.18 a and 1.18 b can be fairly accurately approximated by the following expressions:

$$\tan\varphi \approx 1.01 - 0.53\chi,$$

$$\sqrt{a^2+b^2} \approx 1.16 + 0.19\chi.$$

This leads to the following simple formula for the surface impedance:

$$\eta = (1.16 + 0.19\chi)\sqrt{\frac{\omega\mu_m}{\sigma}}e^{j\tan^{-1}(1.01-0.53\chi)}. \tag{1.250}$$

In the last formula, the dependence of the surface impedance η on the shape of the hysteresis loop is represented by the two parameters only: "squareness" of the loop χ and magnetic permeability μ_m.

Next, consider the penetration depth, δ, in the case of boundary condition (1.235). This depth can be defined as follows:

$$\delta = z_0(\frac{T}{2} + t_0).$$

By using formula (1.209) and expressions (1.223) and taking into account that $H_c \ll H_m, t_0 = 0$, and $B_0(\frac{T}{2} + t_0) = B_c$, we derive

$$\delta = \left[\frac{\int_0^{\frac{T}{2}} H_0(\tau) d\tau}{\sigma B_c} \right]^{\frac{1}{2}}. \tag{1.251}$$

By using boundary condition (1.235) and the definition (1.244) of the "squareness" factor χ, we obtain

$$\delta = \left[\frac{H_m \int_0^{\frac{T}{2}} \sin \omega\tau d\tau}{\chi \sigma B_m} \right]^{\frac{1}{2}}. \tag{1.252}$$

By performing integration in formula (1.252) and recalling that $\mu_m = \frac{B_m}{H_m}$, we finally arrive at

$$\delta = \sqrt{\frac{2}{\chi \omega \sigma \mu_m}}. \tag{1.253}$$

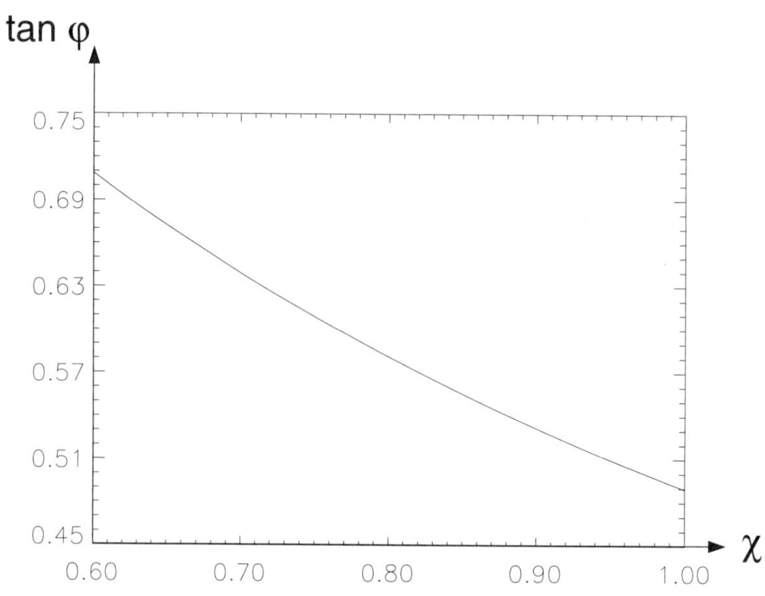

Fig. 1.18 a

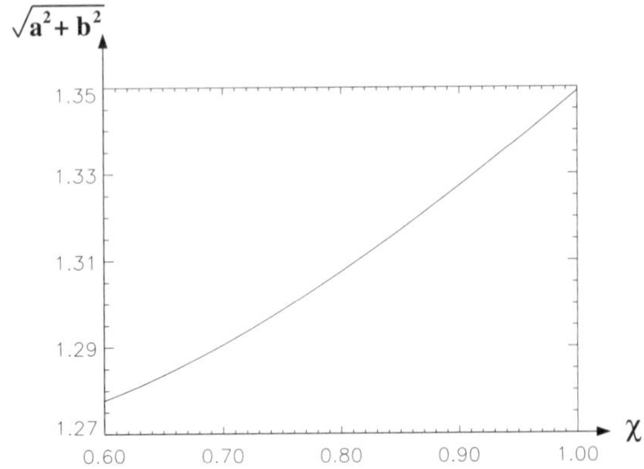

Fig. 1.18 b

As in the case of surface impedance, we can observe that the dependence of penetration depth δ on the shape of the hysteresis loop is represented by two parameters only: "squareness" of the loop χ and magnetic permeability μ_m. The latter is a nonlinear function of H_m, which is determined by a main magnetization curve that passes through vertices of symmetric hysteresis loops. This makes the penetration depth field dependent.

The previous analysis can be easily extended to the important case of magnetically nonlinear conducting laminations. Indeed, during initial stages of positive half-cycles, nonlinear diffusion of magnetic fields at both sides of laminations occurs in the same way as in the case of conducting half-space (see Fig. 1.19 a). The motion of front $z_0(t)$ can be calculated by using formula (1.209). At the instant of time t_Δ such that

$$z_0(t_\Delta) = \frac{\Delta}{2}, \tag{1.254}$$

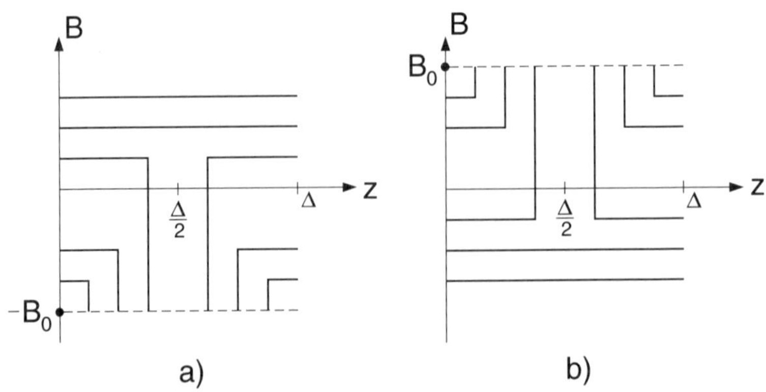

Fig. 1.19

two rectangular fronts are merged together and the distribution of magnetic flux density over a lamination cross-section is uniform. It remains this way until the commencement of negative half-cycle (see the same Fig. 1.19 a). During negative half-cycles, the situation reverses itself (see Fig. 1.19 b). It is apparent from this discussion that in the case of gradual magnetic transitions eddy currents are being induced all the time, whereas in the case of abrupt magnetic transitions eddy currents are limited in time and only induced during initial stages of half-cycles, that is, before the rectangular fronts merge.

1.6 STANDING MODE OF NONLINEAR DIFFUSION

We have seen that in the case of abrupt magnetic transitions magnetic flux density profiles are rectangular ones, while in the case of gradual magnetic transitions actual profiles of magnetic flux density are close to (and can be approximated by) rectangular ones. There is, however, an important difference between these two cases. In the case of abrupt transition, rectangular profiles have heights that are constant with time, whereas in the case of gradual transitions profile heights vary with time. This difference is best illuminated by a "standing" mode of nonlinear diffusion, which may occur in the case of gradual transitions and which is not possible in the case of abrupt transitions.

In the case of the "standing" mode of nonlinear diffusion, the magnetic flux density $b_0(t)$ at the boundary is increased with time, whereas the front, $z_0(t)$, does not change with time. To find how $b_0(t)$ should vary with time for that mode to be realized, we assume that by time t_1 a rectangular front is already formed and that

$$z_0(t) = z_0 = \text{const} \qquad \text{for } t \geq t_1. \tag{1.255}$$

By using the last condition in formula (1.209), we obtain

$$\int_0^t h_0(\tau)d\tau = \frac{z_0^2\sigma}{2}b_0(t) \qquad \text{for } t \geq t_1. \tag{1.256}$$

By differentiating both sides of the last expression with respect to time, we find

$$h_0(t) = \frac{z_0^2\sigma}{2}\frac{db_0}{dt}, \tag{1.257}$$

which leads to

$$\frac{db_0}{h_0} = \frac{2}{z_0^2\sigma}dt. \tag{1.258}$$

By integrating both sides of formula (1.258), we have

$$\int_{t_1}^{t} \frac{db_0}{h_0} = \frac{2}{z_0^2 \sigma}(t - t_1). \tag{1.259}$$

To evaluate the integral in formula (1.259), a relation between b_0 and h_0 should be employed. To be specific and for the sake of computational simplicity, a "power" approximation (see (1.96))

$$h_0 = \frac{b_0^n}{k} \tag{1.260}$$

is adopted below. By substituting relation (1.260) into formula (1.259), after integration we obtain

$$b_0^{1-n}(t) - b_0^{1-n}(t_1) = \frac{2(n-1)(t_1 - t)}{k\sigma z_0^2}. \tag{1.261}$$

The last formula leads to the following expression:

$$b_0(t) = \frac{1}{\left[b_0^{1-n}(t_1) + \frac{2(n-1)(t_1-t)}{k\sigma z_0^2}\right]^{\frac{1}{n-1}}}, \tag{1.262}$$

which can be further transformed into the form

$$b_0(t) = \frac{c}{(t_0 - t)^{\frac{1}{n-1}}}. \tag{1.263}$$

Here, we have adopted the following notations:

$$c = \left(\frac{k\sigma z_0^2}{2(n-1)}\right)^{\frac{1}{n-1}}, \tag{1.264}$$

$$t_0 = \frac{k\sigma z_0^2 b_0^{1-n}(t_1)}{2(n-1)} + t_1. \tag{1.265}$$

Thus, we have established that, if the magnetic flux density on the boundary of conducting half-space varies with time according to expressions (1.263)–(1.265), then the zero front of $b(z,t)$ stands still during the time interval $t_1 \le t \le t_0$. In other words, during this time interval the electromagnetic field diffusion exhibits a standing mode illustrated in Fig. 1.20. It is clear from (1.263) that, for the standing mode to be realized, the magnetic flux density $b_0(t)$ on the boundary should be increased very rapidly and even

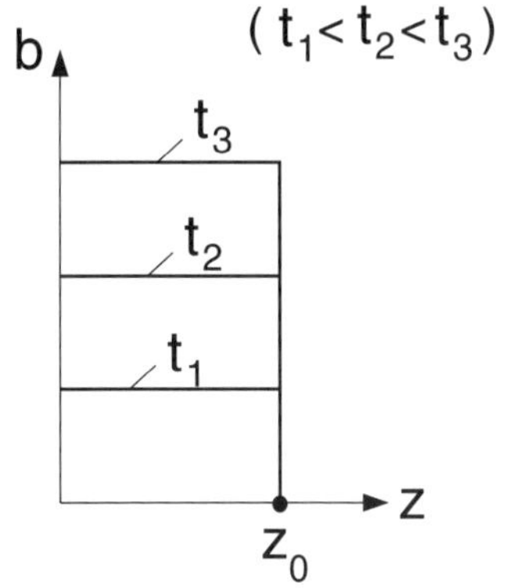

Fig. 1.20

approach infinity as t approaches t_0. In practice, it is impossible to increase $b_0(t)$ up to infinity. However, if $b_0(t)$ is increased in time in accordance with expressions (1.263)–(1.265) during time interval $[t_1, t_2]$ (with $t_2 < t_0$), then for this time interval the standing mode will occur.

The origin of the standing mode can be elucidated on physical grounds as follows. When the magnetic flux density $b_0(t)$ at the boundary is increased in time, then usually two things may happen: (1) the height of the rectangular profile of magnetic flux density is increased and (2) further inward progress of this rectangular profile occurs. These two phenomena require an additional supply of electromagnetic energy for their realizations. It turns out that, under the boundary condition (1.263)–(1.265), the electromagnetic energy entering the conducting material at any instant of time is just enough to affect the uniform (in z) increase in magnetic flux density in the region ($0 \leq z \leq z_0$) already "occupied" by the electromagnetic field, but insufficient to affect the further inward diffusion of the field into the material.

Our previous discussion of the standing mode of nonlinear diffusion has been based on the "rectangular profile" approximation for magnetic flux density. The question can be asked whether this mode is an artifact of the rectangular profile approximation. For this reason, it is interesting to find a standing mode solution without resorting to the above approximation, but rather through analytical solution of nonlinear diffusion Eq. (1.99). It is remarkable that the standing mode solution can be obtained by using the method of separation of variables. Actually, this is the only solution that can be obtained by this method.

According to the method of separation of variables, we look for a solution of nonlinear diffusion Eq. (1.99) in the form

$$b(z,t) = \varphi(z)\psi(t). \tag{1.266}$$

By substituting expression (1.266) into Eq. (1.99), after obvious transformations we obtain

$$\frac{1}{\varphi(z)}\frac{d^2\varphi^n_{(z)}}{dz^2} = \frac{k\sigma}{\psi^n(t)}\frac{d\psi(t)}{dt}. \tag{1.267}$$

Since the left-hand side of the last equation depends only on z, while the right-hand side of the same equation depends only on t, we conclude that these two sides can be equal to one another only if they are equal to the same constant λ. Thus, we have

$$\frac{k\sigma}{\psi^n(t)}\frac{d\psi(t)}{dt} = \lambda, \tag{1.268}$$

$$\frac{1}{\varphi(z)}\frac{d^2\varphi^n(z)}{dz^2} = \lambda. \tag{1.269}$$

First, we shall integrate Eq. (1.268). To this end, we rewrite this equation as follows:

$$\frac{d\psi}{\psi^n} = \frac{\lambda}{k\sigma}dt, \tag{1.270}$$

and after integration we obtain

$$\psi(t) = [\frac{k\sigma}{(n-1)\lambda(t_0 - t)}]^{\frac{1}{n-1}}, \tag{1.271}$$

where t_0 is some constant of integration.

Equation (1.269) is more complicated than Eq. (1.268), and its integration is more involved. To integrate Eq. (1.269), we introduce the following functions:

$$\varphi^n(z) = \theta(z), \tag{1.272}$$

$$\frac{d\theta(z)}{dz} = R(z). \tag{1.273}$$

From the last two expressions and Eq. (1.269), we derive

$$\frac{dR}{dz} = \frac{d^2\theta}{dz^2} = \frac{d^2\varphi^n(z)}{dz^2} = \lambda\varphi(z) = \lambda\theta^{\frac{1}{n}}(z). \tag{1.274}$$

On the other hand, by using formula (1.273), we have

$$\frac{dR}{dz} = \frac{dR}{d\theta}\cdot\frac{d\theta}{dz} = R\frac{dR}{d\theta} = \frac{1}{2}\frac{dR^2}{d\theta}. \tag{1.275}$$

By equating right-hand sides of formulas (1.274) and (1.275), we obtain the first-order differential equation for R:

$$\frac{dR^2}{d\theta} = 2\lambda\theta^{\frac{1}{n}}. \tag{1.276}$$

By integrating the last equation, we obtain

$$R(z) = \sqrt{\frac{2n}{n+1}\lambda} \quad [\theta(z)]^{\frac{n+1}{2n}}. \tag{1.277}$$

In formula (1.277), a constant of integration was set to zero. This can be justified on physical grounds. Indeed, the electric field should vanish at the zero front z_0, that is, at the same point where $b(z,t)$ vanishes. But the electric field is proportional to $R(z)$. Indeed, by using formula (1.266), (1.272), and (1.275), we find:

$$E(z,t) = -\frac{1}{\sigma}\frac{\partial h}{\partial z} = -\frac{1}{k\sigma}\frac{\partial b^n}{\partial z} = -\frac{\psi^n(t)}{k\sigma}R(z). \tag{1.278}$$

The magnetic flux density $b(z,t)$ is proportional to $\varphi(z)$. Thus, functions $R(z)$ and $\varphi(z)$ should vanish simultaneously. According to formula (1.272), this is only possible if the integration constant in (1.277) is set to zero. Next, by substituting expression (1.273) into formula (1.277), we arrive at the differential equation:

$$\frac{d\theta(z)}{dz} = \sqrt{\frac{2n}{n+1}\lambda} \quad [\theta(z)]^{\frac{n+1}{2n}}. \tag{1.279}$$

By integrating the last equation, we derive

$$[\theta(z)]^{\frac{n-1}{n}} = \frac{(n-1)^2}{2(n+1)n}\lambda(z_0 - z)^2, \tag{1.280}$$

where $z_0 > 0$ is some constant of integration.

Now, by using the relation (1.272) between $\varphi(z)$ and $\theta(z)$, from the last formula we obtain

$$\varphi(z) = \left[\frac{(n-1)^2\lambda}{2(n+1)n}(z_0 - z)^2\right]^{\frac{1}{n-1}}. \tag{1.281}$$

Finally, by substituting expressions (1.271) and (1.281) into formula (1.266), we arrive at the following analytical (and exact) solution of nonlinear diffusion Eq. (1.99):

$$b(z,t) = [\frac{(n-1)k\sigma(z_0 - z)^2}{2(n+1)n(t_0 - t)}]^{\frac{1}{n-1}}. \tag{1.282}$$

It is remarkable that, as a result of the above substitution, constant λ cancels out. As a consequence, we end up with the only solution that can be obtained by the method of separation of variables and this solution does not depend on "separation" constant λ at all.

In the form (1.282), the above solution is not physically meaningful. This is because $b(z,t)$ approaches infinity as z is increased. However, the above solution can be "cut off" at $z = z_0$ and then physically reinterpreted as follows. Suppose that at time $t = 0$ the magnetic flux density satisfies the following initial condition:

$$b(z,0) = \begin{cases} \left[\frac{(n-1)k\sigma(z_0-z)^2}{2(n+1)nt_0}\right]^{\frac{1}{n-1}}, & \text{if } 0 \leq z \leq z_0, \\ 0, & \text{if } z \geq z_0. \end{cases} \tag{1.283}$$

Suppose also that the magnetic flux density satisfies the following boundary condition for the time interval $0 \leq t \leq t_0$:

$$b_0(t) = b(0,t) = \left[\frac{(n-1)k\sigma z_0^2}{2(n+1)n(t_0-t)}\right]^{\frac{1}{n-1}}. \tag{1.284}$$

Then, according to formula (1.282), the exact solution to the initial boundary value problem (1.285)–(1.284) for the nonlinear diffusion Eq. (1.99) can be written as follows:

$$b(z,t) = \begin{cases} \left[\frac{(n-1)k\sigma(z_0-z)^2}{2(n+1)n(t_0-t)}\right]^{\frac{1}{n-1}}, & \text{if } 0 \leq z \leq z_0, \\ 0, & \text{if } z \geq z_0. \end{cases} \tag{1.285}$$

This solution is illustrated by Fig. 1.21, and it is apparent that it has the physical meaning of the standing mode. It is also clear from the last formula as well as from Fig. 1.21 that this solution has the self-similarity property. Namely, the profiles of magnetic flux density for different instants of time can be obtained from one another by dilation (or contraction) along the b-axis. In other words, these profiles remain similar to one another. This suggests that the standing mode solution (1.285) can be derived by using dimensional analysis. However, we shall not delve further into this matter. Instead, we shall use the self-similarity property of the above solution in order to give another interpretation (and definition) of the standing mode. To this end, we introduce the normalized profile $b^*(z,t)$ of magnetic flux density:

$$b^*(z,t) = \frac{b(z,t)}{b_0(t)}. \tag{1.286}$$

From formulas (1.285) and (1.286), we find

$$b^*(z,t) = \begin{cases} \left(1 - \frac{z}{z_0}\right)^{\frac{2}{n-1}}, & \text{if } 0 \leq z \leq z_0, \\ 0, & \text{if } z \geq z_0. \end{cases} \tag{1.287}$$

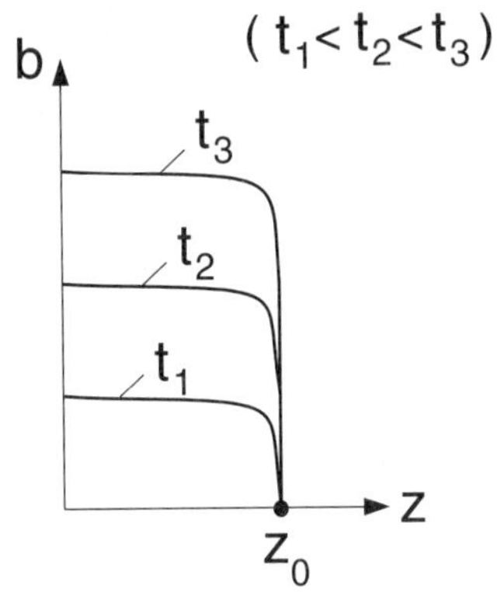

Fig. 1.21

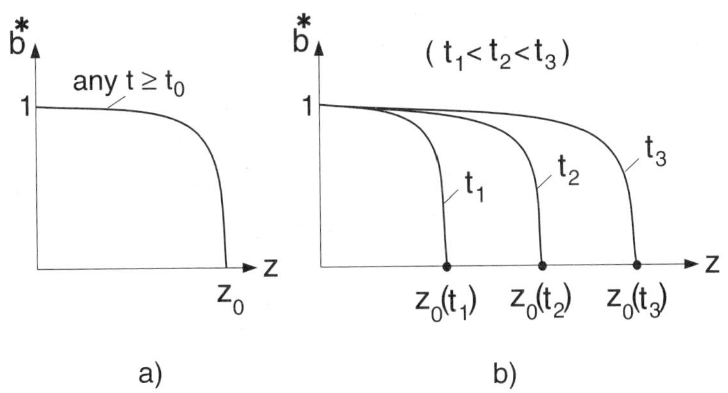

Fig. 1.22

Thus, in the case of the standing mode, the normalized profile of magnetic flux density does not change with time (see Fig. 1.22 a). This property can be used as another definition of the standing mode. It is instructive to note that, in the case of self-similar solutions (1.193), the normalized profile of magnetic flux density is dilated (expanded) with time (see Fig. 1.22 b), which is natural for "forward" diffusion. It is also instructive to point out that the standing mode solution (1.285) is limited both in time and space. This is related to the fact that exponent "n" in the nonlinear diffusion equation is larger than one. In Chapter 3 we shall show that, in the case when this exponent is smaller than one, there exists the self-similar standing mode solution (that is, the solution with the independent of time normalized profile), and this solution is "unlimited" in time and in space. In other words, this solution exists for the semi-infinite interval $0 \le t \le \infty$ and at any instant of time t it asymptotically approaches zero as z is increased to infinity (i.e., there is no zero front). Moreover, it will be shown that there also exist self-similar solutions with normalized profiles contracting in time. These solutions can be physically interpreted as "backward" diffusion.

It is interesting to compare the exact standing mode solution (1.285) with the standing mode expressions derived in this section on the basis of rectangular profile approximation. First, it is clear from formula (1.285) (as well as Fig. 1.21) that, for sufficiently large "n," the actual magnetic flux density profiles are almost rectangular. Second, it is apparent that the boundary condition (1.284) can be written in the form (1.263) with "c" and "t_0" defined as follows:

$$c = \left[\frac{(n-1)k\sigma z_0^2}{2(n+1)n}\right]^{\frac{1}{n-1}}, \tag{1.288}$$

$$t_0 = \frac{(n-1)k\sigma z_0^2 b_0^{1-n}(0)}{2(n+1)n}. \tag{1.289}$$

By comparing formulas (1.288)–(1.289) with formulas (1.264)–(1.265), respectively, we find that for sufficiently large "n" these expressions are practically identical (up to inessential initial time t_1 in (1.265)). This once again suggests that the rectangular profile approximation is a fairly accurate one.

Finally, it should be noted that, in the case of the standing mode of nonlinear diffusion, very peculiar shielding (screening) of internal layers of the conducting material takes place. It is amusing that this self-shielding occurs for rapidly increasing magnetic flux density at the boundary. It would be interesting to find some meaningful practical applications for the standing mode of nonlinear diffusion.

1.7 NONLINEAR DIFFUSION IN A CYLINDER

In previous sections, we dealt with nonlinear diffusion of electromagnetic fields in conductors with plane (flat) boundaries. In this section, we shall extend our study to the case of nonlinear diffusion in a cylinder. This study will shed some light on how the curvature of conducting boundaries may affect the process of nonlinear diffusion. In this section, we shall also discuss the case of magnetically inhomogeneous conducting media.

Consider an infinite conducting cylinder of radius R (see Fig. 1.23) subject to time-varying uniform magnetic field $H_0(t)$ whose direction is parallel to the cylinder axis. First, we assume that this cylinder is magnetically homogeneous with constitutive relation described by Eq. (1.18). In other words, we shall first consider the case of abrupt (sharp) magnetic transition.

Suppose that initial values of magnetic flux density and magnetic field are equal to $-B_m$ and 0, respectively. Next, suppose that the magnetic field at the boundary is increased from its zero value and remains positive. As the magnetic field at the boundary is increased, this increase extends inside the conducting cylinder causing the transition of magnetic flux density from $-B_m$ to $+B_m$. As a result, a rectangular front of magnetic flux density is formed and it moves from the boundary of the cylinder toward its axis (see Fig. 1.24). We intend to derive the expression for the radial coordinate, $r_0(t)$, of this front in terms of σ, B_m, R and the magnetic field, $H_0(t)$, at the boundary. To this end, we shall exploit the circular symmetry of the problem. According to this symmetry, electric field lines and lines of electric current density are circular ones. They exist only for $r \geq r_0(t)$. Consider an electric field line L_r of radius r and let us apply the law

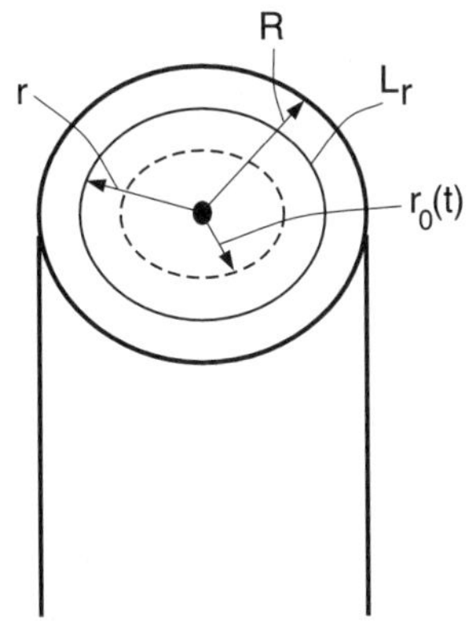

Fig. 1.23

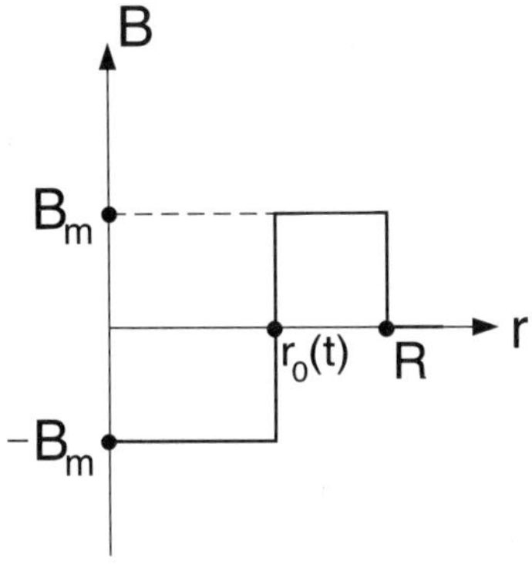

Fig. 1.24

of electromagnetic induction to this line:

$$\oint_{L_r} \mathbf{E} \cdot \mathbf{dl} = -\frac{d\Phi(r,t)}{dt}, \tag{1.290}$$

where $\Phi(r,t)$ is the magnetic flux that links L_r.

Due to the circular symmetry, we have

$$\oint_{L_r} \mathbf{E} \cdot \mathbf{dl} = E(r,t) \cdot 2\pi r. \tag{1.291}$$

By using Fig. 1.24, it is easy to see that the following expression is valid for flux $\Phi(r,t)$:

$$\Phi(r,t) = B_m \pi (r^2 - r_0^2(t)) - B_m \pi r_0^2(t), \tag{1.292}$$

which can be further reduced to the form

$$\Phi(r,t) = B_m \pi (r^2 - 2r_0^2(t)). \tag{1.293}$$

From the last formula, we find

$$-\frac{d\Phi(r,t)}{dt} = 2\pi B_m \frac{dr_0^2(t)}{dt}. \tag{1.294}$$

By substituting expressions (1.291) and (1.294) into formula (1.290), we end up with

$$E(r,t) = \frac{B_m}{r} \frac{dr_0^2(t)}{dt}. \tag{1.295}$$

By using the last expression, we obtain the following equation for electric current density:

$$j(r,t) = \sigma E(r,t) = \frac{\sigma B_m}{r} \frac{dr_0^2(t)}{dt}. \tag{1.296}$$

Now, we recall that

$$j(r,t) = -\frac{\partial H(r,t)}{\partial r}. \tag{1.297}$$

By integrating the last expression from $r_0(t)$ to R and taking into account that $H(r_0(t),t) = 0$, we obtain

$$H_0(t) = H(R,t) = -\int_{r_0(t)}^{R} j(r,t)dr. \tag{1.298}$$

By substituting formula (1.296) into (1.298) and performing the integration, we arrive at

$$H_0(t) = \sigma B_m \left(\ln \frac{r_0(t)}{R} \right) \frac{dr_0^2(t)}{dt}. \tag{1.299}$$

It can be shown that

$$\ln\left(\frac{r_0(t)}{R}\right)\frac{dr_0^2(t)}{dt} = \frac{d}{dt}\left[r_0^2(t)\ln\left(\frac{r_0(t)}{R}\right)\right] - \frac{1}{2}\frac{dr_0^2(t)}{dt}. \tag{1.300}$$

Indeed,

$$\begin{aligned} \frac{d}{dt}\left[r_0^2(t)\ln\left(\frac{r_0(t)}{R}\right)\right] &= \ln\left(\frac{r_0(t)}{R}\right)\frac{dr_0^2(t)}{dt} + r_0(t)\frac{dr_0(t)}{dt} \\ &= \ln\left(\frac{r_0(t)}{R}\right)\frac{dr_0^2(t)}{dt} + \frac{1}{2}\cdot\frac{dr_0^2(t)}{dt}, \end{aligned} \tag{1.301}$$

which justifies equality (1.300).

By substituting this equality into formula (1.299), then performing integration from 0 to t and taking into account that $r_0(0) = R$, we obtain

$$\frac{\int_0^t H_o(\tau)d\tau}{\sigma B_m} = r_0^2(t)\ln\frac{r_0(t)}{R} - \frac{r_0^2(t) - R^2}{2}. \tag{1.302}$$

This is a nonlinear equation for $r_0(t)$. It is convenient to transform this equation as follows:

$$\boxed{\frac{2\int_0^t H_0(\tau)d\tau}{\sigma R^2 B_m} = \frac{r_0^2(t)}{R^2}\ln\frac{r_0^2(t)}{R^2} - \frac{r_0^2(t)}{R^2} + 1.} \tag{1.303}$$

We shall next introduce the variable

$$\lambda(t) = \frac{r_0^2(t)}{R^2}, \tag{1.304}$$

and the function

$$F(\lambda) = \lambda(\ln\lambda - 1) + 1. \tag{1.305}$$

By using the above function, Eq. (1.303) can be represented in the form

$$\frac{2\int_0^t H_0(\tau)d\tau}{\sigma R^2 B_m} = F(\lambda(t)). \tag{1.306}$$

This form is convenient for the graphical solution of Eq. (1.303). Indeed, the function, $F(\lambda)$, can be precomputed and its graph can be constructed (see Fig. 1.25). Then, for every instant of time, the left-hand side of Eq. (1.306) can be evaluated and plotted along the vertical axis in Fig. 1.25. Finally, by drawing the horizontal line until it intersects with the graph of the function $F(\lambda)$ and the vertical line until it intersects with the horizontal axis, we determine the value of $\lambda(t)$ corresponding to the left-hand side of Eq. (1.306). By using this value and formula (1.304), we find $r_0(t)$.

Equation (1.306) can also be solved numerically by using, for instance, the Newton iterative technique. This equation can also be solved analytically, albeit approximately. The analytical solution is based on the following approximation of $F(\lambda)$:

$$F(\lambda) \approx \frac{(\lambda - 1)^4 + (\lambda - 1)^2}{2}. \tag{1.307}$$

To appreciate the accuracy of approximation (1.307), the graphs of function $F(\lambda)$ and its approximation are plotted in Fig. 1.26. This figure suggests that approximation (1.307) is quite accurate. By substituting formula (1.307) into Eq. (1.306), we end up with the quadratic equation for $(\lambda - 1)^2$. By solving this equation and taking into account expression (1.304) for λ, we arrive at the following approximate formula for $r_0(t)$:

$$r_0(t) \approx R\sqrt{1 - \sqrt{-\frac{1}{2} + \sqrt{\frac{1}{4} + \frac{4\int_0^t H_0(\tau)d\tau}{\sigma R^2 B_m}}}} \tag{1.308}$$

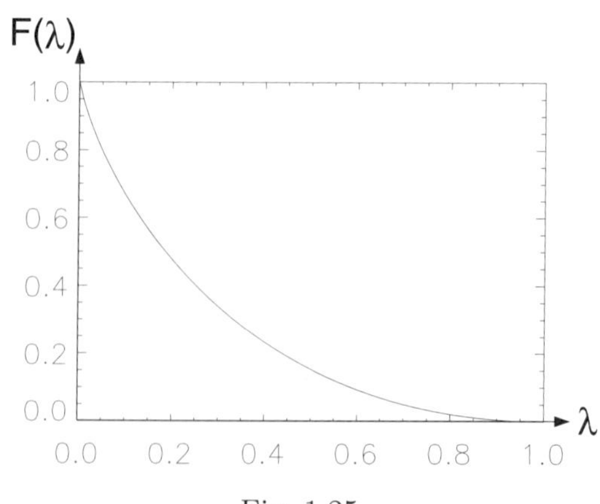

Fig. 1.25

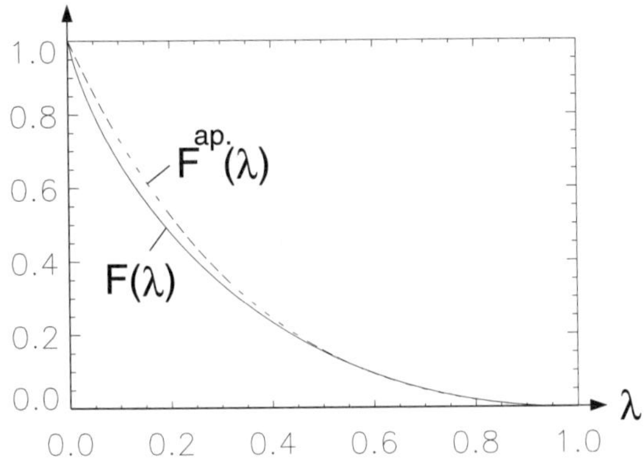

Fig. 1.26

The value of $r_0(t)$ given by formula (1.308) can be used as an initial guess in Newton iterations. However, this may be unnecessary, because it is quite conceivable that the accuracy of formula (1.308) is higher than the accuracy of abrupt magnetic transition assumption, which was used in the derivation of Eq. (1.306).

Having determined the radial coordinate, $r_0(t)$, of the front, we can fully describe the distribution of magnetic flux density. We can also find the electric field by invoking formula (1.295). However, there is some inconvenience in using this formula. This is because this formula requires differentiation of $r_0(t)$, which is found graphically or numerically. This difficulty can be circumvented by recalling formula (1.299) and expressing $\frac{dr_0^2}{dt}$ as follows:

$$\frac{dr_0^2}{dt} = \frac{H_0(t)}{\sigma B_m \ln \frac{r_0(t)}{R}}. \tag{1.309}$$

By substituting the last expression into Eq. (1.295), we end up with

$$E(r,t) = \frac{H_0(t)}{\sigma r \ln \frac{r_0(t)}{R}}. \tag{1.310}$$

The last formula does not contain the derivative of $r_0(t)$ and, for this reason, it is convenient for calculations. By using formula (1.310), we can also find an important relationship between electric and magnetic fields at the cylinder boundary:

$$E_0(t) = \frac{H_0(t)}{\sigma R \ln \frac{r_0(t)}{R}}, \tag{1.311}$$

where the notation $E(R,t) = E_0(t)$ has been used.

Up to this point, we have discussed the situation when the magnetic field $H_0(t)$ at the boundary is assumed to be positive. If the magnetic field $H_0(t)$ is reduced to zero and then becomes negative, the motion of the "positive" rectangular front of magnetic flux density is terminated and the "negative" rectangular front is formed and it moves from the cylinder boundary toward its axis (see Fig. 1.27). By repeating the same line of reasoning that was used in the derivation of formula (1.303), we arrive at the similar expression:

$$\frac{-2\int_0^t H_0(\tau)d\tau}{\sigma R^2 B_m} = \frac{r_0^2(t)}{R^2}\ln\frac{r_0^2(t)}{R^2} - \frac{r_0^2(t)}{R^2} + 1. \tag{1.312}$$

This equation can be used for the determination of the radial coordinate $r_0(t)$ of the "negative" rectangular front of magnetic flux density in the same way as we have used Eq. (1.303).

The previous discussion can be easily extended to the case when the abrupt (sharp) magnetic transition is described by the rectangular hysteresis loop shown in Fig. 1.7. In this case, the "positive" rectangular fronts are formed and they move inward when $H_0(t) - H_c$ is positive. The "negative" rectangular fronts are formed and they move inward when $H_0(t) + H_c$ is negative. The radial coordinates of "positive" and "negative" fronts can be determined by solving modified Eqs. (1.303) and (1.312), respectively. Modification of these equations consists in the replacement of the integral $\int_0^t H_0(\tau)d\tau$ by integrals $\int_0^t (H_0(\tau) - H_c)d\tau$ and $\int_0^t (H_0(t) + H_c)d\tau$, correspondingly. When $H_0(t)$ is between $-H_c$ and $+H_c$, everything is still and there is no movement of rectangular fronts.

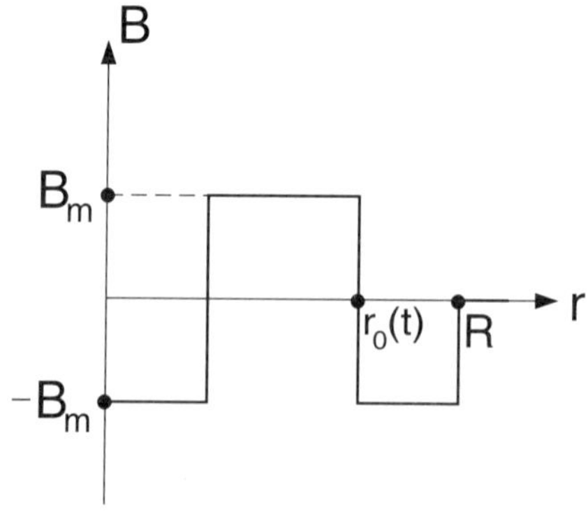

Fig. 1.27

So far, we have only considered magnetically homogeneous conducting media. However, in many applications magnetic properties of surface layers are different from magnetic properties of interior layers. To account for these inhomogeneities, we consider magnetically layered (piecewise homogeneous) media. Namely, we shall assume that a conducting cylinder consists of several cylindrical layers ($R_{k+1} \leq r \leq R_k, k = 1, 2, \ldots, n$), which exhibit abrupt magnetic transitions with different saturation values $\pm B_{mk}$ (see Fig. 1.28). When rectangular fronts move through the first layer, $r_0(t)$ can be evaluated by using Eqs. (1.303) and (1.312). Next, we shall derive similar equations for $r_0(t)$ in the case when rectangular fronts move through cylindrical layer number $k+1$, that is, when $R_{k+1} < r_0(t) < R_k$. To this end, we shall first establish some relation (boundary condition) between $H(R_k, t)$ and $E(R_k, t)$. Consider a circular line L_r with $R_k \leq r \leq R$ and let us apply the law of electromagnetic induction to this line:

$$E(r,t)2\pi r = -\frac{d\Phi(r,t)}{dt}. \tag{1.313}$$

Similarly, when $r = R_k$, we have

$$E(R_k,t)2\pi r = -\frac{d\Phi(R_k,t)}{dt}. \tag{1.314}$$

When the rectangular front moves through the layer number $k+1$, the magnetic flux density changes only within this layer. For this reason, we conclude that

$$\frac{d\Phi(r,t)}{dt} = \frac{d\Phi(R_k,t)}{dt}. \tag{1.315}$$

By using the last expression in Eqs. (1.313) and (1.314), we find

$$E(r,t)2\pi r = E(R_k,t)2\pi R_k, \tag{1.316}$$

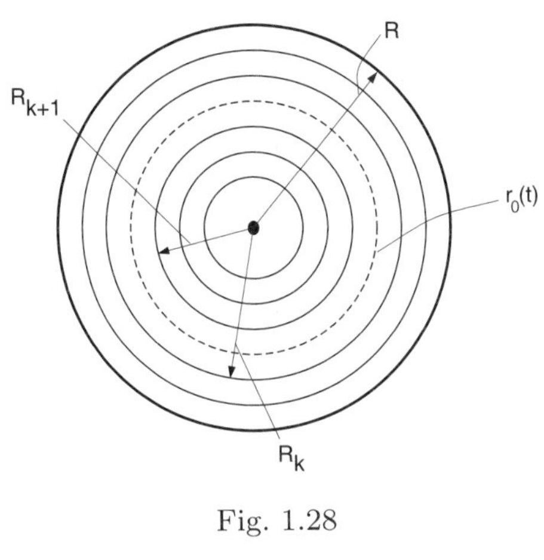

Fig. 1.28

which leads to

$$E(r,t) = \frac{R_k}{r} E(R_k,t). \tag{1.317}$$

From the last formula, we obtain

$$j(r,t) = \frac{\sigma R_k}{r} E(R_k,t). \tag{1.318}$$

By combining expressions (1.297) and (1.318), we derive

$$H(R_k,t) - H_0(t) = \int_{R_k}^{R} j(r,t)dr = (\sigma R_k \ln \frac{R}{R_k}) E(R_k,t), \tag{1.319}$$

which can be rewritten as

$$H(R_k,t) - (\sigma R_k \ln \frac{R}{R_k}) E(R_k,t) = H_0(t). \tag{1.320}$$

The last relation can be construed as the boundary condition that holds at $r = R_k$ for any time $t > t_k$, where t_k is the time when the front moved through the interface boundary $r = R_k$. It is remarkable that this boundary condition does not depend on values of B_{mk}, that is, on the specific nature of abrupt magnetic transitions. Boundary condition (1.320) can be converted into a nonlinear equation for $r_0(t)$. To do this, we recall formulas (1.295) and (1.303), which for the layer number $k+1$ can be rewritten as follows:

$$E(R_k,t) = \frac{B_{mk}}{R_k}\frac{dr_0^2(t)}{dt},$$

$$H(R_k,t) = \frac{\sigma B_{mk}}{2}\frac{d}{dt}\left[r_0^2(t)\left(\ln\frac{r_0^2(t)}{R_k^2} - 1\right)\right].$$

By substituting these expressions into formula (1.320), we arrive at

$$\frac{\sigma B_{mk}}{2}\frac{d}{dt}\left[-\left(\ln\frac{R^2}{R_k^2} + 1\right) r_0^2(t) + r_0^2(t)\ln\frac{r_0^2(t)}{R_k^2}\right] = H_0(t).$$

By integrating the last equation from t_k to t, we derive

$$\frac{2\int_{t_k}^{t} H_0(\tau)d\tau}{\sigma R_k^2 B_{mk}} = \frac{r_0^2(t)}{R_k^2}\ln\frac{r_0^2(t)}{R_k^2} + \left(1 + \ln\frac{R^2}{R_k^2}\right)\left(1 - \frac{r_0^2(t)}{R_k^2}\right). \tag{1.321}$$

The last equation is similar to Eq. (1.303) and, actually, it is reducible to it for $R_k = R$. For this reason, the last equation can be solved in the same manner as was discussed for Eq. (1.303). By solving Eq. (1.321) for $k = 1, 2, \dots n$, we can completely describe nonlinear diffusion in magnetically inhomogeneous conducting media.

Next, we consider nonlinear diffusion in a cylinder in the case of gradual magnetic transitions described by hysteresis loops exemplified by Fig. 1.9. In this case, we shall use the rectangular profile approximation introduced in Section 1.5. According to this approximation, as soon as $h_0(t) = H_0(t) - H_c$ becomes positive, a rectangular profile of magnetic flux density is formed and it moves inward (see Fig. 1.29). We intend to derive the expression for the radial coordinate, $r_0(t)$, of this profile. This derivation, in many respects, closely parallels the derivation of formula (1.303). We start with the law of electromagnetic induction:

$$E(r,t)2\pi r = -\frac{d\Phi(r,t)}{dt}. \tag{1.322}$$

By using Fig. 1.29, the flux, $\Phi(r,t)$, can be evaluated as follows:

$$\Phi(r,t) = (b_0(t) - B_c)\pi(r^2 - r_0^2(t)) - B_c\pi r_0^2(t), \tag{1.323}$$

which can be further reduced to the form

$$\Phi(r,t) = \pi r^2 b_0(t) - \pi b_0(t) r_0^2(t) - \pi B_c r^2. \tag{1.324}$$

By substituting the last expression into formula (1.322), we end up with

$$E(r,t) = -\frac{r}{2}\frac{db_0(t)}{dt} + \frac{1}{2r}\frac{d}{dt}[b_0(t)r_0^2(t)]. \tag{1.325}$$

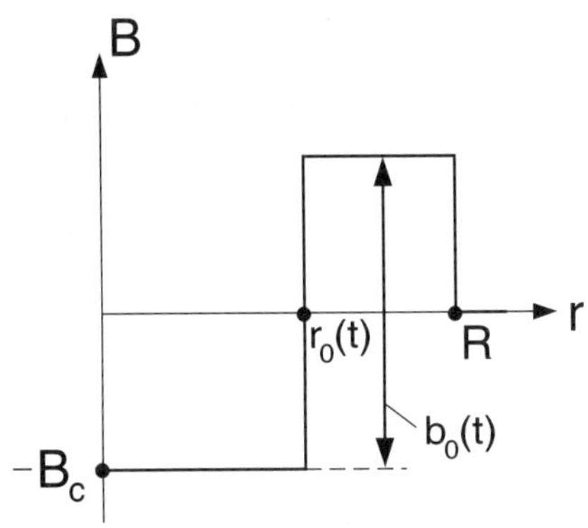

Fig. 1.29

Equation (1.325) leads to the following expression for the electric current density:

$$j(r,t) = -\frac{\sigma r}{2}\frac{db_0(t)}{dt} + \frac{\sigma}{2r}\frac{d}{dt}[b_0(t)r_0^2(t)]. \tag{1.326}$$

By using formulas (1.297) and (1.326) and performing integration with respect to r, we obtain

$$h_0(t) = \frac{\sigma}{4}(R^2 - r_0^2(t))\frac{db_0(t)}{dt} + \left(\frac{\sigma}{2}\ln\frac{r_0(t)}{R}\right)\frac{d}{dt}\left[b_0(t)r_0^2(t)\right], \tag{1.327}$$

where, as before, $h_0(t) = H_0(t) - H_c$. By using simple calculus, we find

$$\begin{aligned}\left(\ln\frac{r_0(t)}{R}\right)\frac{d}{dt}\left[b_0(t)r_0^2(t)\right] = &\frac{d}{dt}\left[b_0(t)r_0^2(t)\ln\frac{r_0(t)}{R}\right]\\ &-\frac{d}{dt}\left[\frac{b_0(t)r_0^2(t)}{2}\right] + \frac{r_0^2(t)}{2}\frac{db_0(t)}{dt}.\end{aligned} \tag{1.328}$$

By substituting formula (1.328) into Eq. (1.327), we arrive at

$$\begin{aligned}h_0(t) =&\frac{\sigma}{4}R^2\frac{db_0(t)}{dt} + \frac{\sigma}{2}\frac{d}{dt}\left[b_0(t)r_0^2(t)\ln\frac{r_0(t)}{R}\right]\\ &-\frac{\sigma}{4}\frac{d}{dt}\left[b_0(t)r_0^2(t)\right].\end{aligned} \tag{1.329}$$

By integrating the last equation with respect to time from 0 to t and taking into account that $b_0(0) = 0$ and $r_0(0) = R$, we obtain

$$\begin{aligned}\int_0^t h_0(\tau)d\tau =&\frac{\sigma}{4}R^2 b_0(t) - \frac{\sigma}{4}r_0^2(t)b_0(t)\\ &+\frac{\sigma}{2}b_0(t)r_0^2(t)\ln\frac{r_0(t)}{R}.\end{aligned} \tag{1.330}$$

The last expression can be rewritten as follows:

$$\frac{4\int_0^t h_0(\tau)d\tau}{\sigma R^2 b_0(t)} = \frac{r_0^2(t)}{R^2}\ln\frac{r_0^2(t)}{R^2} - \frac{r_0^2(t)}{R^2} + 1. \tag{1.331}$$

The last equation is a generalization of Eq. (1.303) to the case of gradual magnetic transitions. It is actually reduced to Eq. (1.303) when $b_0(t) = 2B_m$ and $H_c = 0$. Thus, we can use the same solution techniques for Eq. (1.331) as we used for Eq. (1.303). Namely, the following approximate formula (similar to formula (1.308)) can be used for the calculation of $r_0(t)$:

$$r_0(t) = R\sqrt{1 - \sqrt{-\frac{1}{2} + \sqrt{\frac{1}{4} + \frac{\int_0^t h_0(\tau)d\tau}{\sigma R^2 b_0(t)}}}}. \tag{1.332}$$

In Eqs. (1.331) and (1.332), $h_0(t)$ and $b_0(t)$ are related through actual shapes of hysteresis loops.

Finally, we shall use Eq. (1.331) to evaluate how the curvature of conducting boundary affects the process of nonlinear diffusion. To this end, we introduce the variable

$$\delta(t) = R - r_0(t), \tag{1.333}$$

which has the meaning of instantaneous penetration depth. By using definition (1.333), we find

$$\frac{r_0^2(t)}{R^2} = \frac{[R-\delta(t)]^2}{R^2} = 1 - 2\frac{\delta(t)}{R} + \frac{\delta^2(t)}{R^2}. \tag{1.334}$$

Similarly,

$$\ln\frac{r_0^2(t)}{R^2} = 2\ln\frac{R-\delta(t)}{R} = 2\ln[1 - \frac{\delta(t)}{R}]. \tag{1.335}$$

Next, by using the power series expansion for $\ln(1-x)$, we obtain

$$\ln\frac{r_0^2(t)}{R^2} = 2[-\frac{\delta(t)}{R} - \frac{\delta^2(t)}{2R^2} - \frac{\delta^3(t)}{3R^3} - \ldots]. \tag{1.336}$$

By using formulas (1.334) and (1.336) on the right-hand side of Eq. (1.331), we derive:

$$\begin{aligned}
&\frac{r_0^2(t)}{R^2}\ln\frac{r_0^2(t)}{R^2} - \frac{r_0^2(t)}{R^2} + 1 \approx 2\frac{\delta(t)}{R} - \frac{\delta^2(t)}{R^2} \\
&+ \left(1 - 2\frac{\delta(t)}{R} + \frac{\delta^2(t)}{R^2}\right)\left(-2\frac{\delta(t)}{R} - \frac{\delta^2(t)}{R^2} - \frac{2\delta^3(t)}{3R^3} - \ldots\right) \\
&\approx 2\left[\frac{\delta^2(t)}{R^2} - \frac{\delta^3(t)}{3R^3} - \ldots\right].
\end{aligned} \tag{1.337}$$

By substituting formula (1.337) into Eq. (1.331) and retaining only the first two terms of the above power series expansion, we obtain

$$\sqrt{\frac{2\int_0^t h_0(t)}{\sigma b_0(t)}} \approx \delta(t)\sqrt{1 - \frac{\delta(t)}{3R}}. \tag{1.338}$$

By using the approximation

$$\sqrt{1 - \frac{\delta(t)}{3R}} \approx 1 - \frac{\delta(t)}{6R}, \tag{1.339}$$

the formula (1.338) can be written in the form

$$\sqrt{\frac{2\int_0^t h_0(t)}{\sigma b_0(t)}} \approx \delta(t)\left(1 - \frac{\delta(t)}{6R}\right). \tag{1.340}$$

In the case when $\delta(t)$ is small in comparison with $6R$ (note factor 6) the second term on the right-hand side of (1.340) can be neglected and the last expression leads to

$$\delta^{(0)}(t) \approx \sqrt{\frac{2\int_0^t h_0(\tau)d\tau}{\sigma b_0(t)}}. \tag{1.341}$$

This formula coincides with Eq. (1.209) derived for flat (plane) boundaries of conducting media, that is, for zero curvature. This justifies the use of superscript "(0)" for $\delta(t)$.

Formula (1.340) can also be treated as a quadratic equation for $\delta(t)$. This leads to the following formula for $\delta(t)$:

$$\delta(t) = 3R\left(1 - \sqrt{1 - \frac{2\delta^{(0)}(t)}{3R}}\right). \tag{1.342}$$

In the case when $\delta^{(0)}(t)$ is small in comparison with R, we can use the power series expansion for $\sqrt{1-x}$ and retain only the first two terms of this expansion. This leads to the following formula:

$$\delta(t) = \delta^{(0)}(t) + \frac{(\delta^{(0)}(t))^2}{6R}. \tag{1.343}$$

The second term in last expression gives the correction for finite curvature of conducting boundaries. It also suggests that, for small penetration depths, nonlinear diffusion in conducting bodies with curvilinear boundaries occurs almost in the same way as in the case of plane (flat) boundaries. This fact will be extensively used in Chapter 5, where the results obtained for plane boundaries will be interpreted as impedance boundary conditions.

1.8 APPLICATIONS TO CIRCUIT ANALYSIS

In our previous discussions, it has been tacitly assumed that the magnetic field (or the electric field) at the boundary of conducting media is known as a function of time. However, in applications such a situation is rarely encountered. Typically, magnetic fields are created by electric currents through coils that are placed around magnetic conductors (laminated cores, for instance). The electric currents through such coils are usually not known in advance. These currents should be determined from the analysis of electric circuits that contain these coils. It is apparent that the electric currents in such circuits are affected by the voltages induced in the coils placed around conducting magnetic cores. These voltages, in turn, are determined by the process of nonlinear diffusion of magnetic fields in the above cores. Thus, we are back full circle: we need to know an electric current through a coil for the analysis of nonlinear diffusion in a conducting magnetic core, however, in order to find this current the analysis of nonlinear diffusion in the magnetic core is needed. There are two ways out of this predicament. The first way is to treat the electric circuit equations as the boundary conditions for the analysis of nonlinear diffusion. This approach is convenient for relatively simple electric circuits. The second way is to use the nonlinear diffusion analysis for the derivation of terminal (current-voltage) relations for coils placed around conducting magnetic cores. When these terminal relations are found, they can be used along with other circuit equations in the analysis of electric circuits with the above coils. This approach is preferable for complicated electric circuits.

Next, we proceed with the discussion of the first approach. As an example, consider a toroidal conducting magnetic core of circular cross-section. We assume that a coil that has N turns is uniformly wound around this core (see Fig. 1.30). Suppose the "source" voltage $v_s(t)$ across the coil terminals is known. Then, we can write the following circuit equation:

$$v_s(t) = R_e i(t) + N\frac{d\Phi(t)}{dt}, \tag{1.344}$$

where R_e is the electrical resistance of the coil, $i(t)$ is the coil current and $\Phi(t)$ is the magnetic flux that links each turn of the coil.

In the case when the magnetic core is assumed to be linear and the distribution of magnetic flux density over the core cross-section is assumed to be uniform, we can introduce the inductance L_e of the coil:

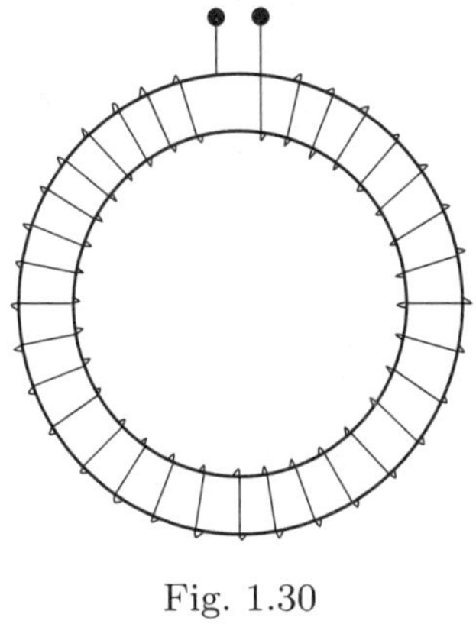

Fig. 1.30

$$N\Phi(t) = L_e i(t). \tag{1.345}$$

By substituting formula (1.345) into expression (1.344), we arrive at the conventional electric circuit differential equation:

$$v_s(t) = R_e i(t) + L_e \frac{di(t)}{dt}. \tag{1.346}$$

Now, we shall remove the assumption that the core is magnetically linear but shall still assume that the magnetic flux distribution over the core cross-section is uniform. Then, the flux, $\Phi(t)$, can be expressed as follows:

$$\Phi(t) = AB(H(t)), \tag{1.347}$$

where A is the cross-sectional area and $B(H)$ stands for the notation for nonlinear relation between B and H.

By using the Ampére Law, we find

$$H(t) = \frac{i(t)N}{\ell}, \tag{1.348}$$

where ℓ is some average length of magnetic field lines within the core.

From formulas (1.347) and (1.348), we derive

$$\frac{d\Phi}{dt} = A\frac{dB}{dH} \cdot \frac{dH}{di} \cdot \frac{di}{dt}, \tag{1.349}$$

which leads to

$$\frac{d\Phi}{dt} = \frac{AN}{\ell}\frac{dB}{dH}(i(t)) \cdot \frac{di(t)}{dt}. \tag{1.350}$$

By combining formulas (1.344) and (1.350), we arrive at the following circuit equation:

$$v_s(t) = R_e i(t) + L_e(i(t))\frac{di(t)}{dt}, \tag{1.351}$$

where we introduced the nonlinear (current dependent) inductance $L_e(i(t))$ defined as follows:

$$L_e(i(t)) = \frac{AN^2}{\ell}\frac{dB}{dH}(i(t)) = \frac{AN^2}{\ell}\mu_d(i(t)). \tag{1.352}$$

In the case of linear media, $B = \mu H$ and expression (1.352) is reduced to the well-known formula:

$$L_e = \frac{\mu AN^2}{\ell}. \tag{1.353}$$

Now, we shall remove the assumption that the distribution of magnetic flux density over the core cross-section is uniform. In other words, we shall try to take into account the screening effect of eddy currents induced in the core during the process of nonlinear diffusion. In this case, we shall use the law of electromagnetic induction and express $\frac{d\Phi}{dt}$ in terms of electric field at the core boundary:

$$\frac{d\Phi}{dt} = -\oint_L \mathbf{E}_0 \cdot \mathbf{dl} = -E_0(t)L = -E_0 2\pi R. \tag{1.354}$$

Here L is the circular boundary of toroidal cross-section and R is its radius.

The second equality in formula (1.354) implies circular symmetry of electric field, which holds fairly accurately when L is small in comparison with ℓ, that is, when $L \ll \ell$. By substituting formulas (1.348) and (1.354) into the circuit Eq. (1.344), we end up with

$$\boxed{\frac{R_e \ell}{N} H_0(t) - NLE_0(t) = v_s(t).} \tag{1.355}$$

Equation (1.355) relates the boundary values of magnetic and electric fields to the known source voltage, $v_s(t)$, applied to the coil terminals. This equation can be construed as a boundary condition, and it can be used for the analysis of nonlinear diffusion of magnetic field in the conducting magnetic core. To illustrate this, consider first the case when the magnetic properties of the core can be modelled as abrupt magnetic transitions. Since it is also assumed that $L \ll \ell$, then the nonlinear diffusion in the toroid occurs almost in the same way as the nonlinear diffusion in the circular cylinder of radius R. For this reason, we can use expressions (1.295) and (1.302), which can be represented in these forms:

$$E_0(t) = \frac{B_m}{R} \frac{dr_0^2(t)}{dt}, \tag{1.356}$$

$$H_0(t) = \frac{\sigma B_m}{2} \frac{d}{dt} \left[r_0^2(t) \ln \frac{r_0^2(t)}{R^2} - r_0^2(t) \right]. \tag{1.357}$$

By using the last two formulas, we can convert the boundary condition (1.355) into a nonlinear equation for $r_0(t)$:

$$\frac{\sigma B_m}{2} \frac{d}{dt} \left[\frac{R_e \ell}{N} r_0^2(t) \ln \frac{r_0^2(t)}{R^2} - \left(\frac{R_e \ell}{N} + \frac{2NL}{\sigma R} \right) r_0^2(t) \right] = v_s(t). \tag{1.358}$$

By integrating in the last formula and taking into account that $r_0(0) = R$, we arrive at

$$\frac{2\int_0^t v_s(\tau)d\tau}{\sigma R^2 B_m} = \frac{R_e\ell}{N}\frac{r_0^2(t)}{R^2}\ln\frac{r_0^2(t)}{R^2} - \left(\frac{R_e\ell}{N} + \frac{2NL}{\sigma R}\right)\left(\frac{r_0^2(t)}{R^2} - 1\right). \tag{1.359}$$

The last equation is very similar to Eq. (1.303). For this reason, the technique that was used for the solution of the former equation can also be used for the solution of Eq. (1.359).

After $r_0(t)$ is found, we can determine the current, $i(t)$, through the coil. To this end, we shall invoke formula (1.311) and transform the boundary condition (1.355) as follows:

$$v_s(t) = \left(\frac{R_e\ell}{N} - \frac{NL}{\sigma R\ln\frac{r_0(t)}{R}}\right)H_0(t). \tag{1.360}$$

By using expression (1.348) in the last equation, we obtain the following formula for $i(t)$:

$$i(t) = \frac{v_s(t)}{R_e - \frac{N^2L}{\sigma R\ell\ln\frac{r_0(t)}{R}}}. \tag{1.361}$$

For $t = 0$, $r_0(0) = R$ and, from the last expression, we find that

$$i(0) = 0, \tag{1.362}$$

as it must.

Let T be the time when $r_0(T) = 0$. Then, for $t \geq T$ from formula (1.361), we obtain

$$i(t) = \frac{v_s(t)}{R_e}, \tag{1.363}$$

as expected.

It is also clear from formula (1.361) that

$$i(t) < \frac{v_s(t)}{R_e} \quad \text{for} \quad 0 < t < T. \tag{1.364}$$

Thus, time T can be physically interpreted as a rise time. This interpretation is especially transparent in the case when $v_s(t) = V_0 = \text{const}$. In this case, for $t \geq T$ we have

$$i_s(t) = I_0 = \frac{V_0}{R_e}. \tag{1.365}$$

The previous analysis can be extended to the important case when the conducting magnetic core has a small air gap. This case exhibits some peculiarities, which we proceed to discuss. First, Eq. (1.348) should be modified. Indeed, by using the Ampére Law, we arrive at

$$H_0(t) \cdot \ell + H_\delta \cdot \delta = Ni(t), \tag{1.366}$$

where H_δ is the magnetic field in the air gap region, while δ is the (effective) length of the air gap.

By invoking the continuity of normal component of magnetic flux density across a media interface, we find

$$H_\delta = \pm \frac{B_m}{\mu_0}, \tag{1.367}$$

where the "$\pm$" signs account for two possible states of magnetic flux density within the core.

Let us assume that, before the source voltage $v_s(t)$ is turned on ($i(t) = 0$), the initial magnetic flux density within the core has been equal to $-B_m$. Then, from Eqs. (1.366) and (1.367), we obtain

$$H_0(0) = \frac{B_m \delta}{\mu_0 \ell} > 0. \tag{1.368}$$

This suggests that the abrupt magnetic transition of the core should be described by a rectangular loop with $H_c > H_0(0)$. Otherwise, the "demagnetizing" field $H(0)$ is too strong for the state of $B = -B_m$ to be sustainable.

Now, suppose that the voltage source, $v_s(t)$, is turned on and the current, $i(t)$, in the coil has reached the value at which the magnetic field, $H_0(t)$, at the boundary of the conducting magnetic core exceeds the value of H_c. Then, the process of nonlinear diffusion of the magnetic field in the core begins and the magnetic flux density at the core boundary is equal to $+B_m$. By using this fact and Eqs. (1.366) and (1.367), we arrive at

$$H_0(t) \cdot \ell + \frac{B_m \delta}{\mu_0} = Ni(t), \tag{1.369}$$

which leads to

$$i(t) = H_0(t) \frac{\ell}{N} + \frac{B_m \delta}{\mu_0 N}. \tag{1.370}$$

By substituting the last equation as well as the Eq. (1.354) into electric circuit Eq. (1.346), we end up with the following modified boundary condition:

$$\boxed{\frac{R_e\ell}{N}H_0(t) - NLE_0(t) = v_s(t) - \frac{R_e B_m \delta}{\mu_0 N}.} \tag{1.371}$$

As before, this boundary condition can be converted into a nonlinear equation for the radial coordinate, $r_0(t)$, of the rectangular front. To this end, we rewrite Eq. (1.357) in the form:

$$H_0(t) - H_c = \frac{\sigma B_m}{2}\frac{d}{dt}\left[r_0^2(t)\ln\frac{r_0^2(t)}{R^2} - r_0^2(t)\right], \tag{1.372}$$

which takes into account the hysteretic nature of the abrupt transition.

Now, by substituting Eqs. (1.356) and (1.372) into the boundary condition (1.371), we obtain

$$\frac{\sigma R_e \ell B_m}{2N}\frac{d}{dt}\left[r_0^2(t)\ln\frac{r_0^2(t)}{R^2} - r_0^2(t)\right] - \frac{NLB_m}{R}\frac{dr_0^2(t)}{dt} = \tilde{v}_s(t), \tag{1.373}$$

where

$$\tilde{v}_s(t) = v_s(t) - \frac{R_e B_m \delta}{\mu_0 N} - \frac{R_e \ell H_c}{N}. \tag{1.374}$$

By integrating Eq. (1.373) and taking into account that $r_0(t_1) = R$, we arrive at the following nonlinear equation:

$$\begin{gathered}\frac{2\int_{t_1}^{t}\tilde{v}_s(\tau)d\tau}{\sigma R^2 B_m} = \\ \frac{R_e\ell}{N}\cdot\frac{r_0^2(t)}{R^2}\ln\frac{r_0^2(t)}{R^2} - \left(\frac{R_e\ell}{N} + \frac{2NL}{\sigma R}\right)\left(\frac{r_0^2(t)}{R^2} - 1\right),\end{gathered} \tag{1.375}$$

where t_1 is the instant of time at which the magnetic field $H_0(t)$ exceeds the value of H_c. This instant of time can be determined from Eqs. (1.344) and (1.369). Indeed, before time t_1, there are no changes in the magnetic flux through the core, and from Eq. (1.344) we find

$$i(t) = \frac{v_s(t)}{R_e}, \qquad \text{for } 0 \le t \le t_1. \tag{1.376}$$

Now, by substituting formula (1.376) into Eq. (1.369) and taking into account that $H_0(t_1) = H_c$, we obtain

$$\frac{R_e \ell H_c}{N} + \frac{R_e B_m \delta}{\mu_0 N} = v_s(t_1), \tag{1.377}$$

which is the desired equation for t_1.

Now, we shall turn back to Eq. (1.375). This equation is very similar to Eq. (1.303) from Section 1.7, and it can be solved by using the techniques described therein. After $r_0(t)$ is found, the current, $i(t)$, through the coil can be determined. To this end, we shall use formula (1.299) in the form

$$H_0(t) - H_c = \sigma B_m \left(\ln \frac{r_0(t)}{R} \right) \frac{dr_0^2(t)}{dt}, \tag{1.378}$$

which accounts for the hysteretic nature of the core.

From formulas (1.356) and (1.378), we find

$$E_0(t) = \frac{H_0(t) - H_c}{\sigma R \ln \frac{r_0(t)}{R}}. \tag{1.379}$$

By substituting the last formula into expression (1.354) and then into Eq. (1.344), we obtain

$$v_s(t) = R_e i(t) - \frac{NL(H_0(t) - H_c)}{\sigma R \ln \frac{r_0(t)}{R}}. \tag{1.380}$$

Finally, by combining the last equation with Eq. (1.369) and by solving these two equations for $i(t)$, we derive the formula:

$$i(t) = \frac{v_s(t) - \frac{B_m \delta N L + \mu_0 N \ell L H_c}{\mu_0 \ell \sigma R \ln(r_0(t)/R)}}{R_e - \frac{N^2 L}{\sigma R \ell \ln(r_0(t)/R)}}, \tag{1.381}$$

which can be used for the calculation of electric current.

For $t = t_1$, $r_0(t_1) = R$ and, from the last expression, we find that

$$i(t_1) = \frac{B_m \delta}{\mu_0 N} + \frac{\ell H_c}{N}, \tag{1.382}$$

which is consistent with formula (1.377).

At $t = T$, we have $r_0(T) = 0$, and, from formula (1.381), we obtain expression (1.363).

The approach that we have thus far discussed is convenient when an electric circuit is described by a relatively simple equation such as Eq. (1.344). If a coil with a magnetic conducting core is a part of a complicated electric circuit described by many differential equations, this approach does not work very well. This is because complicated circuit equations cannot be reduced to a simple boundary condition. Under those circumstances, another approach can be pursued. In this approach, the nonlinear diffusion analysis is used in order to derive a terminal voltage-current relation for a coil placed around a conducting magnetic core. In general, this terminal relation can be written in the form

$$\hat{L}_c(v(t), i(t)) = 0, \tag{1.383}$$

where $v(t)$ is the voltage induced in the coil by a time varying magnetic flux through the core, $i(t)$ is the current through the coil, and $\hat{L}_c$ stands for some (integral-differential) operator. The actual form of this operator depends on the geometry and magnetic characteristics of the core. If the relation (1.383) is found, it can then be used to complement the electric circuit equations. In other words, it can be discretized and solved numerically jointly with these circuit equations.

It is apparent that the essence of the above approach is in the derivation of terminal relation (1.383). We shall first demonstrate how this type of relation can be derived for a gapless, circular cross-section toroid. For this toroid, according to formula (1.354), voltage $v(t)$ is related to the electric field $E_0(t)$ at the boundary by the expression

$$v(t) = -LNE_0(t), \tag{1.384}$$

while the magnetic field $H_0(t)$ at the boundary is related to the electric current by formula (1.348):

$$i(t) = \frac{\ell H_0(t)}{N}. \tag{1.385}$$

On the other hand, $H_0(t)$ and $E_0(t)$ are related to one another by expression (1.311):

$$H_0(t) = \frac{\sigma R}{2} E_0(t) \ln \frac{r_0^2(t)}{R^2}. \tag{1.386}$$

By using formulas (1.384) and (1.385) in the last expression, we arrive at

$$\frac{N}{\ell} i(t) = -\frac{\sigma R}{2LN} v(t) \ln \frac{r_0^2(t)}{R^2}. \tag{1.387}$$

The last equation can be solved for $\frac{r_0^2(t)}{R^2}$, which yields

$$\frac{r_0^2(t)}{R^2} = \exp\left(-\frac{2N^2 Li(t)}{\sigma \ell R v(t)}\right) \tag{1.388}$$

Next, from formulas (1.356) and (1.384), we obtain

$$v(t) = -NRLB_m \frac{d}{dt}\left(\frac{r_0^2(t)}{R^2}\right). \tag{1.389}$$

Finally, by using expression (1.388) in the last formula, it can be transformed as follows:

$$v(t) + NRLB_m \frac{d}{dt}\left[\exp\left(-\frac{2N^2 Li(t)}{\sigma \ell R v(t)}\right)\right] = 0. \tag{1.390}$$

The last expression can be construed as a terminal relation of the type (1.383), and it can be used in circuit analysis in conjunction with other circuit equations.

As another example, consider a core that consists of one thin magnetic conducting lamination (or a stack of such laminations). Suppose also that the core has an air gap of length δ and that magnetic properties of the core exhibit gradual magnetic transitions described by hysteresis loops as exemplified by Fig. 1.9. In this case, we can use "rectangular profile" approximation for magnetic flux density, which leads to the relation (1.224) between electric ($E_0(t)$) and magnetic ($H_0(t)$) fields at the boundary. We shall next express this relation in terms of $v(t)$ and $i(t)$. First, we remark that the Ampére Law leads to the expression (1.366). In this expression, H_δ can be related to $B_0(t)$ by the formula similar to (1.367):

$$H_\delta = \frac{B_0(t)}{\mu_0}. \tag{1.391}$$

By substituting the last equation into expression (1.366), we obtain

$$H_0(t)\ell + \frac{B_0(t)\delta}{\mu_0} = Ni(t). \tag{1.392}$$

Magnetic flux density $B_0(t)$ is a nonlinear function of $H_0(t)$:

$$B_0(t) = f(H_0(t)), \tag{1.393}$$

which describes an ascending branch of hysteresis loop.

By using formula (1.393) in Eq. (1.392), we have

$$H_0(t)\ell + \frac{f(H_0(t))\delta}{\mu_0} = Ni(t). \tag{1.394}$$

Next, by using formulas (1.384) and (1.393) in relation (1.224), we arrive at

$$v(t) + LN\frac{d}{dt}\left[\frac{2}{\sigma}\left(f(H_0(t)) + B_c\right)\int_{t_0}^{t}(H_0(\tau) - H_c)d\tau\right]^{\frac{1}{2}} = 0. \tag{1.395}$$

Formulas (1.394) and (1.395) considered together can be construed as the terminal relation between $v(t)$ and $i(t)$. Indeed, formula (1.394) is a nonlinear equation for $H_0(t)$. By solving this equation, $H_0(t)$ can be found as a nonlinear function of $i(t)$:

$$H_0(t) = F(i(t)). \tag{1.396}$$

By substituting this function into formula (1.395), we arrive at the terminal relation between $v(t)$ and $i(t)$. However, in practical applications, instead of finding function F and excluding $H_0(t)$ from the coupled Eqs. (1.394) and (1.395), we can use the above coupled equations as the terminal relation between $v(t)$ and $i(t)$. In other words, we can discretize these coupled equations and solve them jointly with the rest of electric circuit equations. This has the advantage that $H_0(t)$ and, consequently, $B_0(t)$ are determined at every instant of time as we solve these equations. By knowing $H_0(t)$ and $B_0(t)$, we can find $z_0(t)$ (see formulas (1.209) and (1.223)). Knowledge of $z_0(t)$ is very important. This is because, as soon as $z_0(t)$ reaches the value of $\frac{\Delta}{2}$ (see formula (1.254)), the distribution of magnetic flux density over the lamination cross-section becomes uniform (within the framework of "rectangular profile" approximation). As a result, at subsequent instances of time the relation between $v(t)$ and $i(t)$ can be expressed in terms of current dependent inductance $L_e(i(t))$:

$$v(t) = L_e(i(t))\frac{di(t)}{dt}, \tag{1.397}$$

where $L_e(i(t))$ is given by formula (1.352).

Thus, coupled Eqs. (1.394) and (1.395) are used as the terminal relation between $v(t)$ and $i(t)$ before the instant of time the two rectangular fronts are merged together at the middle of the lamination, while the terminal relation (1.397) is used after that instant of time.

1.9 EDDY CURRENT HYSTERESIS AND THE PREISACH MODEL

In Section 1.8, we discussed how the electric circuit analysis and nonlinear diffusion analysis can be coupled together. However, it was tacitly assumed in our previous discussion that the current $i(t)$ through a coil (and magnetic field $H_0(t)$ at the boundary) are monotonically increased (or decreased) in time. Under this assumption, we were able to derive the terminal relation between voltage $v(t)$ induced in the coil and current $i(t)$ through the coil. In the general case when current through the coil is a piecewise monotonic function of time, the derivation of the above terminal relation becomes quite complicated. This is because of the phenomenon

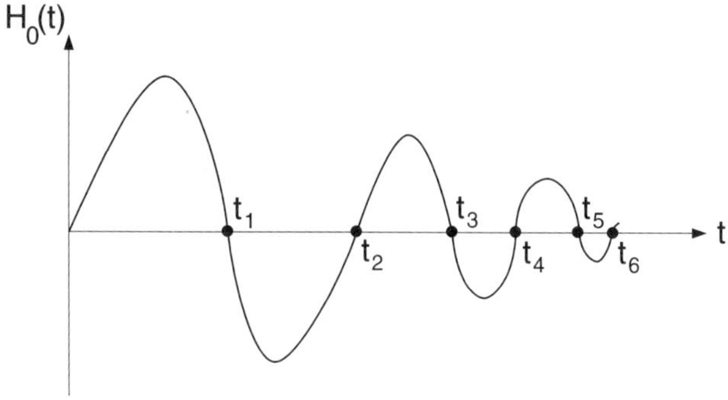

Fig. 1.31

of eddy current hysteresis. To explain this phenomenon, consider, as an example, a time variation of magnetic field $H_0(t)$ shown in Fig. 1.31. It will be assumed that the following inequalities are valid:

$$\int_0^{t_1} H_0(\tau)d\tau > -\int_{t_1}^{t_2} H_0(\tau)d\tau > \int_{t_2}^{t_3} H_0(\tau)d\tau > \cdots > -\int_{t_5}^{t_6} H_0(\tau)d\tau. \tag{1.398}$$

It will also be assumed that magnetic properties of conducting material are described by rectangular magnetization curves (abrupt magnetic transitions) and that the initial value of magnetic flux density is equal to $-B_m$.

According to the previous assumptions, we conclude that during the time interval $0 < t \le t_1$ a "positive" rectangular front of magnetic flux density is formed and moves toward the axis of a cylinder (see Fig. 1.32a). The radial coordinate $r_0(t)$ of this front can be determined by solving Eq. (1.303) or by using the approximate formula (1.308). At time $t = t_1$, the motion of the positive rectangular front is terminated and a negative rectangular front of magnetic flux density is formed. During the time interval $t_1 < t \le t_2$, the latter front moves toward the cylinder axis (see Fig. 1.32 b). The radial coordinate of this front can be determined by solving nonlinear Eq. (1.312). At time $t = t_2$, the motion of this negative rectangular front is terminated and a new positive rectangular front of magnetic flux density is formed. This front moves toward the cylinder axis during the time interval $t_2 < t \le t_3$, and the distribution of magnetic flux density for this time interval is shown in Fig. 1.32 c. At subsequent time intervals ($t_3 < t \le t_4, t_4 < t \le t_5$, and $t_5 < t \le t_6$), new negative and positive rectangular fronts of magnetic flux density are formed, and the distribution of magnetic flux density at time $t = t_6$ looks like the one shown in Fig. 1.32 d. It is important to note that radial coordinates $r_0^{(k)}$ of still (motionless) rectangular fronts form a

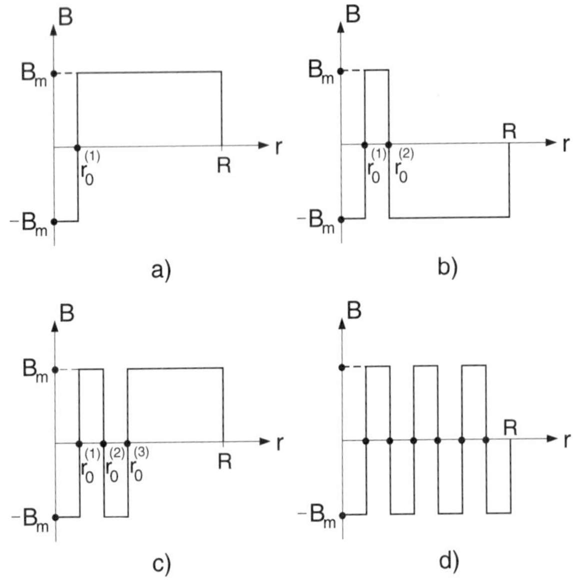

Fig. 1.32

monotonically increasing sequence:

$$r_0^{(1)} < r_0^{(2)} < r_0^{(3)} < r_0^{(4)} < r_0^{(5)} < r_0^{(6)} < \cdots < R. \tag{1.399}$$

This follows directly from inequality (1.398) and Eq. (1.303) and (1.312). In other words, the past time variations of magnetic field $H_0(t)$ at the conductor boundary leave their mark upon future magnetic flux distributions over the conductor cross-section. This suggests that there is a hysteretic relation between the magnetic flux through the conductor cross-section and the magnetic field at the boundary. To clearly understand this hysteretic relation, we introduce the magnetic flux:

$$\Phi(t) = 2\pi \int_0^R B(r,t) r dr, \tag{1.400}$$

and the function:

$$w_0(t) = \int_0^t H_0(\tau) d\tau. \tag{1.401}$$

Next, we shall plot $\Phi(t)$ versus $w_0(t)$. It is evident from Eq. (1.303) and Fig. 1.32 a that as $w_0(t)$ is monotonically increased during the time interval $0 < t < t_1$, the flux $\Phi(t)$ is also monotonically increased starting from its initial value $-\Phi_m = -\pi R^2 B_m$. Thus, the branch "1" is traced in Fig. 1.33 during the above time interval. For the time interval $t_1 < t \leq t_2$, the magnetic field, $H_0(t)$, is negative and the function $w_0(t)$ is monotonically decreased. It is clear from Fig. 1.32 b that for the same interval the magnetic flux is monotonically decreased as well. As a result, the branch "2" is traced in Fig. 1.33. During the time interval $t_2 < t \le t_3$, the magnetic field $H_0(t)$ is positive and the function $w(t)$ is monotonically increased. It is obvious from Fig. 1.32c that for the same time interval the magnetic flux is monotonically increased as well. This results in the branch "3" in Fig. 1.33. By using the same line of reasoning, it is easy to see that new branches "4," "5," and "6" will be formed during the time intervals $t_3 < t \le t_4, t_4 < t \le t_5$ and $t_5 < t \le t_6$. Thus, the relation between $\Phi(t)$ and $w_0(t)$ is a multibranch nonlinearity. It is also clear that branch-to-branch transitions occur after each extremum value of $w_0(t)$. Indeed, the function $w_0(t)$ assumes its (local) maximum values at times $t = t_1$, $t = t_3$, $t = t_5$, and its (local) minimum values at times $t = t_2, t = t_4$, and $t = t_5$, and at all these time instants transitions to new branches occur. It is important to stress that the $\Phi(t)$ vs. $w_0(t)$ relation is rate independent. This means that the value of $\Phi(t)$ depends on the past extremum values of $w_0(t)$ as well as the current value of $w_0(t)$, however, it does not depend on the rate of time variations of $w_0(t)$. The last statement is obvious from the fact that $\Phi(t)$ is fully determined by radial coordinates of rectangular fronts of magnetic flux density, and these "front" coordinates depend only on the values of $w_0(t)$ and do not depend on the rate of its time variations.

It is apparent that the branching described above occurs inside some major ("limiting") hysteresis loop shown in Fig. 1.34. This major loop is formed when for two subsequent monotonic variations of $w_0(t)$ the corresponding fronts of magnetic flux density reach the cylinder axis. Beyond the major loop magnetic flux $\Phi(t)$ may assume only two values: $+\Phi_m$ or $-\Phi_m$.

The major hysteresis loop as well as the branching inside this loop

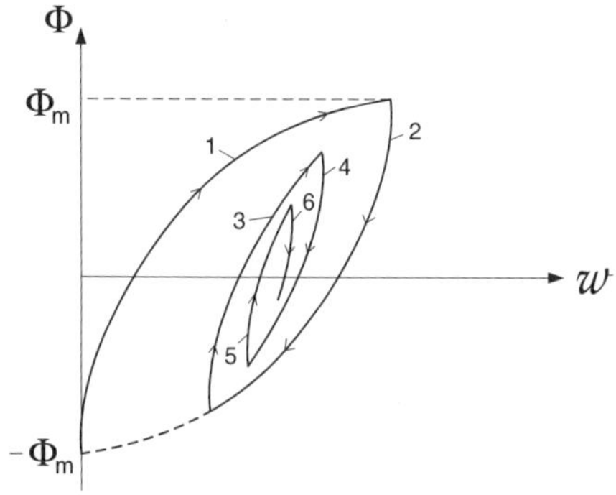

Fig. 1.33

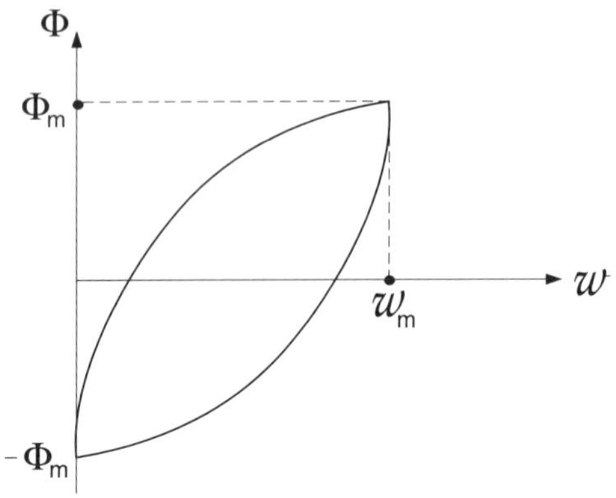

Fig. 1.34

exhibit some asymmetry. This asymmetry can be completely removed by redefining the function $w_0(t)$ as

$$\tilde{w}_0(t) = w_0(t) - \frac{w_m}{2}, \qquad (1.402)$$

where w_m is specified in Fig. 1.34. In our subsequent discussion, it will be tacitly assumed that the above shifting of $w(t)$ is performed when it is needed.

Now, we can summarize our previous discussion by stating that the essence of eddy current hysteresis is the multibranch rate independent nonlinear relation between the magnetic flux $\Phi(t)$ and the function $w_0(t) = \int_0^t H_0(\tau)d\tau$. It is interesting to explore the use of mathematical models of hysteresis for the description of eddy current hysteresis. One of the most powerful hysteresis models is the Preisach model. The origin of this model can be traced back to the landmark paper of F. Preisach [19]. Initially, this model was developed for the description of magnetic hysteresis. However, it was gradually realized that the Preisach model is a general mathematical tool that can be used for the description of hysteresis of various physical natures. Presently, there exists an extensive monographic literature on the Preisach model [5], [6], [9], [16], [20]. To make our discussion self-contained, the main facts related to the Preisach model are summarized in Appendix A. Our presentation of these facts closely follows the book [16]. These facts also will be used in Chapter 4 when superconducting hysteresis will be discussed.

The Preisach model is given by the following expression:

$$f(t) = \iint_{\alpha \geq \beta} \mu(\alpha, \beta)\hat{\gamma}_{\alpha\beta}u(t)d\alpha d\beta, \qquad (1.403)$$

where $u(t)$ is a physical quantity called input, $f(t)$ is a physical quantity called output, $\hat{\gamma}_{\alpha\beta}$ are rectangular loop operators with α and β being "up" and "down" switching values, respectively, while $\mu(\alpha, \beta)$ is a weight function. Details related to formula (1.403) can be found in Appendix A.

The Preisach model (1.403) describes a rate-independent hysteretic relation between input $u(t)$ and output $f(t)$. In Appendix A, the theorem is proven, which states that wiping-out and congruency properties consistute the necessary and sufficient conditions for a hysteretic nonlinearity to be represented by the Preisach model. The last theorem is very instrumental in establishing the connection between eddy current hysteresis and the Preisach model. To do this, we recall that in the case of eddy current hysteresis, $\Phi(t)$ vs. $w_0(t)$ relation is a rate-independent hysteretic nonlinearity. Now, we shall demonstrate that this nonlinearity exhibits wiping-out and congruency properties. Indeed, each time $w_0(t)$ is monotonically increased (or decreased), a rectangular front of magnetic flux density is formed and it moves toward the cylinder axis. This moving front will wipe out those previous rectangular fronts of magnetic flux density if they correspond to the previous extremum values of $w_0(t)$, which are exceeded by its new extremum value. In this way, the effect of those previous extremum values of $w_0(t)$ on the future values of magnetic flux $\Phi(t)$ is completely eliminated. This means that the wiping-out property holds. Next, we shall demonstrate the validity of the congruency property. Consider two variations of $w_0(t)$: $w_0^{(1)}(t)$ and $w_0^{(2)}(t)$. Suppose that $w_0^{(1)}(t)$ and $w_0^{(2)}(t)$ have different past histories (different past extrema) but, starting from some instant of time, they vary monotonically back-and-forth between the same reversal (extremum) values. It is apparent from the mechanism of nonlinear diffusion described at the beginning of this section that these back-and-forth variations of $w_0^{(1)}(t)$ and $w_0^{(2)}(t)$ will affect in the identical way the same surface layers of the conducting cylinder. Consequently, these variations will result in equal increments of magnetic flux $\Phi(t)$, which is tantamount to the congruency of the corresponding minor loops. Since the wiping-out and congruency properties are established for the $\Phi(t)$ vs. $w_0(t)$ relation, this relation can be represented by the Preisach model. Thus, by taking formula (1.401) into account, we find

$$\Phi(t) = \iint\limits_{\alpha \geq \beta} \mu(\alpha, \beta)\hat{\gamma}_{\alpha\beta}\left(\int_0^t H_0(\tau)d\tau\right) d\alpha d\beta. \tag{1.404}$$

By using the following relations between $\Phi(t)$ and $v(t)$ as well as between $H_0(t)$ and $i(t)$:

$$\Phi(t) = \frac{1}{N}\int_0^t v(\tau)d\tau + \Phi_0, \tag{1.405}$$

$$H_0(t) = \frac{N}{\ell} i(t), \tag{1.406}$$

expression (1.404) can be written in the form:

$$\frac{1}{N} \int_0^t v(\tau) d\tau + \Phi_0 = \iint\limits_{\alpha \geq \beta} \mu(\alpha, \beta) \hat{\gamma}_{\alpha\beta} \left(\frac{N}{\ell} \int_0^t i(\tau) d\tau \right) d\alpha d\beta. \tag{1.407}$$

The last expression can be interpreted as a terminal voltage-current relation for a coil placed around a conducting magnetic cylinder. It is important to note that this terminal relation is valid for arbitrary time variations of current and voltage.

Formula (1.404) (as well as (1.407)) has been derived for a conducting magnetic cylinder of circular cross-section. However, this formula can be generalized for a conducting magnetic cylinder of "arbitrary" cross-section. For such a cylinder, the nonlinear diffusion equation has the form:

$$\frac{\partial^2 H}{\partial x^2} + \frac{\partial^2 H}{\partial y^2} = \sigma \frac{\partial B(H)}{\partial t}, \tag{1.408}$$

where for the case of abrupt magnetic transitions

$$B(H(t)) = B_m \text{ sign } H(t). \tag{1.409}$$

In Eq. (1.408), x and y are coordinates in the cylinder cross-section plane, while the magnetic field is always normal to this plane.

Let us now assume that the initial value of the magnetic flux density in the cylinder is equal to $-B_m$. Let us also assume that $H_0(t)$ varies with time as is shown in Figure 1.31. By using the same line of reasoning as before, we conclude that positive rectangular fronts of magnetic flux density are formed and moved inwards for odd time intervals ($t_{2k} < t < t_{2k+1}$), while negative fronts are formed and moved inwards for even time intervals ($t_{2k+1} < t < t_{2k}$). Next, we shall transform nonlinear diffusion Eqs. (1.408–1.409) into rate independent forms for odd and even time intervals. To this end, we introduce the function

$$w_{2k+1}^{+} = \int_{t_{2k}}^{t} H(\tau) d\tau, \quad \left(H(t) = \frac{\partial w_{2k}^{+}}{\partial t} \right). \tag{1.410}$$

By integrating Eq. (1.408) with respect to time from t_{2k} to t and by using formula (1.409), we derive:

$$\nabla^2 w_{2k+1}^{+} = \sigma \left[B_m \text{ sign} \left(\frac{\partial w_{2k}^{+}}{\partial t} \right) - B(t_{2k}) \right]. \tag{1.411}$$

The last equation is valid within the region $\Omega_{2k+1}^{+}(t)$ occupied by a newly formed positive front. In this region, function w_{2k}^{+} is monotonically increased with time and, consequently, sign $\left(\frac{\partial w_{2k}^{+}}{\partial t}\right) = 1$. In the same region, we also have $B(t_{2k}) = -B_m$. As a result, Eq. (1.411) takes the form of the Poisson equation:

$$\nabla^2 w_{2k+1}^{+} = 2\sigma B_m. \tag{1.412}$$

The solution of the last equation is subject to the following boundary conditions:

$$w_{2k+1}^{+}(t)|_L = w_{0,2k+1}^{+}(t) = \int_{t_{2k}}^{t} H_0(\tau)d\tau, \tag{1.413}$$

$$w_{2k+1}^{+}(t)|_{L_{2k+1}^{+}(t)} = 0, \tag{1.414}$$

$$\frac{\partial w_{2k+1}^{+}}{\partial \nu}|_{L_{2k+1}^{+}(t)} = 0, \tag{1.415}$$

where ν is a normal to the moving boundary $L_{2k+1}^{+}(t)$ of the region $\Omega_{2k+1}^{+}(t)$.

Boundary conditions (4.414) and (4.415) at the moving boundary $L_{2k+1}^{+}(t)$ follow from the fact that magnetic field and tangential component of electric field are equal to zero at the points of $L_{2k+1}^{+}(t)$ for the time interval $t_{2k} < \tau < t$, that is, before the arrival of the positive front.

During even time intervals, $H_0(t) < 0$ and negative fronts of the magnetic flux density are formed and they extend inwards with time. By introducing the function

$$w_{2k}^{-} = \int_{t_{2k-1}}^{t} H(\tau)d\tau, \tag{1.416}$$

and by literally repeating the same line of reasoning as before, we end up with the following boundary value problem:

$$\nabla^2 w_{2k}^{-} = -2\sigma B_m, \tag{1.417}$$

$$w_{2k}^{-}(t)|_L = w_{0,2k}(t) = \int_{t_{2k-1}}^{t} H_0(\tau)d\tau, \tag{1.418}$$

$$w_{2k}^{-}(t)|_{L_{2k}^{-}(t)} = 0, \tag{1.419}$$

$$\frac{\partial w_{2k}^{-}}{\partial \nu}|_{L_{2k}^{-}(t)} = 0. \tag{1.420}$$

It is interesting to note that nonlinear diffusion Eq. (1.408) is transformed into linear Poisson Eqs. (1.412) and (1.417). However, nonlinearity of the problem did not disappear; it is present in boundary conditions (1.414)–(1.415) and (1.419)–(1.420), which should be satisfied at moving boundaries $L^+_{2k+1}(t)$ and $L^-_{2k}(t)$, respectively. Locations of these boundaries are not known beforehand and should be determined from the fact that zero Dirichlet and Neumann boundary conditions must be simultaneously satisfied at these boundaries. In other words, formulas (1.412)–(1.415) and (1.417)–(1.420) define boundary value problems with moving boundaries, and these moving boundaries are the source of nonlinearity.

The following properties can be inferred by inspecting boundary-value problems (1.412)–(1.415) and (1.417)–(1.420).

Rate Independence Property.

Boundary value problems (1.412)–(1.415) and (1.417)–(1.420) are rate independent because there are no time derivatives in the formulations of these boundary value problems. Consequently, the instantaneous positions and shapes of moving boundaries $L^+_{2k+1}(t)$ and $L^-_{2k}(t)$ are determined by instantaneous boundary values of $w^+_{0,2k+1}(t)$ and $w^-_{0,2k}(t)$, respectively.

Symmetry Property.

Boundary value problems (1.412)–(1.415) and (1.417)–(1.420) have identical (up to a sign) mathematical structures. This suggests that, if $|w^-_{0,2k}| = |w^+_{0,2k+1}|$, then the corresponding boundaries L^-_{2k} and L^+_{2k+1} are identical. In other words, there is complete **symmetry** between inward motions of positive and negative fronts.

Now we introduce the function

$$w_0(t) = \int_0^t H_0(\tau)d\tau. \tag{1.421}$$

It is clear that function $w_0(t)$ is a sum of the appropriate functions $w^+_{0,2k}(t)$ and $w^-_{0,2k}(t)$. It is also clear that $w_0(t)$ achieves local maxima at $t = t_{2k+1}$ and local minima at $t = t_{2k}$. Next, we intend to show that $\Phi(t)$ vs. $w_0(t)$ is a rate-independent hysteretic relation. The rate independence of the above relation directly follows from the previously stated rate independence property. It is also true that $\Phi(t)$ vs. $w_0(t)$ is a hysteretic relation. Indeed, the current value of $\Phi(t)$ depends not only on the current value of $w_0(t)$ but on the past extremum values of $w_0(t)$ as well. This is because the past extremum values of $w_0(t)$ determine the final locations and shapes of positive and negative rectangular fronts of B that were generated in the past. These past and motionless rectangular fronts affect current values of $\Phi(t)$. It is also apparent that there are reversals of $\Phi(t)$ at extremum values of $w_0(t)$. In other words, new branches of Φ vs. w_0 relation are formed after local extrema of $w_0(t)$. The previous discussion clearly suggests that Φ vs. w_0 is a rate independent hysteretic relation. Next, we shall demonstrate that this hysteretic relation exhibits the wiping-out and congruency properties. Indeed, every monotonic increase (or decrease) of $w_0(t)$ results in the formation of a positive (or negative) rectangular front of the magnetic flux density, which extends inwards. This moving front will wipe out those previous rectangular fronts if they correspond to those previous extremum values of $w_0(t)$, which are exceeded by a new extremum value of $w_0(t)$. In this way, the effect of those previous extremum values of $w_0(t)$ on the future values of magnetic flux $\phi(t)$ is completely eliminated. This means that the wiping-out property holds. Now we shall demonstrate the validity of the congruency property. Consider two different boundary conditions: $w_0^{(1)}(t)$ and $w_0^{(2)}(t)$. Suppose that $w_0^{(1)}(t)$ and $w_0^{(1)}(t)$ have different past histories (different past extrema) but, starting from some instant of time, they vary monotonically back-and-forth between the same two extremum (reversal) values. It is apparent that these back-and-forth variations of $w_0^{(1)}(t)$ and $w_0^{(2)}(t)$ will affect in the identical way the same surface layers of the conducting cylinder. Consequently, those variations will result in equal increments of the magnetic flux, which is tantamount to the congruency of the corresponding minor loops. Since the wiping-out and congruency properties constitute necessary and sufficient conditions for applicability of the Preisach model, we conclude that the Φ vs. w_0 relation can be represented by the Preisach model. As a result, we arrive at the following representation of eddy current hysteresis:

$$\Phi(t) = \iint\limits_{\alpha \geq \beta} \mu(\alpha, \beta)\hat{\gamma}_{\alpha\beta}\left(\int_0^t H_0(\tau)d\tau\right) d\alpha d\beta, \tag{1.422}$$

which is valid for cylinders of arbitrary cross-sections.

It is worthwhile to stress two remarkable points related to the above result. First, memory effects and dynamic effects of eddy current hysteresis are clearly separated. The memory effects are taken into account by the structure of the Preisach model, while the dynamic effects are accounted for by the nature of the input $\left(\int_0^t H_0(\tau)d\tau\right)$ to this model. Second, the last formula suggests that the Preisach model can be useful for the description of hysteresis exhibited by spatially distributed systems. This is in contrast with the traditionally held point of view that the Preisach model describes only local hysteretic effects in magnetic materials.

Next, we turn to the discussion of properties of function $\mu(\alpha, \beta)$ in formula (1.422). By using the symmetry property, it can be inferred **that the same increments of** $w_0(t)$, which **occurred after different extremum values of** $w_0(t)$, **result in the same increments of** $\Phi(t)$. This fact implies that the integral

$$F(\alpha,\beta) = \iint\limits_{T(\alpha,\beta)} \mu(\alpha',\beta')d\alpha' d\beta' \tag{1.423}$$

over a triangle $T(\alpha,\beta)$, defined by inequalities $\alpha' < \alpha, \beta' > \beta, \alpha' - \beta' \geq 0$, does not depend on α and β separately but rather on the difference $\alpha - \beta$. In other words, the value of the above integral is invariant with respect to parallel translations of the triangle $T(\alpha,\beta)$ along the line $\alpha = \beta$. This is only possible if

$$\mu(\alpha,\beta) = \mu(\alpha - \beta). \tag{1.424}$$

This means that function μ assumes constant values along the lines $\alpha - \beta$ = const. By using this fact, it can be easily observed that function μ can be found by measuring only the ascending (or descending) branch of the major loop of Φ vs. w_0 hysteretic nonlinearity. It can also be seen that any path traversed on the (w_0, Φ) plane is piecewise congruent to the ascending branch of the major loop. (See Fig. 1.33.) Thus, Φ vs. w_0 hysteretic nonlinearity is completely characterized by the ascending branch of the major loop. This branch can be found experimentally by measuring the step response of eddy current hysteresis. Indeed, by assuming initial condition $B(0) = -B_m$, we apply the field $H_0(t) = s(t)$, where $s(t)$ is the unit step function. We can then measure flux $\Phi_s(t)$, which corresponds to $w_0(t) = t$. By excluding time t, we find the function $\Phi_s(w_0)$, which describes the ascending branch of the major loop. Thus, we arrive at the remarkable conclusion that **nonlinear (and dynamic) eddy current hysteresis can be fully characterized by a step response.**

It is clear from the previous discussions that the ascending branch of the major loop can also be experimentally found by measuring response $\Phi(t)$ to any monotonically increasing function $w_0 = X(t)$, that is, to any positive and sufficiently large current $i(t)$. Indeed, for any monotonically increasing function $w_0 = X(t)$, we can find the inverse function $t = X^{-1}(w_0)$. By substituting the latter function into response $\Phi(t)$, we find the ascending branch $\Phi(X^{-1}(w_0))$ of the major loop. By using this branch, we can predict eddy current hysteresis for arbitrary time variations of current $i(t)$.

Boundary value problems (1.412)–(1.415) and (1.417)–(1.420) can be used for a very elegant derivation of the formula for the front $z_0(t)$ in the case of plane boundary, that is, in the 1D case. In that case, the boundary value problem (1.412)–(1.415) is reduced to:

$$\frac{d^2w}{dz^2} = 2\sigma B_m \qquad if\ 0 < z < z_0(t), \tag{1.425}$$

$$w(0,t) = w_0(t), \tag{1.426}$$

$$w(z_0(t), t) = 0, \tag{1.427}$$

$$\frac{dw(z,t)}{dz}|_{z_0(t)} = 0. \tag{1.428}$$

The solution to Eq. (1.425) that satisfies the boundary conditions (1.426) and (1.428) has the form

$$w(z,t) = \sigma B_m z^2 - 2\sigma B_m z z_0(t) + w_0(t). \tag{1.429}$$

To find $z_0(t)$, the boundary condition (1.427) is used, which leads to

$$-\sigma B_m z_0^2(t) + w_0(t) = 0. \tag{1.430}$$

The last expression yields

$$z_0(t) = \sqrt{\frac{w_0(t)}{\sigma B_m}} = \left(\frac{\int_0^t H_0(\tau) d\tau}{\sigma B_m} \right)^{\frac{1}{2}}, \tag{1.431}$$

which is identical to formula (1.42).

To find $H(z,t)$, we differentiate both sides of Eq. (1.429) with respect to time and recall (1.410), which yields

$$H(z,t) = -2\sigma B_m z \frac{dz_0(t)}{dt} + H_0(t). \tag{1.432}$$

From formula (1.431) we find

$$\frac{dz_0(t)}{dt} = \frac{H_0(t)}{2\sigma B_m z_0(t)}. \tag{1.433}$$

By substituting the last expression into formula (1.432), we arrive at

$$H(z,t) = H_0(t) \left(1 - \frac{z}{z_0(t)} \right), \tag{1.434}$$

which is consistent with formula (1.37).

By using formula (1.431), we can derive the expression for the ascending branch of the major loop of eddy current hysteresis in the case of a magnetically nonlinear conducting lamination. In the above case, we have:

$$\Phi = -\Phi_m + 2B_m z_0(t). \tag{1.435}$$

By substituting formula (1.431) in the last expression, we find:

$$\Phi = -\Phi_m + 2 \left[\frac{B_m w_0}{\sigma} \right]^{\frac{1}{2}}, \tag{1.436}$$

which is the equation for the ascending branch. It is clear from this equation that this branch has a vertical (infinite) initial slope, that is, the slope at $w_0 = 0$ (see Fig. 1.33). For this reason, weight function $\mu(\alpha, \beta)$ is singular and must be understood as a distribution. Although we have arrived at this conclusion for the case of lamination, it is of a general nature. This is because at initial stages (i.e., for small penetration depths) nonlinear diffusion in conducting bodies with curvilinear boundaries occurs almost (i.e., asymptotically) in the same way as in the case of plane (flat) boundaries. As a result, the ascending branches always have vertical initial slopes and functions $\mu(\alpha, \beta)$ are always singular.

The above difficulty can be completely circumvented if we consider the inverse w_0 vs. Φ hysteretic relation. This relation is shown in Fig. 1.35. It can be mathematically shown that this inverse hysteretic relation can also be represented by the Preisach model

$$w_0(t) = \iint_{\alpha \geq \beta} \nu(\alpha, \beta) \hat{\gamma}_{\alpha\beta} \Phi(t) d\alpha d\beta \tag{1.437}$$

with the weight function $\nu(\alpha, \beta)$, which has the following property:

$$\nu(\alpha, \beta) = \nu(\alpha - \beta). \tag{1.438}$$

The last property suggests that function $\nu(\alpha, \beta)$ can be fully determined by using the ascending branch of the major loop. In the case of lamination, this branch can be found analytically. Indeed, by using formula (1.436), we derive

$$w_0 = \frac{\sigma}{B_m} \left[\frac{\Phi + \Phi_m}{2} \right]^2, \tag{1.439}$$

which is the equation for the ascending branch.

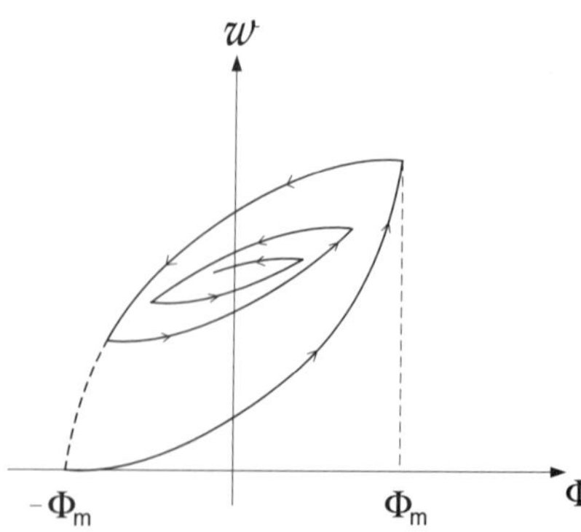

Fig. 1.35

To find function ν, we shall invoke formula (A.22) from Appendix A. For the case when property (1.438) is valid, function $\mathcal{F}(\alpha, \beta)$ also depends on $\alpha - \beta$, and formula (A.22) takes the form:

$$\nu(\alpha) = \frac{d^2 F}{d\alpha^2}. \tag{1.440}$$

For the ascending branch (1.439), function $\mathcal{F}$ and argument α can be identified as follows:

$$\mathcal{F} = w_o, \qquad \alpha = \Phi. \tag{1.441}$$

By substituting the last equalities into expression (1.439) and then by using formula (1.440), we derive:

$$\nu(\alpha - \beta) = \frac{\sigma}{2B_m} = \text{const.} \tag{1.442}$$

In the general case, function $\nu(\alpha - \beta)$ can be found experimentally. The best way to see how it can be accomplished is to write formula (1.437) in terms of current $i(t)$ and voltage $v(t)$. To this end, we employ formulas (1.428), (1.406), and (1.405) and after simple transformations we arrive at the following expression:

$$\frac{N}{\ell}\int_0^t i(\tau)d\tau = \iint\limits_{\alpha \geq \beta} \nu(\alpha - \beta)\hat{\gamma}_{\alpha\beta}\left[\frac{1}{N}\int_0^t v(\tau)d\tau + \Phi_0\right] d\alpha d\beta. \tag{1.443}$$

The last expression can be construed as a terminal voltage-current relation for a coil placed around a conducting magnetic cylinder. The difference of this relation from the one given by formula (1.407) is that terminal relation (1.443) is in a "voltage controlled" form. This suggests that by applying any positive voltage $v(t)$ (for instance $v(t) = s(t)$) and by measuring $i(t)$, we can find corresponding functions $\Phi(t)$ and $w_0(t)$. By excluding time t from those functions, we can find a relation $w_0(\Phi)$, which represents the ascending branch. This relation can be used for determination of $\nu(\alpha - \beta)$, or it can be directly used to predict current $i(t)$ for arbitrary variations of voltage $v(t)$.

Finally, we remark that formula (1.422) can be generalized to the case when abrupt magnetic transitions are described by rectangular hysteresis loops (see Fig. 1.7). It can be easily shown that in this case formula (1.442) can be modified as follows:

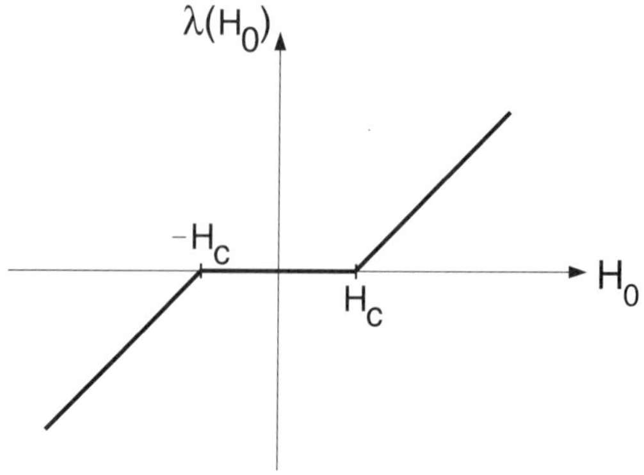

Fig. 1.36

$$\Phi(t) = \iint\limits_{\alpha\geq\beta} \mu(\alpha,\beta)\hat{\gamma}_{\alpha\beta}\left(\int_0^t \lambda(H_0(\tau))d\tau\right) d\alpha d\beta, \tag{1.444}$$

where function $\lambda(H_0)$ is defined as (see Fig. 1.36):

$$\lambda(H_0) = (H_0 - H_c)s(H_0 - H_c) + (H_0 + H_c)s(-H_0 - H_c), \tag{1.445}$$

and $s(\cdot)$ is the unit step function.

As approximations, the last two formulas can be used in situations when actual hysteresis loops are close to rectangular ones. In those situations, the same simple experiments as described before can be used for identification of function μ.

REFERENCES

[1] P. Agarwal, AIEE Trans. Commun. Electron. **78**, pp. 169–179, (1959).
[2] V. Arkad'ev, in the book "Practical Problems of Electromagnetism," published by the Division of Technical Sciences, USSR National Academy of Sciences, pp. 19–42, (1939).
[3] G. Barenblatt, "Similarity, Self-Simularity and Intermediate Asymptotics," Consultants Bureau, New York, (1979).
[4] G. Barenblatt, Prikl. Math. Mech., **16**, 11, pp. 67–78, (1952).
[5] G. Bertotti, "Hysteresis in Magnetism," Academic Press (forthcoming).
[6] M. Brokate and J. Sprekels, "Hysteresis and Phase Transitions," Springer, (1996).

[7] N. Goldenfeld, "Lectures on Phase Transitions and the Renormalization Group," Addison-Wesley, (1992).
[8] J.A. Kong, "Electromagnetic Wave Theory," John Wiley, (1986).
[9] M.A. Krasnoselskii and A. Pokrovskii, "Systems with Hysteresis," Nauka, Moscow, (1983).
[10] W. MacLean, Journal of Applied Physics, **25**, pp. 1267–1270, (1954).
[11] I.D. Mayergoyz, Izvestia USSR Academy of Sciences, Energetika and Transport, No. 5, pp. 135–141, (1967).
[12] I.D. Mayergoyz, Automatic Control and Remote Sensing, No. 10, pp. 137–146 (English Translation pp. 1670–1678), (1969).
[13] I.D. Mayergoyz, Archiv für Electrotechnik, **64**, No. 314, pp. 153–162, (1981).
[14] I.D. Mayergoyz, IEEE Transactions on Magnetics, **18**, pp. 1716–1718, (1982).
[15] I.D. Mayergoyz, Physical Review Letters, **56**, No. 15, pp. 1518–1521, (1986).
[16] I.D. Mayergoyz, "Mathematical Models of Hysteresis," Springer, (1991).
[17] H.M. McConnell, AIEE Transactions, **73**, pp. 226–235, (1954).
[18] L.R. Neumann, "Skin Effect in Ferromagnetics," Gosenergoisdat, Moscow, (1949).
[19] F.Z. Preisach, Zeitschrift für Physik, **94**, pp. 227–302, (1935).
[20] A. Visintin, "Differential Models of Hysteresis," Springer, (1994).
[21] W. Wolman and H. Kaden, Z. Techn. Phys. **13**, pp. 330–345, (1932).
[22] Ya. Zeldovich and A. Kompaneyets, in the book "Collection of Papers Dedicated to A.F. Ioffe on the Occasion of his Seventieth Birthday," Publication of USSR National Academy of Sciences, (1950).

CHAPTER 2

Diffusion of Electromagnetic Fields in Magnetically Nonlinear Conducting Media (Vector Polarization)

2.1 NONLINEAR DIFFUSION OF CIRCULARLY POLARIZED ELECTROMAGNETIC FIELDS IN ISOTROPIC MEDIA

In this chapter, we shall continue our discussion of nonlinear diffusion of plane electromagnetic waves in the conducting half-space. In the previous chapter, our discussion of this problem was carried out for the case of linearly polarized magnetic fields. In that case, the analysis was reduced to the solution of a **scalar** nonlinear diffusion equation. In many applications, the magnetic field is not linearly polarized. For this reason, it is of importance to consider nonlinear diffusion of arbitrary polarized electromagnetic fields. As was shown in Section 1.1, the above problem is reduced to the solution of the **vector** nonlinear diffusion equation:

$$\frac{\partial^2 \mathbf{H}}{\partial z^2} = \sigma \frac{\partial \mathbf{B}(\mathbf{H})}{\partial t}, \tag{2.1}$$

or two coupled scalar nonlinear diffusion equations:

$$\frac{\partial^2 H_x}{\partial z^2} = \sigma \frac{\partial B_x(H_x, H_y)}{\partial t}, \tag{2.2}$$

$$\frac{\partial^2 H_y}{\partial z^2} = \sigma \frac{\partial B_y(H_x, H_y)}{\partial t}. \tag{2.3}$$

This obviously raises the level of mathematical difficulties. However, these difficulties can be completely circumvented in the case of isotropic media and circular polarization of electromagnetic fields. This is due to the high degree of **symmetry** of the problem in the above case. The analysis of nonlinear diffusion for the circular polarization of an electromagnetic field is of interest for the following two reasons. First, linear and circular polarizations can be viewed as two limiting cases for other types of polarization. Therefore, the solution of the problem for these two limiting cases may provide some insights in how the surface impedance of magnetically nonlinear conducting media depends on the type of polarization. Second, elliptical polarizations of electromagnetic fields can be treated as perturbations of the circular polarization. This will allow us to use extensively the perturbation technique for the solution of the vector nonlinear diffusion Eq. (2.1) in the case of elliptical polarizations of electromagnetic fields.

However, we shall first proceed with the discussion of nonlinear diffusion in the case of circular polarization and isotropic media. For isotropic media, the Cartesian components of the magnetic flux density are related to the Cartesian components of the magnetic field by the formulas

$$B_x(H_x, H_y) = \mu\left(\sqrt{H_x^2 + H_y^2}\right) H_x, \tag{2.4}$$

$$B_y(H_x, H_y) = \mu\left(\sqrt{H_x^2 + H_y^2}\right) H_y, \tag{2.5}$$

where $\mu(|\mathbf{H}|) = \mu\left(\sqrt{H_x^2 + H_y^2}\right)$ is the magnetic permeability of isotropic conducting media.

It is clear that we deal with the case of unhysteretic media. The case of isotropic hysteretic media will be treated at the end of this section.

By substituting formulas (2.4) and (2.5) into Eqs. (2.2) and (2.3), respectively, we end up with the following coupled nonlinear diffusion equations:

$$\frac{\partial^2 H_x}{\partial z^2} = \sigma \frac{\partial}{\partial t}\left[\mu\left(\sqrt{H_x^2 + H_y^2}\right) H_x\right], \tag{2.6}$$

$$\frac{\partial^2 H_y}{\partial z^2} = \sigma \frac{\partial}{\partial t}\left[\mu\left(\sqrt{H_x^2 + H_y^2}\right) H_y\right]. \tag{2.7}$$

We shall be interested in time-periodic solutions to the above equations subject to the following boundary conditions:

$$H_x(0, t) = H_m \cos(\omega t + \theta_0), \tag{2.8}$$

$$H_y(0,t) = H_m \sin(\omega t + \theta_0), \tag{2.9}$$

$$H_x(\infty,t) = 0, \tag{2.10}$$

$$H_y(\infty,t) = 0. \tag{2.11}$$

The boundary conditions (2.8) and (2.9) correspond to the circular polarization of the magnetic field, whereas the boundary conditions (2.10) and (2.11) reflect the fact that the magnetic field decays to zero.

Now we shall make the following very important observation. The mathematical structure of nonlinear partial differential Eqs. (2.6) and (2.7) as well as of boundary conditions (2.8)–(2.11) is invariant with respect to rotations of x- and y-axes around the z-axis. In other words, the mathematical form of the above equations and boundary conditions will remain the same for any choice of x- and y-axes in the plane $z = 0$. This suggests that the solution of the boundary value problem (2.6)–(2.11) should also be invariant with respect to the rotations of the x- and y-axes*. This, in turn, implies that the magnetic field is circularly polarized everywhere within the conducting media:

$$H_x(z,t) = H(z)\cos(\omega t + \theta(z)), \tag{2.12}$$

$$H_y(z,t) = H(z)\sin(\omega t + \theta(z)). \tag{2.13}$$

Next, we shall formally show that the circularly polarized solution (2.12)–(2.13) is consistent with the mathematical form of the boundary value problem (2.6)–(2.11). First, it is clear from formulas (2.12) and (2.13) that:

$$|\mathbf{H}(z)| = \sqrt{H_x^2(z) + H_y^2(z)} = H(z). \tag{2.14}$$

This means that the magnitude of the magnetic field and, consequently, the magnetic permeability $\mu(|\mathbf{H}|)$ do not change with time at every point within the conducting media.

Next, we represent formulas (2.12) and (2.13) in the phasor form:

$$\hat{H}_x(z) = H(z)e^{j\theta(z)}, \tag{2.15}$$

*Strictly speaking, this statement is valid when the solution to the boundary value problem (2.6)–(2.11) is unique, which is assumed here on physical grounds. In the case when there are many (or infinite number of solutions, the symmetry of equations may not be reflected in the symmetry of each individual solution, but rather in the symmetry of the overall pattern of all solutions. this is the so-called "spontaneous symmetry breaking" phenomenon.

$$\hat{H}_y(z) = -jH(z)e^{j\theta(z)}, \tag{2.16}$$

where, as before, the symbol "ˆ" is used for the notation of phasors, while $j = \sqrt{-1}$.

It is apparent from expressions (2.14), (2.15), and (2.16) that

$$H(z) = |\hat{H}_x(z)| = |\hat{H}_y(z)|, \tag{2.17}$$

and

$$\mu\left(|\mathbf{H}(z)|\right) = \mu\left(|\hat{H}_x(z)|\right) = \mu\left(|\hat{H}_y(z)|\right). \tag{2.18}$$

By using phasors (2.15) and (2.16) as well as the formula (2.18), it is easy to transform the boundary value problem (2.6)–(2.11) into the following boundary value problems:

$$\frac{d^2\hat{H}_x(z)}{dz^2} = j\omega\sigma\mu\left(|\hat{H}_x(z)|\right)\hat{H}_x(z), \tag{2.19}$$

$$\hat{H}_x(0) = H_m e^{j\theta_0}, \tag{2.20}$$

$$\hat{H}_x(\infty) = 0, \tag{2.21}$$

and

$$\frac{d^2\hat{H}_y(z)}{dz^2} = j\omega\sigma\mu\left(|\hat{H}_y(z)|\right)\hat{H}_y(z), \tag{2.22}$$

$$\hat{H}_y(0) = -jH_m e^{j\theta_0}, \tag{2.23}$$

$$\hat{H}_y(\infty) = 0. \tag{2.24}$$

This **exact** transformation of the boundary value problem (2.6)–(2.11) into boundary value problems (2.19)–(2.21) and (2.22)–(2.24) can be construed as a mathematical proof that the circular polarization of the incident wave is preserved everywhere within the magnetically nonlinear conducting media. This also proves the remarkable fact that there are no higher-order time-harmonics of the magnetic field anywhere within the media despite its nonlinear magnetic properties.

From the purely mathematical point of view, the achieved simplification of the boundary value problem (2.6)–(2.11) is tremendous. First, partial differential Eqs. (2.6) and (2.7) are exactly reduced to ordinary differential Eqs. (2.19) and (2.22), respectively. Second, the boundary value problem (2.6)–(2.11) for coupled equations is reduced to two completely decoupled boundary value problems (2.19)–(2.21) and (2.22)–(2.24). Finally, the decoupled boundary value problems have identical mathematical structures. As a result, the same solution technique can be applied to both of them.

It turns out that simple analytical solutions to Eqs. (2.19) and (2.22) can be found in the case of a power law approximation of a magnetization curve. This approximation is given by the expression:

$$H = \left(\frac{B}{k}\right)^n, \qquad (n > 1), \tag{2.25}$$

which can also be rewritten as follows:

$$B = kH^{\frac{1}{n}}. \tag{2.26}$$

This approximation implies the following formula for the magnetic permeability:

$$\mu(H) = kH^{\frac{1}{n}-1}. \tag{2.27}$$

Sketches of the B vs. H and μ vs. H relations corresponding to the power law approximation are shown in Figs. 2.1 and 2.2, respectively. It is clear that for the above approximation the magnetic permeability is decreased as the magnetic field is increased. Thus, this approximation takes magnetic saturation of media into account. However, this approximation idealizes the actual magnetic properties of media for very small values of magnetic field. Namely, the permeability approaches infinity as the field approaches zero. The physical implications of this idealization will be discussed later.

By using formula (2.27), we find the following expression for the magnetic permeability μ_m at the boundary of media:

$$\mu_m = kH_m^{\frac{1}{n}-1}. \tag{2.28}$$

By combining formulas (2.27) and (2.28), we can exclude the coefficient k from expression (2.27):

$$\mu(H) = \mu_m \left(\frac{H}{H_m}\right)^{\frac{1}{n}-1}. \tag{2.29}$$

From formulas (2.18) and (2.29), we obtain:

$$\mu\left(|\hat{H}_x(z)|\right) = \mu_m \left|\frac{\hat{H}_x(z)}{H_m}\right|^{\frac{1}{n}-1}, \tag{2.30}$$

$$\mu\left(|\hat{H}_y(z)|\right) = \mu_m \left|\frac{\hat{H}_y(z)}{H_m}\right|^{\frac{1}{n}-1}. \tag{2.31}$$

By substituting expressions (2.30) and (2.31) into Eqs. (2.19) and (2.22), respectively, we arrive at the following boundary value problems:

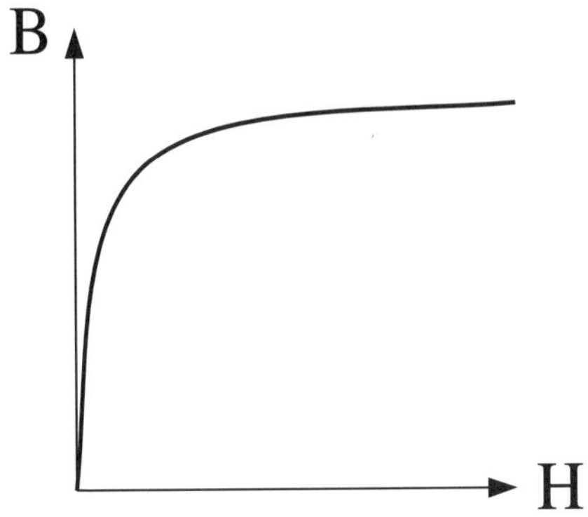

Fig. 2.1

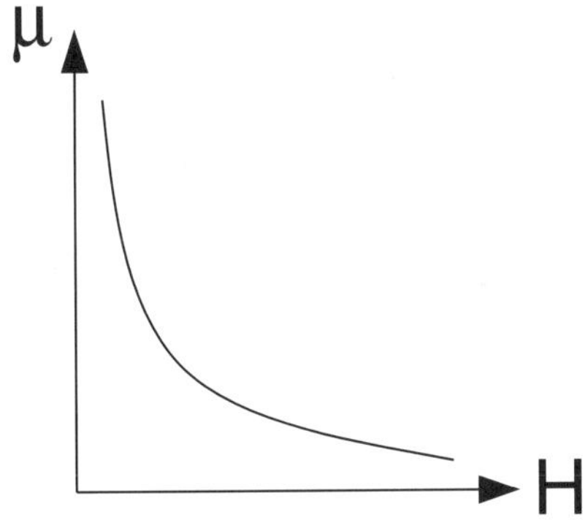

Fig. 2.2

$$\frac{d^2\hat{H}_x(z)}{dz^2} = j\omega\sigma\mu_m \left|\frac{\hat{H}_x(z)}{H_m}\right|^{\frac{1}{n}-1} \hat{H}_x(z), \tag{2.32}$$

$$\hat{H}_x(0) = H_m e^{j\theta_0}, \tag{2.33}$$

$$\hat{H}_x(\infty) = 0, \tag{2.34}$$

and

$$\frac{d^2\hat{H}_y(z)}{dz^2} = j\omega\sigma\mu_m \left|\frac{\hat{H}_y(z)}{H_m}\right|^{\frac{1}{n}-1} \hat{H}_y(z), \tag{2.35}$$

$$\hat{H}_y(0) = -jH_m e^{j\theta_0}, \tag{2.36}$$

$$\hat{H}_y(\infty) = 0. \tag{2.37}$$

First, we find the solution to the boundary value problem (2.32)–(2.34). This solution will be sought in the form:

$$\hat{H}_x(z) = \begin{cases} \hat{H}_m \left(1 - \frac{z}{z_0}\right)^{\alpha}, & \text{if } 0 \le z \le z_0, \\ 0, & \text{if } z \ge z_0, \end{cases} \tag{2.38}$$

where

$$\hat{H}_m = H_m e^{j\theta_0}, \tag{2.39}$$

$$\alpha = \alpha' + j\alpha''. \tag{2.40}$$

It is clear that the function (2.38) satisfies the boundary conditions (2.33) and (2.34). Next, we shall choose parameters z_0, α' and α'' in the way that Eq. (2.32) will be also satisfied. To this end, we shall rewrite formula (2.38) as follows:

$$\frac{\hat{H}_x(z)}{\hat{H}_m} = \left(1 - \frac{z}{z_0}\right)^{\alpha}. \tag{2.41}$$

The last expression is written in an abbreviated form with the tacit understanding that it is valid for $0 \le z \le z_0$, while for $z \ge z_0$ the right-hand side is equal to zero. Similar abbreviations will be tacitly used in subsequent formulas when they are appropriate.

From formulas (2.39)–(2.41), we find:

$$\left|\frac{\hat{H}_x(z)}{H_m}\right| = \left(1 - \frac{z}{z_0}\right)^{\alpha'}, \tag{2.42}$$

which leads to

$$\left|\frac{\hat{H}_x(z)}{H_m}\right|^{\frac{1}{n}-1} = \left(1 - \frac{z}{z_0}\right)^{-\frac{\alpha'(n-1)}{n}}. \tag{2.43}$$

By substituting formulas (2.41) and (2.43) into Eq. (2.32), we arrive at:

$$\alpha(\alpha - 1)\hat{H}_m \left(1 - \frac{z}{z_0}\right)^{\alpha-2} = j\omega\sigma\mu_m \hat{H}_m z_0^2 \left(1 - \frac{z}{z_0}\right)^{\alpha - \frac{\alpha'(n-1)}{n}}. \tag{2.44}$$

It is clear that the last equality will hold, if the following two conditions are satisfied:

$$2 = \frac{\alpha'(n-1)}{n}, \tag{2.45}$$

and:

$$\alpha(\alpha - 1) = j\omega\sigma\mu_m z_0^2. \tag{2.46}$$

From formula (2.45), we find:

$$\alpha' = \frac{2n}{n-1}. \tag{2.47}$$

Next, we shall use the characteristic Eq. (2.46) to determine α'' and z_0. To this end, we shall represent this equation in the form:

$$(\alpha' + j\alpha'')(\alpha' - 1 + j\alpha'') = j\omega\sigma\mu_m z_0^2, \tag{2.48}$$

which is equivalent to the following two equations:

$$\alpha'(\alpha' - 1) - (\alpha'')^2 = 0, \tag{2.49}$$

$$\alpha''(2\alpha' - 1) = \omega\sigma\mu_m z_0^2. \tag{2.50}$$

By using formulas (2.47) and (2.49), we find

$$\alpha'' = \frac{\sqrt{2n(n+1)}}{n-1}. \tag{2.51}$$

Next, by substituting this expression for α'' into Eq. (2.50), we arrive at:

$$(2\alpha' - 1)\sqrt{\alpha'(\alpha' - 1)} = \omega\sigma\mu_m z_0^2. \tag{2.52}$$

By taking formula (2.47) into account in the last equation, we finally obtain:

$$z_0 = \frac{[2n(n+1)(3n+1)^2]^{\frac{1}{4}}}{\sqrt{\omega\sigma\mu_m}(n-1)}. \tag{2.53}$$

Formulas (2.38), (2.47), (2.51), and (2.53) completely define the solution to the boundary value problem (2.32)–(2.34). The boundary value problem (2.35)–(2.37) is identical (up to notations) to the boundary value problem (2.32)–(2.34). For this reason, the solution to the boundary value problem (2.35)–(2.37) can be written in the form:

$$\hat{H}_y(z) = -j\hat{H}_m\left(1 - \frac{z}{z_0}\right)^{\alpha}, \tag{2.54}$$

where α and z_0 are given by the expressions (2.40), (2.47), (2.51), and (2.53).

Solutions (2.38) and (2.54) are written in terms of phasors. We shall next transform them into time-domain forms (2.12) and (2.13). To this end, we shall first use the following transformations:

$$\begin{aligned}\hat{H}_x(z) &= H_m e^{j\theta_0}\left(1-\frac{z}{z_0}\right)^{\alpha'+j\alpha''}\\ &= H_m\left(1-\frac{z}{z_0}\right)^{\alpha'} e^{j\left[\theta_0+\alpha''\ln\left(1-\frac{z}{z_0}\right)\right]}.\end{aligned} \tag{2.55}$$

Now, by using the standard expression

$$H_x(z,t) = Re[\hat{H}_x(z)e^{j\omega t}] \tag{2.56}$$

and formula (2.55), (2.47), and (2.51), we derive

$$H_x(z,t) = H_m\left(1-\frac{z}{z_0}\right)^{\frac{2n}{n-1}}\cos\left[\omega t+\theta_0+\frac{\sqrt{2n(n+1)}}{n-1}\ln\left(1-\frac{z}{z_0}\right)\right]. \tag{2.57}$$

By repeating the same line of reasoning, we arrive at the following expression for $H_y(z,t)$:

$$H_y(z,t) = H_m\left(1-\frac{z}{z_0}\right)^{\frac{2n}{n-1}}\sin\left[\omega t+\theta_0+\frac{\sqrt{2n(n+1)}}{n-1}\ln\left(1-\frac{z}{z_0}\right)\right]. \tag{2.58}$$

In the last two formulas, parameter z_0 is given by expression (2.53).

Formulas (2.57) and (2.58) give the **exact** analytical solution to the boundary value problem (2.6)–(2.11) in the case of power law relation (2.27). Next, we shall analyze this solution. If we fix time t in the last two formulas and consider H_x and H_y as functions of z, then we can easily observe that on the interval $0 \le z \le z_0$ these functions have infinite numbers of zeros (infinite numbers of oscillations). It is also clear that the sequences of H_x-zeros and H_y-zeros converge to z_0. This is the result of logarithmic dependence of phase $\theta(z)$ on z. Indeed, by comparing the last two formulas with expressions (2.12) and (2.13), we find

$$\theta(z) = \theta_0 + \frac{\sqrt{2n(n+1)}}{n-1}\ln\left(1-\frac{z}{z_0}\right), \tag{2.59}$$

$$H(z) = H_m\left(1-\frac{z}{z_0}\right)^{\frac{2n}{n-1}}. \tag{2.60}$$

By using the last equation and formula (2.29), we find the following expression for the magnitude of magnetic flux density as a function of z:

$$B(z) = \mu_m \left(\frac{H(z)}{H_m} \right)^{\frac{1}{n}-1} H(z) = B_m \left(1 - \frac{z}{z_0} \right)^{\frac{2}{n-1}}, \qquad (2.61)$$

where the following notation is introduced:

$$B_m = \mu_m H_m. \qquad (2.62)$$

A typical plot of $B(z)$ as a function of z is shown in Fig. 2.3. It is apparent from this figure (as well as from formula (2.61) and other previous formulas) that there is a finite depth z_0 of penetration of electromagnetic fields into magnetically nonlinear conducting media. This can be explained by the fact that power law approximation (2.26) introduces idealization of magnetic properties of conducting media by allowing for the infinite growth of the magnetic permeability when the magnetic field tends to zero. This infinite growth in μ causes the complete attenuation of the magnetic field at the finite distance z_0. Actual z-variation of magnetic flux density $B(z)$, schematically shown in Fig. 2.4, exhibits a tail "1" at small values of $B(z)$. This tail is usually of no practical significance and can be neglected. As a result, the depth z_0 attains the physical meaning of the penetration depth of the "bulk" part of magnetic flux density.

Now, we proceed to the discussion of surface impedance in the case of circular polarization. To find this impedance, we shall use the following formulas for the electric field phasors:

$$\hat{E}_x(z) = -\frac{1}{\sigma}\frac{d\hat{H}_y(z)}{dz}, \quad \hat{E}_y(z) = \frac{1}{\sigma}\frac{d\hat{H}_x(z)}{dz}. \qquad (2.63)$$

By using the last formulas as well as expressions (2.38) and (2.54), we find:

$$\hat{E}_x(0) = \frac{\alpha}{\sigma z_0}\hat{H}_y(0), \quad \hat{E}_y(0) = -\frac{\alpha}{\sigma z_0}\hat{H}_x(0). \qquad (2.64)$$

From the last equations we obtain the following expression for the surface impedance:

$$\eta = \frac{\hat{E}_x(0)}{\hat{H}_y(0)} = -\frac{\hat{E}_y(0)}{\hat{H}_x(0)} = \frac{\alpha}{\sigma z_0}. \qquad (2.65)$$

Now, by invoking formulas (2.40), (2.47), (2.51), and (2.53), we represent the surface impedance in the form:

$$\eta = (a + jb)\sqrt{\frac{\omega \mu_m}{\sigma}} \qquad (2.66)$$

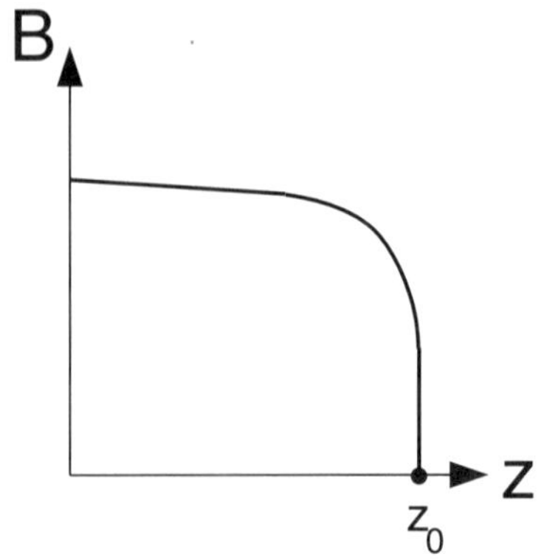

Fig. 2.3

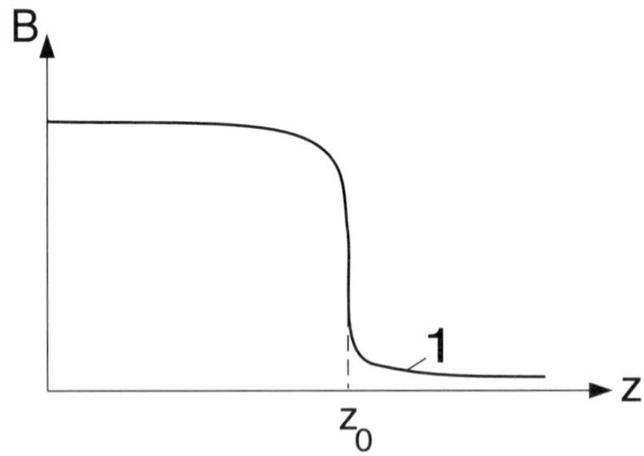

Fig. 2.4

where

$$a = \frac{2n}{[2n(n+1)(3n+1)^2]^{\frac{1}{4}}}, \tag{2.67}$$

$$b = \frac{\sqrt{2n(n+1)}}{[2n(n+1)(3n+1)^2]^{\frac{1}{4}}}. \tag{2.68}$$

It is also convenient to represent the surface impedance in the polar form

$$\eta = |\eta| e^{j\varphi} \tag{2.69}$$

where

$$|\eta| = \left(\frac{2n}{n+1}\right)^{\frac{1}{4}} \sqrt{\frac{\omega\mu_m}{\sigma}}, \qquad \tan\varphi = \frac{b}{a} = \sqrt{\frac{n+1}{2n}}. \tag{2.70}$$

By performing simple calculation in accordance with formulas (2.67), (2.68), and (2.70), we find that as n varies from 7 to ∞, coefficients a and b as well as $\sqrt{a^2+b^2}$ and $\tan\varphi$ vary within the following narrow limits:

$$.92 \leq a \leq .971, \tag{2.71}$$

$$.687 \leq b \leq .694, \tag{2.72}$$

$$1.15 \leq \sqrt{a^2+b^2} \leq 1.19, .707 \leq \tan\varphi \leq .76. \tag{2.73}$$

This suggests that with fair accuracy the surface impedance η can be represented by the formulas

$$\eta = (.945 + j.69)\sqrt{\frac{\omega\mu_m}{\sigma}}, \tag{2.74}$$

or

$$\eta = 1.17\sqrt{\frac{\omega\mu_m}{\sigma}}e^{j36.3^\circ}. \tag{2.75}$$

It is instructive to compare these results with the expression for the surface impedance obtained in the previous chapter (see Section 1.5) for the case of linear polarization of electromagnetic fields. This expression can be written in the same form as (2.69)–(2.70), however, the limits of variations for $\sqrt{a^2+b^2}$ and $\tan\varphi$ are appreciably different and specified here:

$$1.28 \leq \sqrt{a^2+b^2} \leq 1.35, \tag{2.76}$$

$$.49 \leq \tan\varphi \leq .71. \tag{2.77}$$

It was shown that in the case of linear polarization the value of surface impedance is most sensitive to the "squareness" χ of hysteresis loops. There is no parameter like that in formulas (2.67) and (2.68). This makes the above comparison between the cases of circular and linear polarization somewhat ambiguous. This ambiguity can be completely removed in the case of abrupt magnetic transitions described by rectangular magnetization curves. These magnetization curves can be obtained from the power law approximation (2.26) in the limit of n approaching infinity. In this limit, by using formulas (2.67), (2.68) and (2.70), we find

$$\sqrt{a^2+b^2} = 1.19, \tag{2.78}$$

$$\tan\varphi = 0.707. \tag{2.79}$$

In the case of linear polarization, for the same quantities we have (see formula (1.55) from Chapter 1):

$$\sqrt{a^2 + b^2} = 1.34, \tag{2.80}$$

$$\tan\varphi = 0.5. \tag{2.81}$$

The last two expressions have been obtained for sinusoidal variations of magnetic field at the boundary of conducting half-space. In the case of sinusoidal variations of electric field at the boundary, the last two expressions are modified as follows (see formulas (1.71) and (1.72)):

$$\sqrt{a^2 + b^2} = 1.47, \tag{2.82}$$

$$\tan\varphi = .424, \tag{2.83}$$

which makes the discrepancy with the case of circular polarization even more pronounced. This discrepancy suggests the importance of examining how the polarization affects the surface impedance. This study will be carried out in subsequent sections of this chapter.

Next, we shall extend the results of this section to the case of isotropic hysteretic media. In the case of isotropic hysteresis, a uniformly rotating magnetic field results in a uniformly rotating component of magnetic flux density. However, due to hysteresis, this uniformly rotating component of magnetic flux density lags behind the magnetic field (see Fig. 2.5). Apart from the above-mentioned uniformly rotating component of magnetic flux density, there can be a component $\mathbf{B}_0$ of the magnetic flux density that does not change with time. This constant component is usually dependent on past history and it is not essential as far as eddy currents are concerned. The existence of the uniformly rotating component $\mathbf{B}(t)$ of magnetic flux density has been observed in numerous experiments and it can be justified on the symmetry grounds. Indeed, the isotropicity of hysteretic media means that the properties of this media should be invariant with respect to rotations of Cartesian coordinates. In particular, the media properties must be invariant with respect to rotations of x- and y-axes. Since the mathematical form of a uniformly rotating (in x–y plane) magnetic field is also invariant with respect to rotations of the same axes, we conclude that the time-varying component of magnetic flux density should have the mathematical form that is invariant with respect to these rotations as well. This implies that the time-varying component of magnetic flux density is a uniformly rotating vector.

By using Fig. 2.5, the relation between the uniformly rotating vectors $\mathbf{B}(t)$ and $\mathbf{H}(t)$ can be expressed in the mathematical form as follows:

$$\hat{\mathbf{B}} = \mu(|\hat{\mathbf{H}}|)e^{-j\delta}\hat{\mathbf{H}}. \tag{2.84}$$

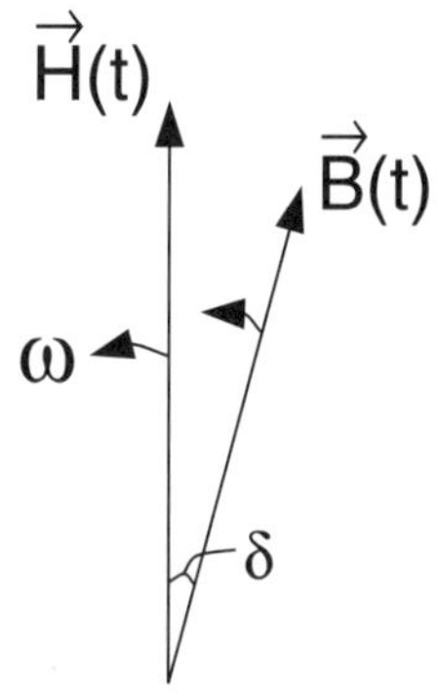

Fig. 2.5

$\hat{\mathbf{B}}$ and $\hat{\mathbf{H}}$ are vector phasors corresponding to $\mathbf{B}(t)$ and $\mathbf{H}(t)$, respectively, while $\mu(|\hat{\mathbf{H}}|)e^{-j\delta}$ can be construed as a complex magnetic permeability with δ being a "loss" angle.

In the sequel, we shall use the approximation (model) (2.27) for μ, which can also be written in the forms (2.30) and (2.31):

$$\mu\left(|\hat{\mathbf{H}}|\right) = \mu_m \left| \frac{\hat{H}_x(z)}{H_m} \right|^{\frac{1}{n}-1} = \mu_m \left| \frac{\hat{H}_y(z)}{H_m} \right|^{\frac{1}{n}-1}. \tag{2.85}$$

In addition, it will be assumed that

$$\delta = \text{const.}$$

The last assumption will be justified (to some extent) after the derivation of the expression for the magnetic flux density.

By using formulas (2.84), (2.85), and the above assumption, we can modify the boundary value problems (2.32)–(2.34) and (2.35)–(2.37) as follows:

$$\frac{d^2\hat{H}_x(z)}{dz^2} = j\omega\sigma\mu_m e^{-j\delta} \left| \frac{\hat{H}_x(z)}{H_m} \right|^{\frac{1}{n}-1} \hat{H}_x(z), \tag{2.86}$$

$$\hat{H}_x(0) = H_m e^{j\theta_0}, \tag{2.87}$$

$$\hat{H}_x(\infty) = 0, \tag{2.88}$$

and

$$\frac{d^2\hat{H}_y(z)}{dz^2} = j\omega\sigma\mu_m e^{-j\delta} \left| \frac{\hat{H}_y(z)}{H_m} \right|^{\frac{1}{n}-1} \hat{H}_y(z), \tag{2.89}$$

$$\hat{H}_y(0) = -jH_m e^{j\theta_0}, \tag{2.90}$$

$$\hat{H}_y(\infty) = 0. \tag{2.91}$$

To solve these boundary value problems, we shall use the same approach as before. Namely, we shall look for the solution of Eq. (2.86) in the form

$$\hat{H}_x(z) = \hat{H}_m \left(1 - \frac{z}{z_0}\right)^{\alpha}, \tag{2.92}$$

where $\hat{H}_m = H_m e^{j\theta_0}$ and $\alpha = \alpha' + j\alpha''$.

It is clear that function (2.92) satisfies the boundary conditions (2.87) and (2.88). From the last formula, we also find

$$\left|\frac{\hat{H}_x(z)}{H_m}\right|^{\frac{1}{n}-1} = \left(1 - \frac{z}{z_0}\right)^{-\frac{\alpha'(n-1)}{n}}. \tag{2.93}$$

Now, by substituting formulas (2.92) and (2.93) into Eq. (2.86), we find that this equation will be satisfied if α is the root of the following characteristic equation:

$$\alpha(\alpha - 1) = j\omega\sigma\mu_m e^{-j\delta} z_0^2, \tag{2.94}$$

and

$$2 = \frac{\alpha'(n-1)}{n}. \tag{2.95}$$

The last relation yields:

$$\alpha' = \frac{2n}{n-1}. \tag{2.96}$$

We shall next proceed to the determination of imaginary part α'' and z_0 from Eq. (2.94). To this end, we shall rewrite the complex Eq. (2.94) as the following two coupled equations:

$$\alpha'(\alpha' - 1) - (\alpha'')^2 = \omega\sigma\mu_m z_0^2 \sin\delta, \tag{2.97}$$

$$\alpha''(2\alpha' - 1) = \omega\sigma\mu_m z_0^2 \cos\delta. \tag{2.98}$$

By using formula (2.96) in the last two equations, we find

$$(\alpha'')^2 - \frac{2n(n+1)}{(n-1)^2} = -\omega\sigma\mu_m z_0^2 \sin\delta, \tag{2.99}$$

$$\alpha'' = \frac{n-1}{3n+1}\omega\sigma\mu_m z_0^2 \cos\delta. \tag{2.100}$$

The last expression yields:

$$\omega\sigma\mu_m z_0^2 = \frac{3n+1}{(n-1)\cos\delta}\alpha''. \tag{2.101}$$

By substituting the last formula into Eq. (2.99), we arrive at

$$(\alpha'')^2 + \alpha''\frac{3n+1}{n-1}\tan\delta - \frac{2n(n+1)}{(n-1)^2} = 0. \tag{2.102}$$

By solving the last quadratic equation, we find

$$\alpha'' = -\frac{3n+1}{2(n-1)}\tan\delta + \sqrt{\frac{(3n+1)^2}{4(n-1)^2}\tan^2\delta + \frac{2n(n+1)}{(n-1)^2}}. \tag{2.103}$$

Finally, by substituting formula(2.103) into Eq. (2.101), we derive

$$z_0 = \frac{\sqrt{\frac{3n+1}{(n-1)\cos\delta}\left(\sqrt{\frac{(3n+1)^2}{4(n-1)^2}\tan^2\delta + \frac{2n(n+1)}{(n-1)^2}} - \frac{3n+1}{2(n-1)}\tan\delta\right)}}{\sqrt{\omega\sigma\mu_m}}. \tag{2.104}$$

It is apparent that in a particular case of $\delta = 0$, formulas (2.103) and (2.104) are reduced to formulas (2.51) and (2.53).

Similar to formula (2.93), we find

$$\hat{H}_y(z) = -jH_m\left(1 - \frac{z}{z_0}\right)^{\alpha}, \tag{2.105}$$

where α and z_0 are specified by the expressions (2.98), (2.103), and (2.104).

By using formulas (2.84), (2.85), (2.92), and (2.105), we derive

$$B(z) = B_m\left(1 - \frac{z}{z_0}\right)^{\frac{2}{n-1}}. \tag{2.106}$$

The last expression suggests that for sufficiently large (and typical) values of n, the magnitude of magnetic flux density is fairly close to its boundary value B_m almost everywhere within the conducting media (see Fig. 2.3). Therefore, the loss angle δ, which is a function of $B(z)$, is also close to its boundary value almost everywhere within the media. This justifies the assumption that δ is constant. Furthermore, the values of δ are usually small (and tend to zero) for sufficiently large values of B_m.

By using formulas (2.92) and (2.105) and the same line of reasoning as before, we derive the following expression for the surface impedance:

$$\eta = \frac{\alpha' + j\alpha''}{\sigma z_0}.$$

By using expressions (2.96), (2.103), and (2.104), this impedance can be written in the form (2.66) with a and b given by the following "messy" expression:

$$a = \frac{2n}{(n-1)\sqrt{\frac{3n+1}{(n-1)\cos\delta}\left(\sqrt{\frac{(3n+1)^2}{4(n-1)^2}\tan^2\delta + \frac{2n(n+1)}{(n-1)^2}} - \frac{3n+1}{2(n-1)}\tan\delta\right)}} \tag{2.107}$$

$$b = \frac{-\frac{3n+1}{2(n-1)}\tan\delta + \sqrt{\frac{(3n+1)^2}{4(n-1)^2}\tan^2\delta + \frac{2n(n+1)}{(n-1)^2}}}{\sqrt{\frac{3n+1}{(n-1)\cos\delta}\left(\sqrt{\frac{(3n+1)^2}{4(n-1)^2}\tan^2\delta + \frac{2n(n+1)}{(n-1)^2}} - \frac{3n+1}{2(n-1)}\tan\delta\right)}}. \tag{2.108}$$

As was pointed out before, coefficients a and b are not very sensitive to particular values of n for $n \geq 7$. Thus, assuming that n is sufficiently large, we can simplify the last two equations:

$$a = \frac{2}{\sqrt{\frac{3}{\cos\delta}\left(\sqrt{\frac{9}{4}\tan^2\delta + 2} - \frac{3}{2}\tan\delta\right)}}, \tag{2.109}$$

$$b = \frac{-\frac{3}{2}\tan\delta + \sqrt{\frac{9}{4}\tan^2\delta + 2}}{\sqrt{\frac{3}{\cos\delta}\left(\sqrt{\frac{9}{4}\tan^2\delta + 2} - \frac{3}{2}\tan\delta\right)}}. \tag{2.110}$$

The results of calculations of $\sqrt{a^2+b^2}$ and $\tan\varphi$ as functions of the loss angle δ are shown in Figs. 2.6 and 2.7.

In concluding this section, we shall make the following historical remark. Many formulas presented in this section were first derived and published by the Russian scientist, L.R. Neumann [9] in 1949. However, there these formulas had entirely different meaning. They were obtained for the problem of nonlinear diffusion of **linearly** polarized electromagnetic fields by using the method of "equivalent sinusoids," that is, by neglecting higher-

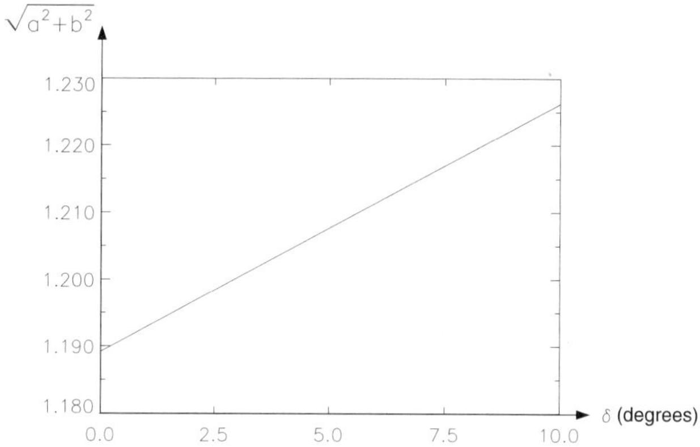

Fig. 2.6

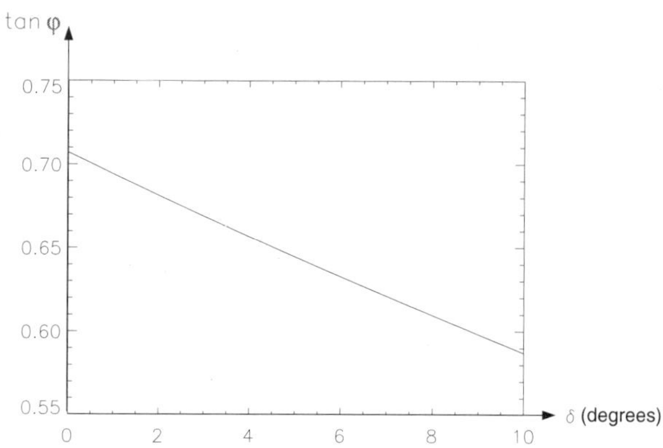

Fig. 2.7

order harmonics. For this reason, these formulas were **approximate** in nature. Many years passed before it was understood and proved (see [4], [5]) that these formulas give the **exact** solution to a different problem, namely, that of nonlinear diffusion of **circularly** polarized electromagnetic fields. This historical fact supports the notion that "scientific progress, considered historically, is not a strictly logical process" (see [10]).

2.2 PERTURBATION TECHNIQUE

In this section, we shall develop the mathematical machinery for the analysis of nonlinear diffusion in the case of noncircular polarizations of electromagnetic fields. We shall treat these polarizations as perturbations of circular polarization. Such an approach requires the development of perturbation technique. This technique will lead to **linear** partial differential equations for perturbations.

To start the discussion, we shall first mathematically state the problem under consideration as the following boundary value problem: find the time-periodic solution to the coupled nonlinear diffusion equations

$$\frac{\partial^2 H_x}{\partial z^2} = \sigma \frac{\partial B_x(H_x, H_y)}{\partial t}, \tag{2.111}$$

$$\frac{\partial^2 H_y}{\partial z^2} = \sigma \frac{\partial B_y(H_x, H_y)}{\partial t}, \tag{2.112}$$

subject to the boundary conditions:

$$H_x(0,t) = H_m \cos \omega t + \epsilon H_m f_x(t), \tag{2.113}$$

$$H_y(0,t) = H_m \sin \omega t + \epsilon H_m f_y(t), \tag{2.114}$$

$$H_x(\infty, t) = H_y(\infty, t) = 0. \tag{2.115}$$

Here ϵ is a small (in some sense) parameter, $f_x(t)$ and $f_y(t)$ are known periodic functions of time with period $T = \frac{2\pi}{\omega}$, while constitutive relations for $B_x(H_x, H_y)$ and $B_y(H_x, H_y)$ are given by

$$B_x(H_x, H_y) = \mu\left(\sqrt{H_x^2 + H_y^2}\right) H_x, \tag{2.116}$$

$$B_y(H_x, H_y) = \mu\left(\sqrt{H_x^2 + H_y^2}\right) H_y. \tag{2.117}$$

It is apparent that boundary conditions (2.113) and (2.114) can be viewed as perturbations of boundary conditions (2.8) and (2.9), which correspond to the circular polarization of the incident electromagnetic field. The only difference is that, for the sake of notational simplicity, initial phase θ_0 has been omitted in boundary conditions (2.113) and (2.114) in comparison with boundary conditions (2.8) and (2.9). This initial phase is not essential, and it can always be removed by the appropriate change of time origin.

Now, by using the general idea of perturbation techniques, we shall consider series expansions for $H_x(z,t), H_y(z,t), B_x(H_x, H_y)$, and $B_y(H_x, H_y)$ with respect to ϵ. By restricting ourselves only to zero- and first-order terms, the above ϵ-expansions can be written as follows:

$$H_x(z,t) = H_x^0(z,t) + \epsilon h_x(z,t), \tag{2.118}$$

$$H_y(z,t) = H_y^0(z,t) + \epsilon h_y(z,t), \tag{2.119}$$

$$B_x(H_x, H_y) = B_x(H_x^0, H_y^0) + \epsilon \left[h_x \frac{\partial B_x}{\partial H_x} (H_x^0, H_y^0) + h_y \frac{\partial B_x}{\partial H_y} (H_x^0, H_y^0) \right], \quad (2.120)$$

$$B_y(H_x, H_y) = B_y(H_x^0, H_y^0) + \epsilon \left[h_x \frac{\partial B_y}{\partial H_x} (H_x^0, H_y^0) + h_y \frac{\partial B_y}{\partial H_y} (H_x^0, H_y^0) \right], \quad (2.121)$$

where $H_x^0(z,t)$, $H_y^0(z,t)$, and $h_x(z,t)$, $h_y(z,t)$ are zero- and first-order terms of $H_x(z,t)$ and $H_y(z,t)$, respectively.

By substituting formulas (2.118)–(2.121) into Eqs. (2.111)–(2.117) and by equating the terms of like powers of ϵ, we end up with the following two boundary value problems:

$$\frac{\partial^2 H_x^0}{\partial z^2} = \sigma \frac{\partial}{\partial t} \left[\mu \left(\sqrt{(H_x^0)^2 + (H_y^0)^2} \right) H_x^0 \right], \quad (2.122)$$

$$\frac{\partial^2 H_y^0}{\partial z^2} = \sigma \frac{\partial}{\partial t} \left[\mu \left(\sqrt{(H_x^0)^2 + (H_y^0)^2} \right) H_y^0 \right], \quad (2.123)$$

$$H_x^0(0,t) = H_m \cos \omega t, \quad (2.124)$$

$$H_y^0(0,t) = H_m \sin \omega t, \quad (2.125)$$

$$H_x^0(\infty,t) = H_y^0(\infty,t) = 0, \quad (2.126)$$

and

$$\frac{\partial^2 h_x}{\partial z^2} = \sigma \frac{\partial}{\partial t} \left[h_x \frac{\partial B_x}{\partial H_x} (H_x^0, H_y^0) + h_y \frac{\partial B_x}{\partial H_y} (H_x^0, H_y^0) \right], \quad (2.127)$$

$$\frac{\partial^2 h_y}{\partial z^2} = \sigma \frac{\partial}{\partial t} \left[h_x \frac{\partial B_y}{\partial H_x} (H_x^0, H_y^0) + h_y \frac{\partial B_y}{\partial H_y} (H_x^0, H_y^0) \right], \quad (2.128)$$

$$h_x(0,t) = H_m f_x(t), \quad (2.129)$$

$$h_y(0,t) = H_m f_y(t), \quad (2.130)$$

$$h_x(\infty,t) = h_y(\infty,t) = 0. \quad (2.131)$$

Thus, we have arrived at the nonlinear boundary value problem (2.122)–(2.126) for zero-order terms $H_x^0(z,t)$ and $H_y^0(z,t)$ and the linear boundary value problem (2.127)–(2.131) for perturbations $h_x(z,t)$ and $h_y(z,t)$.

It is obvious that the boundary value problem (2.122)–(2.126) is identical to the boundary value problem (2.6)–(2.11) studied in the previous section for the case of circular polarization. Consequently, we can use the results from the previous section and write the solution to the boundary value problem (2.122)–(2.126) in the form:

$$H_x^0(z,t) = H_m \left(1 - \frac{z}{z_0}\right)^{\frac{2n}{n-1}} \cos\left[\omega t + \frac{\sqrt{2n(n+1)}}{n-1} \ln\left(1 - \frac{z}{z_0}\right)\right], \tag{2.132}$$

$$H_y^0(z,t) = H_m \left(1 - \frac{z}{z_0}\right)^{\frac{2n}{n-1}} \sin\left[\omega t + \frac{\sqrt{2n(n+1)}}{n-1} \ln\left(1 - \frac{z}{z_0}\right)\right], \tag{2.133}$$

where, as before, n is the exponent in the power law approximation (2.25), while z_0 is given by the formula (2.53).

Our next step is to use the last two expressions to transform Eqs. (2.127) and (2.128) for perturbations. We begin this lengthy transformation by evaluating the derivatives

$$\frac{\partial B_x}{\partial H_x}, \quad \frac{\partial B_x}{\partial H_y}, \quad \frac{\partial B_y}{\partial H_x}, \quad \frac{\partial B_y}{\partial H_y}. \tag{2.134}$$

By using formulas (2.116) and (2.117) and straightforward differentiation, we obtain

$$\frac{\partial B_x}{\partial H_x} = \mu(H) + \frac{d\mu}{dH}(H) \cdot \frac{H_x^2}{H}, \tag{2.135}$$

$$\frac{\partial B_x}{\partial H_y} = \frac{\partial B_y}{\partial H_x} = \frac{d\mu}{dH}(H) \cdot \frac{H_x H_y}{H}, \tag{2.136}$$

$$\frac{\partial B_y}{\partial H_y} = \mu(H) + \frac{d\mu}{dH}(H) \cdot \frac{H_y^2}{H}, \tag{2.137}$$

where we used the following notation:

$$H = \sqrt{H_x^2 + H_y^2}. \tag{2.138}$$

By invoking the power law approximation, we find (see (2.28)–(2.29)):

$$\mu(H) = \mu_m \left(\frac{H}{H_m}\right)^{\frac{1}{n}-1}, \tag{2.139}$$

$$\frac{d\mu}{dH}(H) = \frac{1-n}{n} \mu_m \left(\frac{H}{H_m}\right)^{\frac{1}{n}-1} \cdot \frac{1}{H}. \tag{2.140}$$

By substituting formulas (2.139) and (2.140) into expressions (2.135)–(2.137), we end up with

$$\frac{\partial B_x}{\partial H_x} = \mu_m \left(\frac{H}{H_m}\right)^{\frac{1}{n}-1} \cdot \left[1 + \frac{1-n}{n} \cdot \frac{H_x^2}{H^2}\right], \tag{2.141}$$

$$\frac{\partial B_x}{\partial H_y} = \frac{\partial B_y}{\partial H_x} = \frac{1-n}{n}\mu_m \left(\frac{H}{H_m}\right)^{\frac{1}{n}-1} \cdot \frac{H_x H_y}{H^2}, \tag{2.142}$$

$$\frac{\partial B_y}{\partial H_y} = \mu_m \left(\frac{H}{H_m}\right)^{\frac{1}{n}-1} \cdot \left[1 + \frac{1-n}{n} \cdot \frac{H_y^2}{H^2}\right]. \tag{2.143}$$

It is clear from formulas (2.132) and (2.133) that

$$H^0(z,t) = H_m \left(1 - \frac{z}{z_0}\right)^{\frac{2n}{n-1}} \tag{2.144}$$

From the last expression, we obtain

$$\left(\frac{H^0}{H_m}\right)^{\frac{1}{n}-1} = \left(1 - \frac{z}{z_0}\right)^{-2} \tag{2.145}$$

By using formulas (2.132), (2.133), and (2.144), we derive

$$\left(\frac{H_x^0}{H^0}\right)^2 = \cos^2[\omega t + \theta(z)] = \frac{1}{2} + \frac{1}{2}\cos[2\omega t + 2\theta(z)], \tag{2.146}$$

$$\left(\frac{H_y^0}{H^0}\right)^2 = \sin^2[\omega t + \theta(z)] = \frac{1}{2} - \frac{1}{2}\cos[2\omega t + 2\theta(z)], \tag{2.147}$$

$$\frac{H_x^0 H_y^0}{(H^0)^2} = \cos[\omega t + \theta(z)]\sin[\omega t + \theta(z)] = \frac{1}{2}\sin[2\omega t + 2\theta(z)], \tag{2.148}$$

where, for the sake of brevity, the following notation is used:

$$\theta(z) = \frac{\sqrt{2n(n+1)}}{n-1}\ln\left(1 - \frac{z}{z_0}\right) = \alpha'' \ln\left(1 - \frac{z}{z_0}\right). \tag{2.149}$$

By substituting formulas (2.145)–(2.148) into expressions (2.141)–(2.143) and then into Eqs. (2.127) and (2.128), we can transform these equations as follows:

$$\begin{aligned}\left(1 - \frac{z}{z_0}\right)^2 \frac{\partial^2 h_x(z,t)}{\partial z^2} &= \sigma\mu_m \frac{\partial}{\partial t}\left\{h_x(z,t)\left[\frac{1+n}{2n}\right.\right.\\ &\left.\left.+ \frac{1-n}{2n}\cos(2\omega t + 2\theta(z))\right] + h_y(z,t)\frac{1-n}{2n}\sin\left(2\omega t + 2\theta(z)\right)\right\},\end{aligned} \tag{2.150}$$

$$\left(1-\frac{z}{z_0}\right)^2\frac{\partial^2 h_y(z,t)}{\partial z^2}=\sigma\mu_m\frac{\partial}{\partial t}\Bigg\{h_x(z,t)\frac{1-n}{2n}\sin(2\omega t+2\theta(z))$$
$$+h_y(z,t)\left[\frac{1+n}{2n}-\frac{1-n}{2n}\cos(2\omega t+2\theta(z))\right]\Bigg\}. \tag{2.151}$$

Equations (2.150) and (2.151) are coupled linear partial differential equations of parabolic type with variable (in time and space) coefficients. These equations look messy, however, their structure can be essentially simplified by introducing new state variables:

$$\phi(z,t)=h_x(z,t)+jh_y(z,t), \tag{2.152}$$

$$\psi(z,t)=h_x(z,t)-jh_y(z,t). \tag{2.153}$$

Indeed, by multiplying Eq. (2.151) by j and then adding it to Eq. (2.150), we find

$$\left(1-\frac{z}{z_0}\right)^2\frac{\partial^2\phi(z,t)}{\partial z^2}=\sigma\mu_m\frac{\partial}{\partial t}\Bigg[\frac{1+n}{2n}\phi(z,t)$$
$$+\frac{1-n}{2n}h_x(z,t)e^{j(2\omega t+2\theta(z))}-j\frac{1-n}{2n}h_y(z,t)e^{j(2\omega t+2\theta(z))}\Bigg]. \tag{2.154}$$

By using formula (2.153) and the fact that according to expression (2.149)

$$e^{j2\theta(z)}=\left(1-\frac{z}{z_0}\right)^{j2\alpha''}, \tag{2.155}$$

Eq. (2.154) can be transformed as follows:

$$\left(1-\frac{z}{z_0}\right)^2\frac{\partial^2\phi(z,t)}{\partial z^2}=\sigma\mu_m\frac{1-n}{2n}\frac{\partial}{\partial t}\Bigg[\frac{1+n}{1-n}\phi(z,t)$$
$$+\left(1-\frac{z}{z_0}\right)^{j2\alpha''}e^{j2\omega t}\psi(z,t)\Bigg]. \tag{2.156}$$

Now, by multiplying Eq. (2.151) by $-j$ and then adding it to Eq. (2.150) and by using the transformations that are similar to those just described, we derive:

$$\left(1-\frac{z}{z_0}\right)^2\frac{\partial^2\psi(z,t)}{\partial z^2}=\sigma\mu_m\frac{1-n}{2n}\frac{\partial}{\partial t}\Bigg[\frac{1+n}{1-n}\psi(z,t)$$
$$+\left(1-\frac{z}{z_0}\right)^{-j2\alpha''}e^{-j2\omega t}\phi(z,t)\Bigg]. \tag{2.157}$$

We shall be interested in periodic solutions to Eqs. (2.156) and (2.157) subject to the boundary conditions

$$\phi(0,t) = H_m(f_x(t) + jf_y(t)), \tag{2.158}$$

$$\psi(0,t) = H_m(f_x(t) - jf_y(t)), \tag{2.159}$$

$$\phi(\infty,t) = \psi(\infty,t) = 0, \tag{2.160}$$

which follow from boundary conditions (2.129)–(2.131) and formulas (2.152) and (2.153).

We look for the periodic solution to the boundary value problems (2.156)–(2.160) in terms of Fourier series:

$$\phi(z,t) = \sum_{k=-\infty}^{\infty} \phi_k(z)e^{jk\omega t}, \tag{2.161}$$

$$\psi(z,t) = \sum_{k=-\infty}^{\infty} \psi_k(z)e^{jk\omega t}. \tag{2.162}$$

By substituting Fourier series (2.161) and (2.162) into Eqs. (2.156) and (2.157) and equating the terms with the same exponents, after simple transformations we derive the following coupled ordinary differential equations:

$$\left(1-\frac{z}{z_0}\right)^2 \frac{d^2\phi_k(z)}{dz^2} = j\chi_k \left[a\phi_k(z) + \left(1-\frac{z}{z_0}\right)^{j2\alpha''} \cdot \psi_{k-2}(z)\right], \tag{2.163}$$

$$\left(1-\frac{z}{z_0}\right)^2 \frac{d^2\psi_k(z)}{dz^2} = j\chi_k \left[a\psi_k(z) + \left(1-\frac{z}{z_0}\right)^{-j2\alpha''} \cdot \phi_{k+2}(z)\right], \tag{2.164}$$

where for the sake of brevity the following notations are introduced:

$$\chi_k = k\omega\sigma\mu_m \frac{1-n}{2n}, \tag{2.165}$$

$$a = \frac{1+n}{1-n}. \tag{2.166}$$

Thus, we have reduced the coupled partial differential Eqs. (2.156) and (2.157) to the infinite set of coupled ordinary differential Eqs. (2.163) and (2.164). **The remarkable property of these simultaneous ordinary differential equations is that they are coupled in separate pairs.**

To make this pairing coupling manifest, we change k to $k-2$ in Eq. (2.164) and rewrite Eqs. (2.163) and (2.164) in the form

$$\left(1-\frac{z}{z_0}\right)^2 \frac{d^2\phi_k(z)}{dz^2} - j\chi_k \left[a\phi_k(z) + \left(1-\frac{z}{z_0}\right)^{j2\alpha''} \cdot \psi_{k-2}(z)\right] = 0, \tag{2.167}$$

$$\left(1-\frac{z}{z_0}\right)^2 \frac{d^2\psi_{k-2}(z)}{dz^2} - j\chi_{k-2} \left[a\psi_{k-2}(z) + \left(1-\frac{z}{z_0}\right)^{-j2\alpha''} \cdot \phi_k(z)\right] = 0,$$
$$(k = 0, \pm 2, \pm 3, \ldots). \tag{2.168}$$

Thus, we can solve each pair of coupled Eqs. (2.167)–(2.168) separately. The solution of these equations should be subject to the following boundary conditions:

$$\phi_k(0) = H_m(f_{x,k} + jf_{y,k}), \tag{2.169}$$

$$\psi_{k-2}(0) = H_m(f_{x,k-2} - jf_{y,k-2}), \tag{2.170}$$

$$\phi_k(\infty) = \psi_{k-2}(\infty) = 0, \quad (k = 0, \pm 1, \pm 2, \pm 3, \ldots), \tag{2.171}$$

where $f_{x,k}$ and $f_{y,k}$ are Fourier coefficients of functions $f_x(t)$ and $f_y(t)$, respectively, while the boundary conditions (2.169)–(2.171) follow from boundary conditions (2.158)–(2.160).

After functions $\phi_k(z)$ and $\psi_k(z)$ are determined, from Eqs. (2.152) and (2.153), we can find complex Fourier coefficients for perturbations $h_x(z,t)$ and $h_y(z,t)$:

$$h_{x,k}(z) = \frac{1}{2}(\phi_k(z) + \psi_k(z)), \tag{2.172}$$

$$h_{y,k}(z) = \frac{1}{2j}\left(\phi_k(z) - \psi_k(z)\right), \tag{2.173}$$

and the perturbations themselves:

$$h_x(z,t) = \sum_{k=-\infty}^{\infty} h_{x,k}(z)e^{jk\omega t}, \tag{2.174}$$

$$h_y(z,t) = \sum_{k=-\infty}^{\infty} h_{y,k}(z)e^{jk\omega t}. \tag{2.175}$$

It is apparent that $h_{x,k}$ and $h_{y,k}$ are complex conjugate to $h_{x,-k}$ and $h_{y,-k}$. For this reason, it suffices to solve the boundary value problems (2.167)–(2.171) for nonnegative values of k only.

We will seek a solution to Eqs. (2.167) and (2.168) in the form

$$\phi_k(z) = A_k \left(1 - \frac{z}{z_0}\right)^{\beta_k}, \tag{2.176}$$

$$\psi_{k-2}(z) = B_{k-2} \left(1 - \frac{z}{z_0}\right)^{\beta_k - j2\alpha''}. \tag{2.177}$$

As before, it is tacitly understood that formulas (2.176) and (2.177) are valid for $0 \le z \le z_0$, whereas for $z \ge z_0$ functions $\phi_k(z)$ and $\psi_{k-2}(z)$ are equal to zero. Thus, the boundary conditions (2.171) are satisfied.

By substituting formulas (2.176) and (2.177) into Eqs. (2.167) and (2.168), we find that these equations are satisfied if A_k, B_{k-2}, and β_k are constrained by the relations

$$\left(\beta_k^2 - \beta_k - j\chi_k a z_0^2\right) A_k - j\chi_k z_0^2 B_{k-2} = 0, \tag{2.178}$$

$$-j\chi_{k-2} z_0^2 A_k + \left[(\beta_k - j2\alpha'')^2 - (\beta_k - j2\alpha'') - j\chi_{k-2} a z_0^2\right] B_{k-2} = 0. \tag{2.179}$$

Relations (2.178) and (2.179) can be construed as linear simultaneous equations with respect to A_k and B_{k-2}. Since these equations are homogeneous, they will have nonzero solution if and only if the corresponding determinant is equal to zero. This yields the following characteristic (dispersion) equation for β_k:

$$\begin{aligned}(\beta_k^2 - \beta_k - j\chi_k a z_0^2)[(\beta_k - j2\alpha'')^2 - (\beta_k - j2\alpha'') - j\chi_{k-2} a z_0^2] \\ + \chi_k \chi_{k-2} z_0^4 = 0.\end{aligned} \tag{2.180}$$

The coefficients of this characteristic equation depend only on k and exponent n of the power law approximation. Indeed, by using formulas (2.53), (2.165), and (2.166), we find

$$\chi_k a z_0^2 = \frac{k(3n+1)(n+1)}{(n-1)^2}\sqrt{\frac{n+1}{2n}}, \tag{2.181}$$

$$\chi_k \chi_{k-2} z_0^4 = \frac{k(k-2)(3n+1)^2(n+1)}{2n(n-1)^2}, \tag{2.182}$$

while α'' is given by (2.51).

From the above fact we conclude that the roots of the characteristic Eq. (2.180) depend only on k and n.

It can be shown that, for any k and n, the characteristic Eq. (2.180) has two roots (two solutions) β'_k and β''_k with positive real parts:

$$\mathrm{Re}(\beta'_k) > 0, \qquad \mathrm{Re}(\beta''_k) > 0. \tag{2.183}$$

By using these roots and formulas (2.176) and (2.177), the solution $\phi_k(z)$ and $\psi_{k-2}(z)$ to differential Eqs. (2.167) and (2.168) can be represented in the form

$$\phi_k(z) = A'_k \left(1 - \frac{z}{z_0}\right)^{\beta'_k} + A''_k \left(1 - \frac{z}{z_0}\right)^{\beta''_k}, \tag{2.184}$$

$$\psi_{k-2}(z) = B'_{k-2} \left(1 - \frac{z}{z_0}\right)^{\beta'_k - j2\alpha''} + B''_{k-2} \left(1 - \frac{z}{z_0}\right)^{\beta''_k - j2\alpha''}. \tag{2.185}$$

The unknown coefficients A'_k, A''_k, B'_{k-2}, and B''_{k-2} can be found by satisfying the boundary conditions (2.169) and (2.170) as well as Eq. (2.178). This results in the following four equations:

$$A'_k + A''_k = H_m \left(f_{x,k} + j f_{y,k}\right), \tag{2.186}$$

$$B'_{k-2} + B''_{k-2} = H_m \left(f_{x,k-2} - j f_{y,k-2}\right), \tag{2.187}$$

$$\left[(\beta'_k)^2 - \beta'_k - j\chi_k a z_0^2\right] A'_k - j\chi_k z_0^2 B'_{k-2} = 0, \tag{2.188}$$

$$\left[(\beta''_k)^2 - \beta''_k - j\chi_k a z_0^2\right] A''_k - j\chi_k z_0^2 B''_{k-2} = 0. \tag{2.189}$$

As far as Eq. (2.179) is concerned, it will be automatically satisfied. This is because it is equivalent to Eq. (2.178) when β_k is the solution to the characteristic Eq. (2.180).

Equations (2.186)–(2.189) can be easily solved and their solution is given by

$$A'_k = H_m \left[\frac{C''_k f_{x,k} - f_{x,k-2}}{C''_k - C'_k} + j\frac{C''_k f_{y,k} + f_{y,k-2}}{C''_k - C'_k}\right], \tag{2.190}$$

$$A''_k = H_m \left[\frac{C'_k f_{x,k} - f_{x,k-2}}{C'_k - C''_k} + j\frac{C'_k f_{y,k} + f_{y,k-2}}{C'_k - C''_k}\right], \tag{2.191}$$

$$B'_{k-2} = C'_k A'_k, \quad B''_{k-2} = C''_k A''_k, \tag{2.192}$$

where the constants C'_k and C''_k are defined as follows:

$$C'_k = \frac{(\beta'_k)^2 - \beta'_k - j\chi_k a z_0^2}{j\chi_k z_0^2}, \tag{2.193}$$

$$C_k'' = \frac{(\beta_k'')^2 - \beta_k'' - j\chi_k a z_0^2}{j\chi_k z_0^2}. \tag{2.194}$$

Thus, the algorithm of calculation of perturbations $h_x(z,t)$ and $h_y(z,t)$ consists of the following steps: (1) solve the characteristic Eq. (2.180) and find the roots that satisfy conditions (2.183); (2) use formulas (2.190)–(2.194) to find the coefficients A_k', A_k'', B_k' and B_k'' and plug them into expressions (2.184) and (2.185) to determine the functions $\phi_k(z)$ and $\psi_k(z)$; (3) use formulas (2.172) and (2.173) to find Fourier coefficients $h_{x,k}$ and $h_{y,k}$ and then the perturbations themselves according to Eqs. (2.174) and (2.175). It is useful to stress that this algorithm is valid for any periodic perturbations $f_x(t)$ and $f_y(t)$.

Up to this point, we have discussed only first-order perturbations in ϵ. For higher-order perturbations in ϵ, calculations become much more convoluted. However, the structure of partial differential equations for the higher-order perturbations remains almost the same as for the first-order perturbations. Indeed, if we use the following ϵ-expansions:

$$H_x(z,t) = H_x^0(z,t) + \epsilon h_x(z,t) + \epsilon^2 \tilde{h}_x(z,t) + \ldots, \tag{2.195}$$

$$H_y(z,t) = H_y^0(z,t) + \epsilon h_y(z,t) + \epsilon^2 \tilde{h}_y(z,t) + \ldots, \tag{2.196}$$

and repeat the same line of reasoning as in the derivation of Eqs. (2.127) and (2.128), then we arrive at the following equations and boundary conditions for the second-order perturbations $\tilde{h}_x(z,t)$ and $\tilde{h}_y(z,t)$:

$$\begin{aligned}
&\frac{\partial^2 \tilde{h}_x(z,t)}{\partial z^2} - \sigma \frac{\partial}{\partial t}\left[\tilde{h}_x(z,t)\frac{\partial B_x}{\partial H_x}\left(H_x^0, H_y^0\right)\right. \\
&\left. + \tilde{h}_y(z,t)\frac{\partial B_x}{\partial H_y}(H_x^0, H_y^0)\right] = \frac{\sigma}{2}\frac{\partial}{\partial t}\left[h_x^2(z,t)\frac{\partial^2 B_x}{\partial H_x^2}\left(H_x^0, H_y^0\right)\right. \\
&\left. + 2h_x(z,t)h_y(z,t)\frac{\partial^2 B_x}{\partial H_x \partial H_y}\left(H_x^0, H_y^0\right) + h_y^2(z,t)\frac{\partial^2 B_x}{\partial H_y^2}\left(H_x^0, H_y^0\right)\right],
\end{aligned} \tag{2.197}$$

$$\begin{aligned}
&\frac{\partial^2 \tilde{h}_y(z,t)}{\partial z^2} - \sigma \frac{\partial}{\partial t}\left[\tilde{h}_x(z,t)\frac{\partial B_y}{\partial H_x}\left(H_x^0, H_y^0\right)\right. \\
&\left. + \tilde{h}_y(z,t)\frac{\partial B_y}{\partial H_y}\left(H_x^0, H_y^0\right)\right] = \frac{\sigma}{2}\frac{\partial}{\partial t}\left[h_x^2(z,t)\frac{\partial^2 B_y}{\partial H_x^2}\left(H_x^0, H_y^0\right) + \right. \\
&\left. 2h_x(z,t)h_y(z,t)\frac{\partial^2 B_y}{\partial H_x \partial H_y}\left(H_x^0, H_y^0\right) + h_y^2(z,t)\frac{\partial^2 B_y}{\partial H_y^2}\left(H_x^0, H_y^0\right)\right],
\end{aligned} \tag{2.198}$$

$$\tilde{h}_x(0,t) = \tilde{h}_y(0,t) = 0, \tag{2.199}$$

$$\tilde{h}(\infty, t) = \tilde{h}_y(\infty, t) = 0. \tag{2.200}$$

By comparing partial differential Eqs. (2.197) and (2.198) with partial differential Eqs. (2.127) and (2.128), we observe that these equations have almost identical structures with the only difference that the equations for the first-order perturbations are homogeneous, whereas the equations for the second-order perturbations are inhomogeneous. The right-hand sides of the later equations are determined by the zero-and first-order terms. On the other hand, the boundary conditions for the second-order perturbations are homogeneous, whereas this is not the case for the first-order perturbations. Since the equations for the second-order perturbations have the same mathematical structure as in the case of the first-order perturbations, the same analytical technique can be used to find their solution. Namely, these partial differential equations can be reduced to the infinite set of ordinary differential equations coupled in separate pairs. However, the solution of these ordinary differential equations cannot be carried out explicitly because the first-order perturbations should be found first. Finally, we shall remark that partial differential equations similar to Eqs. (2.197) and (2.198) can be derived for perturbations of any order. However, the right-hand sides of these equations become more complex as the order of perturbation is increased.

2.3 NONLINEAR DIFFUSION OF ELLIPTICALLY POLARIZED ELECTROMAGNETIC FIELDS IN ISOTROPIC MEDIA

In this section, we shall apply the perturbation technique developed in the previous section to the analysis of nonlinear diffusion of elliptically polarized electromagnetic fields. We begin with a brief review of how elliptical polarizations can be characterized. There are two commonly used ways to describe elliptical polarizations. The first way is to specify the equations for x- and y-components of the field

$$H_x(0, t) = H_{mx} \cos(\omega t + \theta_{0x}), \tag{2.201}$$

$$H_y(0, t) = H_{my} \cos(\omega t + \theta_{0y}), \tag{2.202}$$

with

$$\theta_0 = \theta_{0x} - \theta_{0y} \tag{2.203}$$

being an initial phase difference. It is well known that the endpoint of the magnetic field vector $\mathbf{H}(0, t)$ specified by Eqs. (2.201)–(2.202) traces an ellipse. The semimajor and the semiminor of this ellipse as well as its orientation with respect to x- and y-axes are completely determined by three parameters H_{mx}, H_{my}, and θ_0. In this sense, these three parameters completely characterize elliptical polarizations. However, these parameters do not have the same dimension. For this reason, it is sometimes more convenient to characterize elliptical polarizations by the Stokes parameters. These parameters are defined as follows:

$$I = H_{mx}^2 + H_{my}^2, \tag{2.204}$$

$$Q = H_{mx}^2 - H_{my}^2, \tag{2.205}$$

$$U = 2H_{mx}H_{my}\cos\theta_0, \tag{2.206}$$

$$V = -2H_{mx}H_{my}\sin\theta_0. \tag{2.207}$$

These parameters have the same dimension and they are related by the equation

$$I^2 = Q^2 + U^2 + V^2. \tag{2.208}$$

The last equation leads to the notion of the Poincaré sphere. The radius of this sphere is I, and points on this sphere are uniquely defined by the Stokes parameters Q, U, and V. These three Stokes parameters also uniquely characterize various elliptical polarizations. Thus, we conclude that the elliptical polarizations can be represented by the points on the Poincaré sphere (2.208). In this representation, the north pole ($V = I, U = 0, Q = 0$) and the south pole ($V = -I, U = 0, Q = 0$) correspond to left- and right-handed circular polarizations, respectively, while the points of the equator $V = 0$ correspond to linear polarizations of different orientations. The points of the northern and southern hemispheres represent left- and right-handed elliptical polarizations.

The Stokes parameters and the Poincaré sphere are frequently used for characterization of elliptical polarization in optics where parameter I has the physical meaning of light intensity. However, these parameters are not convenient for the development of the perturbation technique, because of nonlinear relations between the Stokes parameters and magnetic field components. For this reason, we shall use the characterization of elliptical polarizations based on parameters H_{mx}, H_{my}, and θ_0. This characterization allows one to take explicitly into account the isotropicity of media and reduce the number of parameters used for specification of elliptical polarizations. Indeed, for fixed H_{mx} and H_{my}, the orientation of an ellipse traced by the endpoint of $\mathbf{H}(0,t)$ is determined by the phase difference θ_0. However, for isotropic media all these orientations are equivalent. Consequently, one can choose x- and y-axes as major and minor axes of the ellipse, which results in the following expressions for the field components:

$$H_x(0,t) = H_{mx}\cos\omega t, \tag{2.209}$$

$$H_y(0,t) = H_{my}\sin\omega t, \tag{2.210}$$

with

$$H_{mx} \geq H_{my}. \tag{2.211}$$

In this way, elliptical polarizations can be characterized by only two parameters H_{mx} and H_{my} of the same dimension. This characterization is also convenient for the introduction of perturbation parameter ϵ:

$$\epsilon = \frac{H_{mx} - H_{my}}{H_{mx} + H_{my}}. \tag{2.212}$$

When $\epsilon = 1$, one deals with linear polarizations, whereas for $\epsilon = 0$ one deals with circular polarizations. By using parameter ϵ, the boundary conditions (2.209) and (2.210) can be written in the form

$$H_x(0,t) = H_m\cos\omega t + \epsilon H_m\cos\omega t, \tag{2.213}$$

$$H_y(0,t) = H_m\sin\omega t - \epsilon H_m\sin\omega t, \tag{2.214}$$

where

$$H_m = \frac{H_{mx} + H_{my}}{2}. \tag{2.215}$$

By comparing boundary conditions (2.213)–(2.214) with boundary conditions (2.113)–(2.114), we can identify functions $f_x(t)$ and $f_y(t)$ as follows:

$$f_x(t) = \cos\omega t, \quad f_y(t) = -\sin\omega t. \tag{2.216}$$

Now, it is clear that the Fourier coefficients of functions $f_x(t)$ and $f_y(t)$ are given by

$$f_{x,-1} = \frac{1}{2}, \quad f_{x,1} = \frac{1}{2}, \quad f_{x,k} = 0 \quad \text{if } |k| \neq 1, \tag{2.217}$$

$$f_{y,-1} = \frac{1}{2j}, \quad f_{y,1} = -\frac{1}{2j}, \quad f_{y,k} = 0 \quad \text{if} |k| \neq 1. \tag{2.218}$$

By using the last expressions and formulas (2.169) and (2.170), we find

$$\phi_{-1}(0) = H_m, \quad \phi_k(0) = 0 \quad \text{if } k \neq -1, \tag{2.219}$$

$$\psi_1(0) = H_m, \quad \psi_k(0) = 0 \quad \text{if } k \neq 1. \tag{2.220}$$

Thus, in the case of elliptical polarizations the boundary value problems (2.167)–(2.171) can be written as follows:

$$\left(1 - \frac{z}{z_0}\right)^2 \frac{d^2\phi_k(z)}{dz^2} - j\chi_k\left[a\phi_k(z) + \left(1 - \frac{z}{z_0}\right)^{j2\alpha''} \cdot \psi_{k-2}(z)\right] = 0, \tag{2.221}$$

$$\left(1-\frac{z}{z_0}\right)^2 \frac{d^2\psi_{k-2}(z)}{dz^2} - j\chi_{k-2}\left[a\psi_{k-2}(z) + \left(1-\frac{z}{z_0}\right)^{-j2\alpha''} \cdot \phi_k(z)\right] = 0, \tag{2.222}$$

$$\phi_k(0) = \psi_{k-2}(0) = 0, \tag{2.223}$$

$$\phi_k(\infty) = \psi_{k-2}(\infty) = 0, \tag{2.224}$$

if $k \geq 0$ and $k \neq 3$.

For $k = 3$, we have

$$\left(1-\frac{z}{z_0}\right)^2 \frac{d^2\phi_3(z)}{dz^2} - j\chi_3\left[a\phi_3(z) + \left(1-\frac{z}{z_0}\right)^{j2\alpha''} \cdot \psi_1(z)\right] = 0, \tag{2.225}$$

$$\left(1-\frac{z}{z_0}\right)^2 \frac{d^2\psi_1(z)}{dz^2} - j\chi_1\left[a\psi_1(z) + \left(1-\frac{z}{z_0}\right)^{-j2\alpha''} \cdot \phi_3(z)\right] = 0, \tag{2.226}$$

$$\phi_3(0) = 0, \quad \psi_1(0) = H_m, \tag{2.227}$$

$$\phi_3(\infty) = \psi_1(\infty) = 0. \tag{2.228}$$

Because all the equations and boundary conditions of the boundary value problems (2.221)–(2.224) are homogeneous, we conclude that these boundary value problems have zero solutions. Consequently, we have

$$\phi_k(z) = 0 \quad \text{if } k \geq 0 \text{ and } k \neq 3, \tag{2.229}$$

$$\psi_k(z) = 0 \quad \text{if } k \geq 0 \text{ and } k \neq 1. \tag{2.230}$$

From formulas (2.229)–(2.230) and (2.172)–(2.173) we find that

$$h_{x,k}(z) = h^*_{x,-k}(z) = h_{y,k}(z) = h^*_{y,-k}(z) = 0 \quad \text{if } k \neq 1 \text{ and } k \neq 3, \tag{2.231}$$

where "$*$" means a complex conjugate quantity.

Thus, we conclude that only the first and the third harmonics of the magnetic field are not equal to zero, while all other harmonics are equal to zero. We have obtained this result because we have restricted ourselves only to first-order perturbations. If we take into account higher-order perturbations, we shall recover higher-order harmonics of the electromagnetic field.

From the purely mathematical point of view, it is quite interesting that the solution to the coupled partial differential Eqs. (2.150)–(2.151) subject to the boundary conditions (2.129)–(2.131) can be represented as a sum of first- and third-order harmonics only. This suggests that the coupled Eqs. (2.150)–(2.151) may have inherited some symmetry properties from the unperturbed problem (2.122)–(2.126) describing nonlinear diffusion of circularly polarized electromagnetic fields.

According to formulas (2.172)–(2.173) and (2.229)–(2.230), we find

$$h_{x,1}(z) = h^*_{x,-1}(z) = \frac{1}{2}\psi_1(z), \tag{2.232}$$

$$h_{y,1}(z) = h^*_{y,-1}(z) = \frac{j}{2}\psi_1(z), \tag{2.233}$$

$$h_{x,3}(z) = h^*_{x,-3}(z) = \frac{1}{2}\phi_3(z), \tag{2.234}$$

$$h_{y,3}(z) = h^*_{y,-3}(z) = -\frac{j}{2}\phi_3(z). \tag{2.235}$$

Thus, if we solve the boundary value problem (2.225)–(2.228) and find the functions $\phi_3(z)$ and $\psi_1(z)$, we can then determine the first and third harmonics of magnetic field perturbations. The solution of the boundary value problem (2.225)–(2.228) proceeds along the lines discussed in the previous section. Namely, we seek the solution to this boundary value problem in the form

$$\phi_3(z) = A'\left(1-\frac{z}{z_0}\right)^{\beta'} + A''\left(1-\frac{z}{z_0}\right)^{\beta''}, \tag{2.236}$$

$$\psi_1(z) = B'\left(1-\frac{z}{z_0}\right)^{\beta'-j2\alpha''} + B''\left(1-\frac{z}{z_0}\right)^{\beta''-j2\alpha''}, \tag{2.237}$$

where β' and β'' are the roots of the following characteristic equation (see (2.180)):

$$\left(\beta^2-\beta-j\chi_3 z_0^2 a\right)\left[(\beta-j2\alpha'')^2-(\beta-j2\alpha'')-j\chi_1 z_0^2 a\right]+\chi_1\chi_3 z_0^4=0, \tag{2.238}$$

which satisfy the conditions (2.183). The coefficients of this characteristic equation depend only on n and they are given by the formulas

$$\chi_3 z_0^2 a = \frac{3(3n+1)(n+1)}{(n-1)^2}\sqrt{\frac{n+1}{2n}}, \tag{2.239}$$

$$\chi_1 z_0^2 a = \frac{(3n+1)(n+1)}{(n-1)^2}\sqrt{\frac{n+1}{2n}}, \tag{2.240}$$

$$\chi_1\chi_3 z_0^4 = \frac{3(3n+1)^2(n+1)}{2n(n-1)^2}. \tag{2.241}$$

Consequently, roots β' and β'' are only functions of n. These functions have been computed and presented in Figs. 2.8 and 2.9, respectively. By knowing roots β' and β'', coefficients A', A'', B', and B'' in formulas (2.236) and (2.237) can be found by using the expressions

$$A' = -A'' = \frac{j\chi_3 z_0^2}{((\beta')^2 - \beta') - ((\beta'')^2 - \beta'')} H_m, \tag{2.242}$$

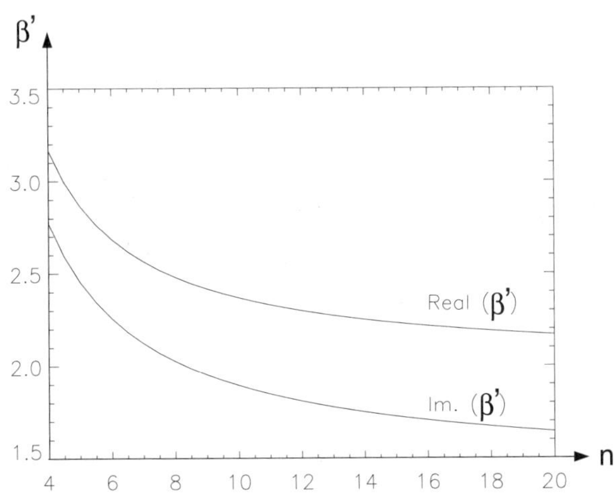

Fig. 2.8

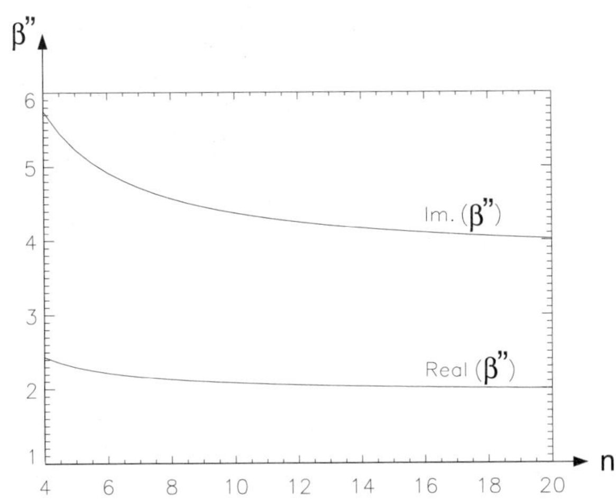

Fig. 2.9

$$B' = \frac{(\beta')^2 - \beta' - j\chi_3 z_0^2 a}{((\beta')^2 - \beta') - ((\beta'')^2 - \beta'')} H_m, \tag{2.243}$$

$$B'' = -\frac{(\beta'')^2 - \beta'' - j\chi_3 z_0^2 a}{((\beta')^2 - \beta') - ((\beta'')^2 - \beta'')} H_m. \tag{2.244}$$

The last expressions are derived in a straightforward way from formulas (2.190)–(2.194) and (2.217)–(2.218).

Knowing the coefficients A', A'', B', and B'', we can now proceed to the calculation of surface impedance. First, we shall find the expressions for phasors of the first and third harmonics of the magnetic field. According to formula (2.174), for the first-order harmonic of the magnetic field perturbation $h_x^{(1)}(z,t)$ we have

$$h_x^{(1)}(z,t) = h_{x,1}(z)e^{j\omega t} + h_{x,1}^*(z)e^{-j\omega t} = \text{Re}\left[2h_{x,1}(z)e^{j\omega t}\right]. \tag{2.245}$$

This means that $2h_{x,1}(z)$ is the phasor of $h_x^{(1)}(z,t)$. By using this fact, the phasor of the first harmonic of the magnetic field can be written as follows:

$$\hat{H}_{x,1}(z) = \hat{H}_x^{(0)}(z) + 2\epsilon h_{x,1}(z). \tag{2.246}$$

Similarly, we find

$$\hat{H}_{y,1}(z) = \hat{H}_y^{(0)}(z) + 2\epsilon h_{y,1}(z). \tag{2.247}$$

Now, by using formulas (2.38), (2.54), (2.232), (2.233), and (2.237) in the last two expressions, we obtain

$$\begin{aligned}\hat{H}_{x,1}(z) =& H_m\left(1-\frac{z}{z_0}\right)^{\alpha} + \epsilon\left[B'\left(1-\frac{z}{z_0}\right)^{\beta'-j2\alpha''}\right. \\ &\left. + B''\left(1-\frac{z}{z_0}\right)^{\beta''-j2\alpha''}\right],\end{aligned} \tag{2.248}$$

$$\begin{aligned}\hat{H}_{y,1}(z) =& -jH_m\left(1-\frac{z}{z_0}\right)^{\alpha} + j\epsilon\left[B'\left(1-\frac{z}{z_0}\right)^{\beta'-j2\alpha''}\right. \\ &\left. + B''\left(1-\frac{z}{z_0}\right)^{\beta''-j2\alpha''}\right].\end{aligned} \tag{2.249}$$

By using the same line of reasoning as before, we derive the following expressions for the phasors of the third harmonic of the magnetic field:

$$\hat{H}_{x,3}(z) = \epsilon\left[A'\left(1-\frac{z}{z_0}\right)^{\beta'} + A''\left(1-\frac{z}{z_0}\right)^{\beta''}\right], \tag{2.250}$$

$$\hat{H}_{y,3}(z) = -j\epsilon\left[A'\left(1-\frac{z}{z_0}\right)^{\beta'} + A''\left(1-\frac{z}{z_0}\right)^{\beta''}\right]. \tag{2.251}$$

It is interesting to note that, in the first-order approximation (with respect to ϵ), the third harmonic of the magnetic field is circularly polarized, and its sense of polarization is identical to that of the incident field at the boundary.

By invoking formulas (2.63), we find the following expressions for the phasors of the first harmonics of the electric field at the boundary:

$$\hat{E}_{x,1}(0) = -\frac{j}{\sigma z_0}\left[\alpha H_m - \epsilon(B'(\beta' - j2\alpha'') + B''(\beta'' - j2\alpha''))\right], \tag{2.252}$$

$$\hat{E}_{y,1}(0) = -\frac{1}{\sigma z_0}\left[\alpha H_m + \epsilon(B'(\beta' - j2\alpha'') + B''(\beta'' - j2\alpha''))\right]. \tag{2.253}$$

Now, we can define surface impedances:

$$\eta_{xy} = \frac{\hat{E}_{x,1}(0)}{\hat{H}_{y,1}(0)}, \quad \eta_{yx} = -\frac{\hat{E}_{y,1}(0)}{\hat{H}_{x,1}(0)}. \tag{2.254}$$

By using the last two expressions and formulas (2.213), (2.214), (2.252), (2.253), as well as (2.65), we derive

$$\frac{\eta_{xy}}{\eta} = \frac{\alpha H_m - \epsilon[B'(\beta' - 2j\alpha'') + B''(\beta'' - j2\alpha'')]}{\alpha(1-\epsilon)H_m}, \tag{2.255}$$

$$\frac{\eta_{yx}}{\eta} = \frac{\alpha H_m + \epsilon[B'(\beta' - 2j\alpha'') + B''(\beta'' - j2\alpha'')]}{\alpha(1+\epsilon)H_m}, \tag{2.256}$$

where η is the surface impedance in the case of circular polarization of the magnetic field.

Formulas (2.255) and (2.256) allow one to evaluate to what extent the surface impedance is affected by deviations from circular polarizations. It is important to note that the right-hand sides of formulas (2.255) and (2.256) do not depend on H_m. This is because, according to formulas (2.243) and (2.244), coefficients B' and B'' are directly proportional to H_m. Thus, the right-hand sides of (2.255) and (2.256) are functions only of n and ϵ. These functions have been computed for various values of n and ϵ and the results of computations are shown in Figs. 2.10, 2.11, 2.12, and 2.13.

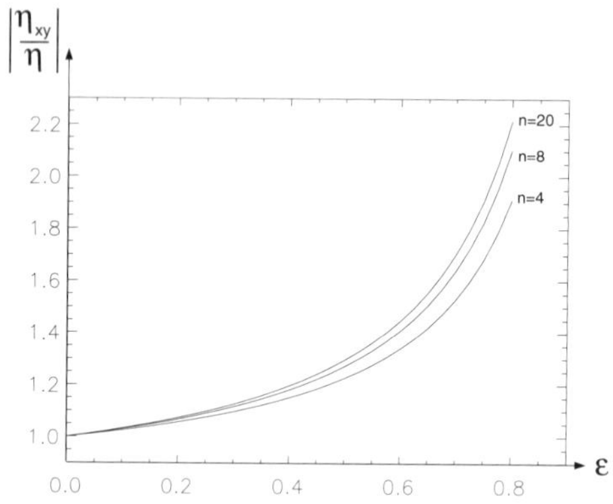

Fig. 2.10

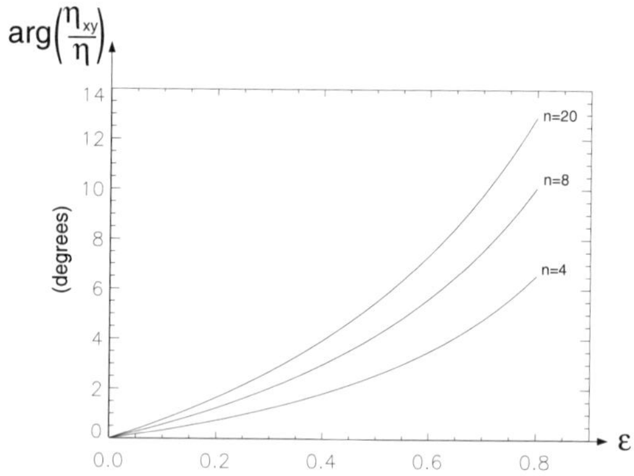

Fig. 2.11

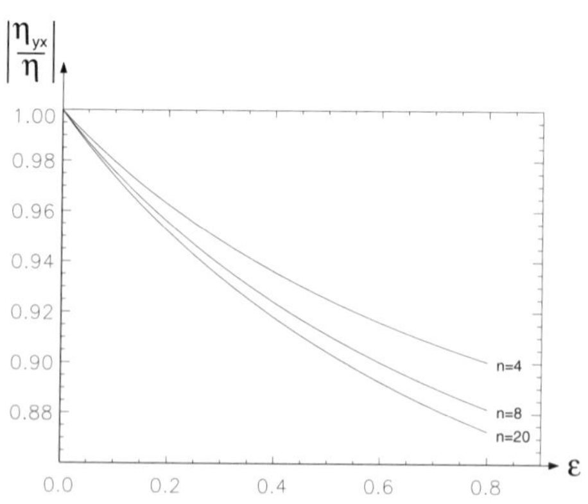

Fig. 2.12

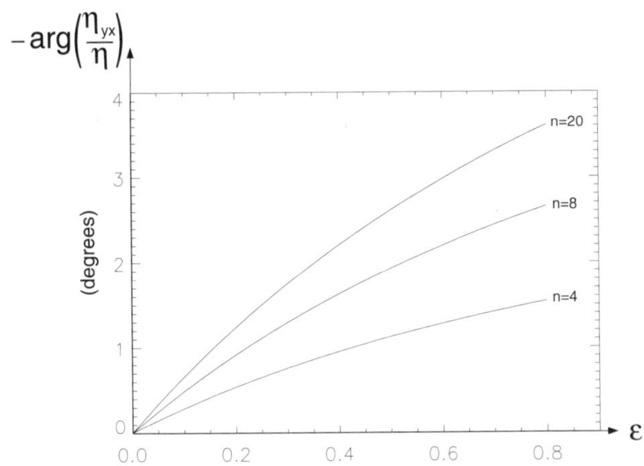

Fig. 2.13

Our calculations are based on the perturbation technique. For this reason, they are inevitably approximate in nature. Therefore, it is desirable to evaluate the accuracy of the perturbation technique. The rigorous mathematical evaluation of the accuracy of the perturbation technique is beyond the scope of this book. However, some insights concerning this accuracy still can be gained. This can be done by applying the perturbation technique to the case of circular polarization. For this case, the exact analytical solution is available (see Section 2.1). The comparison of this exact solution with the solution obtained by using the perturbation technique will provide some insights into the accuracy of the perturbation technique.

Consider nonlinear diffusion of the magnetic field in the case of the following boundary conditions:

$$H_x(0,t) = H_m(\cos\omega t - \epsilon \sin\omega t), \tag{2.257}$$

$$H_y(0,t) = H_m(\sin\omega t + \epsilon \cos\omega t), \tag{2.258}$$

where ϵ is some small parameter.

It is easy to see that the perturbed and unperturbed magnetic fields are circularly polarized. First, we shall use the perturbation technique developed in the previous section. By comparing formulas (2.257)–(2.258) with formulas (2.113)–(2.114), we observe that

$$f_x(t) = -\sin\omega t, \quad f_y(t) = \cos\omega t. \tag{2.259}$$

The Fourier coefficient of the functions $f_x(t)$ and $f_y(t)$ are given by:

$$f_{x,-1} = \frac{1}{2j}, \quad f_{x,1} = \frac{-1}{2j}, \quad f_{x,k} = 0 \quad \text{if } |k| \neq 1, \tag{2.260}$$

$$f_{y,-1} = \frac{1}{2}, \quad f_{y,1} = \frac{1}{2}, \quad f_{y,k} = 0 \quad \text{if } |k| \neq 1. \tag{2.261}$$

By using the last expressions and formulas (2.169) and (2.170), we find

$$\phi_1(0) = jH_m, \quad \phi_k(0) = 0 \quad \text{if } k > 1, \tag{2.262}$$

$$\psi_k(0) = 0 \quad \text{for all } k \geq 0. \tag{2.263}$$

By using last equations and the same reasoning as before, we easily conclude that

$$h_{x,k}(z) = h^*_{x,-k}(z) = h_{y,k}(z) = h^*_{y,-k}(z) = 0 \quad \text{if } k > 1. \tag{2.264}$$

Thus, as expected, there are no higher-order harmonics of the magnetic field perturbation. To compute the first harmonic of the magnetic field perturbation, we invoke Eq. (2.167), which for $k = 1$ can be written as follows:

$$\left(1 - \frac{z}{z_0}\right)^2 \frac{d^2\phi_1(z)}{dz^2} - j\chi_1 \left[a\phi_1(z) + \left(1 - \frac{z}{z_0}\right)^{j2\alpha''} \cdot \psi_{-1}(z)\right] = 0. \tag{2.265}$$

From formulas (2.172), (2.173), (2.262), and (2.263) we find

$$\psi_{-1}(z) = h_{x,-1}(z) - jh_{y,-1}(z) = h^*_{x,1}(z) - jh^*_{y,1}(z) = \phi^*_1(z). \tag{2.266}$$

By using equality (2.266), Eq. (2.265) can be modified as follows:

$$\left(1 - \frac{z}{z_0}\right)^2 \frac{d^2\phi_1(z)}{dz^2} - j\chi_1 \left[a\phi_1(z) + \left(1 - \frac{z}{z_0}\right)^{j2\alpha''} \cdot \phi^*_1(z))\right] = 0. \tag{2.267}$$

According to formula (2.262), the solution to this equation is subject to the following boundary conditions:

$$\phi_1(0) = jH_m, \qquad \phi_1(\infty) = 0. \tag{2.268}$$

Now it is easy to prove that the solution to the boundary value problem (2.267)–(2.268) is given by

$$\phi_1(z) = jH_m \left(1 - \frac{z}{z_0}\right)^{\alpha}. \tag{2.269}$$

Indeed, by taking into account the fact that $\alpha'' = Im\ (\alpha)$ and by substituting expression (2.269) into Eq. (2.267), we arrive at:

$$j\frac{\alpha^2 - \alpha}{z_0^2} H_m \left(1 - \frac{z}{z_0}\right)^{\alpha} + \chi_1 H_m (a - 1) \left(1 - \frac{z}{z_0}\right)^{\alpha} = 0, \tag{2.270}$$

which is tantamount to

$$\alpha^2 - \alpha - j\chi_1(a-1)z_0^2 = 0. \tag{2.271}$$

According to formulas (2.165) and (2.166), we have

$$\chi_1(a-1)z_0^2 = \omega\sigma\mu_m z_0^2. \tag{2.272}$$

This means that Eq. (2.271) coincides with the characteristic Eq. (2.46). Because α is the root of the latter equation, we conclude that Eq. (2.271) is satisfied. This proves that expression (2.269) is the solution to the boundary value problem (2.267)–(2.268).

Now, by using formulas (2.232), (2.233), (2.245), and (2.269), we find that the perturbation technique leads to the following expression for the phasors of the magnetic field:

$$\hat{H}_x(z) = (1+j\epsilon)H_m\left(1-\frac{z}{z_0}\right)^\alpha, \tag{2.273}$$

$$\hat{H}_y(z) = -j(1+j\epsilon)H_m\left(1-\frac{z}{z_0}\right)^\alpha. \tag{2.274}$$

In the last two formulas, the penetration depth z_0 is determined by the value of μ_m computed for $H = H_m$:

$$\mu_m = kH_m^{\frac{1}{n}-1}. \tag{2.275}$$

Next, we shall use the formulas (2.38) and (2.54) and write the exact expressions for the phasors of the magnetic field. First, we note that, according to the boundary conditions (2.257) and (2.258), we have

$$\hat{H}_m = (1+j\epsilon)H_m. \tag{2.276}$$

Now, by using formulas (2.38) and (2.54), we find

$$\hat{H}_x(z) = (1+j\epsilon)H_m\left(1-\frac{z}{\tilde{z}_0}\right)^\alpha, \tag{2.277}$$

$$\hat{H}_y(z) = -j(1+j\epsilon)H_m\left(1-\frac{z}{\tilde{z}_0}\right)^\alpha. \tag{2.278}$$

In these two formulas, the symbol $\sim$ is used to distinguish between the exact value of penetration depth and the one found through the perturbation technique (see (2.273)–(2.274)). By comparing formulas (2.277)–(2.278) and (2.273)–(2.274), we find that they are identical with the exception that the penetration depth $\tilde{z}_0$ is determined by the value of $\tilde{\mu}_m$ computed for the exact magnitude of the magnetic field $\tilde{H} = \sqrt{1+\epsilon^2}H_m$:

$$\tilde{\mu}_m = k\left(\sqrt{1+\epsilon^2}H_m\right)^{\frac{1}{n}-1}. \tag{2.279}$$

This discrepancy in the penetration depth determines the inaccuracy of the perturbation technique. Let us examine how this discrepancy affects the surface impedance, which is the quantity that is most interesting in many applications. By using the same reasoning as before, we find that the values of the surface impedance obtained by using the perturbation technique and the exact solutions are given by

$$\eta = \frac{\alpha}{\sigma z_0}, \qquad \tilde{\eta} = \frac{\alpha}{\sigma \tilde{z}_0}. \tag{2.280}$$

From these formulas, we conclude that the phase of the surface impedance is not affected by the inaccuracy of the perturbation technique. As far as the magnitude of the surface impedance is concerned, from formulas (2.280) and (2.53) we find

$$\left|\frac{\tilde{\eta}}{\eta}\right| = \sqrt{\frac{\tilde{\mu}_m}{\mu_m}} = \left(1+\epsilon^2\right)^{\frac{1-n}{4n}}. \tag{2.281}$$

The results of computations of the ratio $\left|\frac{\tilde{\eta}}{\eta}\right|$ for various values of n and ϵ are shown in Fig. 2.14. It is clear from this figure that for $\epsilon \leq 0.5$ the error of the perturbation technique does not exceed 5%.

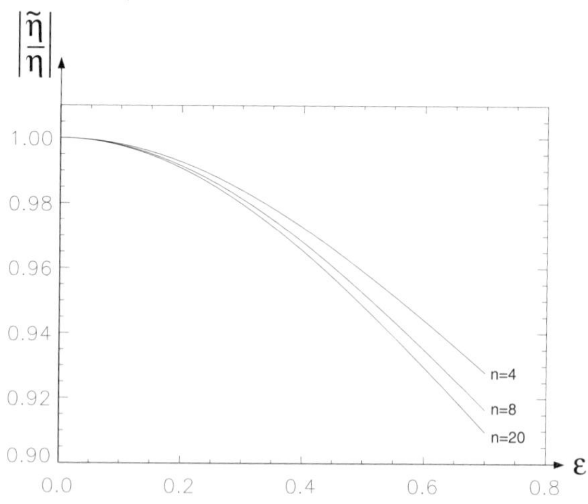

Fig. 2.14

2.4 NONLINEAR DIFFUSION OF CIRCULARLY POLARIZED ELECTROMAGNETIC FIELDS IN ANISOTROPIC MEDIA

In this section, we shall extend the perturbation technique to the analysis of nonlinear diffusion of circularly polarized electromagnetic fields in anisotropic media. We shall mathematically treat constitutive relations for anisotropic media as perturbations of constitutive relations for isotropic media. However, we shall first discuss the question of how the material properties of magnetically nonlinear and anisotropic media should be mathematically described. This question is of independent interest. In the sequel, we assume that magnetic media is unhysteretic. Since hysteresis is neglected, the relation between the magnetic flux density **B** and the magnetic field **H** should be represented by a single-valued function:

$$\mathbf{B} = \mathbf{B}(\mathbf{H}). \tag{2.282}$$

It turns out that not any single-valued function $\mathbf{B}(\mathbf{H})$ can be used for the description of magnetic properties of materials. The absence of hysteresis imposes some restrictions on $\mathbf{B}(\mathbf{H})$ (see [7]). To find these restrictions, let us recall that the absence of hysteresis means that any local cyclic losses are equal to zero. Mathematically, it can be expressed as

$$\oint_L \mathbf{H} \cdot d\mathbf{B} = 0, \tag{2.283}$$

where L is an arbitrary closed path in **H**-space, that is, in the three dimensional space with axes H_x, H_y, and H_z.

On the other hand, we have

$$0 = \oint_L d(\mathbf{H} \cdot \mathbf{B}) = \oint_L \mathbf{H} \cdot d\mathbf{B} + \oint_L \mathbf{B} \cdot d\mathbf{H}. \tag{2.284}$$

By taking into account expression (2.283) in formula (2.284), we find

$$\oint_L \mathbf{B} \cdot d\mathbf{H} = 0. \tag{2.285}$$

The last equation means that the **B**-field is curl-free in **H**-space:

$$\text{curl}_{\mathbf{H}} \mathbf{B}(\mathbf{H}) = 0. \tag{2.286}$$

This implies that:

$$\frac{\partial B_{x_i}}{\partial H_{x_j}} - \frac{\partial B_{x_j}}{\partial H_{x_i}} = 0 \quad \text{if } i \neq j, \tag{2.287}$$

where x_1, x_2, x_3 stand for x, y, and z, respectively.

From formula (2.287), we conclude that the Jacobian matrix

$$\hat{J}(\mathbf{H}) = \left\{ a_{ij} = \frac{\partial B_{x_i}}{\partial H_{x_j}} \right\} \tag{2.288}$$

is symmetric.

This is a mathematical restriction that must be imposed on any constitutive relation $\mathbf{B}(\mathbf{H})$ for nonhysteretic media. However, this is not the only restriction on $\mathbf{B}(\mathbf{H})$. There are two additional (and natural) constraints on $\mathbf{B}(\mathbf{H})$, which can be expressed by the following inequalities (see [7]):

$$c|\Delta\mathbf{H}|^2 \leq \Delta\mathbf{B} \cdot \Delta\mathbf{H} \leq C|\Delta\mathbf{H}|^2, \tag{2.289}$$

where $C \geq c > 0$, $\Delta\mathbf{H}$ is an arbitrary increment of the magnetic field, while $\Delta\mathbf{B}$ is the corresponding increment of magnetic flux density.

Inequalities (2.289) admit the following physical interpretation. The left inequality in (2.289) is valid for **passive** media. For such media, increments $\Delta\mathbf{H}$ of magnetic field and the corresponding increments $\Delta\mathbf{M}$ of magnetization form acute angles. This means that projections of $\Delta\mathbf{M}$ on the directions of $\Delta\mathbf{H}$ are nonnegative. In other words, passivity of media implies that any increment $\Delta\mathbf{H}$ of the magnetic field results in an increment $\Delta\mathbf{M}$ of magnetization whose component, parallel to $\Delta\mathbf{H}$, should have the same direction as $\Delta\mathbf{H}$ (see Fig. 2.15). The property of passivity of media requires that

$$\Delta\mathbf{M} \cdot \Delta\mathbf{H} \geq 0. \tag{2.290}$$

On the other hand, we have

$$\Delta\mathbf{B} = \mu_0 \Delta\mathbf{M} + \mu_0 \Delta\mathbf{H}. \tag{2.291}$$

From the last two formulas, we derive

$$\Delta\mathbf{B} \cdot \Delta\mathbf{H} \geq \mu_0 |\Delta\mathbf{H}|^2, \tag{2.292}$$

which is tantamount to the left inequality in (2.289) with $c \geq \mu_0$.

The right inequality in (2.289) reflects the property of saturation of magnetic media. This property means that, for any fixed $\mathbf{H}$ and $\mathbf{B}$, projections of $\Delta\mathbf{B}$ on the directions of $\Delta\mathbf{H}$ are monotonically increasing functions of $|\Delta\mathbf{H}|$, which exhibit saturation. In other words, these functions are represented by curves that lie below some straight line (see Fig. 2.16). Mathematically, the last fact can be expressed as follows:

$$\Delta \mathbf{B} \cdot \mathbf{u} \leq C|\Delta \mathbf{H}|, \quad \mathbf{u} = \frac{\Delta \mathbf{H}}{|\Delta \mathbf{H}|}, \tag{2.293}$$

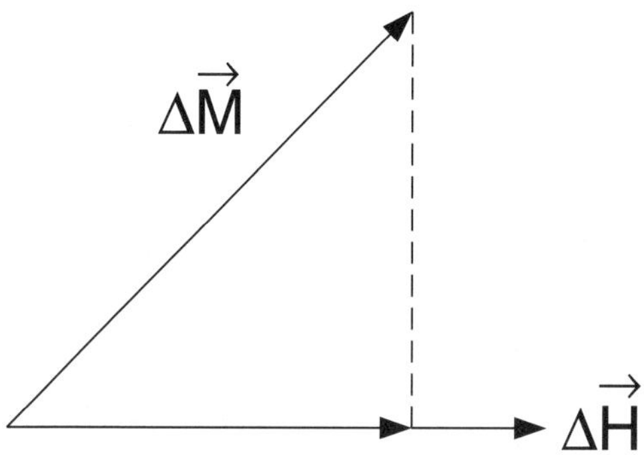

Fig. 2.15

where C is the slope of the above-mentioned straight line.

It is obvious that the last formula is equivalent to the right inequality in (2.289).

Inequalities (2.289) can be expressed in terms of Jacobian matrix (2.288) as follows:

$$c|\vec{\xi}|^2 \leq \hat{J}(\mathbf{H})\vec{\xi} \cdot \vec{\xi} \leq C|\vec{\xi}|^2, \tag{2.294}$$

where $\mathbf{H}$ is an arbitrary magnetic field, while $\vec{\xi}$ is an arbitrary vector.

The proof of equivalence of inequalities (2.289) and (2.294) is based on the following relation:

$$\Delta \mathbf{B} = \left(\int_0^1 \hat{J}(\mathbf{H}_\nu) d\nu \right) \Delta \mathbf{H}, \tag{2.295}$$

where

$$\mathbf{H}_\nu = \mathbf{H} + \nu \Delta \mathbf{H}, \quad (0 \leq \nu \leq 1). \tag{2.296}$$

To establish relation (2.295), we begin with the obvious identity

$$\Delta B_{x_i} = B_{x_i}(\mathbf{H} + \Delta \mathbf{H}) - B_{x_i}(\mathbf{H}) = \int_0^1 \frac{dB_{x_i}(\mathbf{H}_\nu)}{d\nu} d\nu. \tag{2.297}$$

On the other hand, we have

$$\frac{dB_{x_i}(\mathbf{H}_\nu)}{d\nu} = \sum_j \frac{\partial B_{x_i}}{\partial H_{x_j}}(\mathbf{H}_\nu) \cdot \frac{dH_{\nu x_j}}{d\nu} = \sum_j \frac{\partial B_{x_i}}{\partial H_{x_j}}(\mathbf{H}_\nu) \Delta H_{x_j}. \tag{2.298}$$

By substituting the last formula into expression (2.297), we find

$$\Delta B_{x_i} = \sum_j \left(\int_0^1 \frac{\partial B_{x_i}}{\partial H_{x_j}} (\mathbf{H}_\nu) d\nu \right) \cdot \Delta H_{x_j}, \tag{2.299}$$

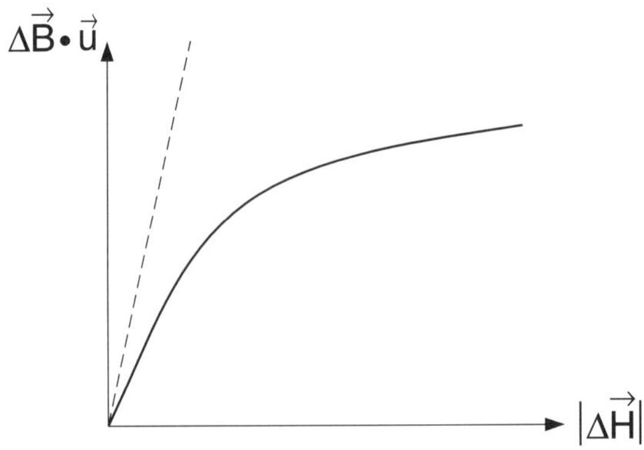

Fig. 2.16

which is the same as formula (2.295).

By using relation (2.295), inequalities (2.289) can be rewritten as follows:

$$c|\Delta \mathbf{H}|^2 \leq \left(\int_0^1 \hat{J}(\mathbf{H}_\nu) \right) \Delta \mathbf{H} \cdot \Delta \mathbf{H} \leq C|\Delta \mathbf{H}|^2, \tag{2.300}$$

or

$$c \leq \left(\int_0^1 \hat{J}(\mathbf{H}_\nu) d\nu \right) \mathbf{u} \cdot \mathbf{u} \leq C, \tag{2.301}$$

where $\mathbf{u}$ is a unit vector defined in formula (2.293).

Now we are ready to prove inequalities (2.294) by using reasoning by contradiction. Suppose that the left inequality (2.294) is not valid. This means that we can find magnetic field $\mathbf{H}$ and unit vector $\mathbf{u}$ such that

$$\hat{J}(\mathbf{H}) \mathbf{u} \cdot \mathbf{u} < c. \tag{2.302}$$

Then, assuming that coefficients of Jacobian matrix are continuous functions of $\mathbf{H}$, we can find such a small value of ΔH that for all $\mathbf{H}_\nu = \mathbf{H} + \nu \Delta H \mathbf{u}$ and $0 \leq \nu \leq 1$ we have

$$\hat{J}(\mathbf{H}_\nu) \mathbf{u} \cdot \mathbf{u} < c. \tag{2.303}$$

By integrating the last inequality with respect to ν, we find

$$\left(\int_0^1 \hat{J}(\mathbf{H}_\nu) d\nu \right) \mathbf{u} \cdot \mathbf{u} < c, \tag{2.304}$$

which contradicts to the left inequality (2.301). This reasoning by contradiction proves the left inequality (2.294). By using the same line of reasoning, the right inequality (2.294) can be established.

Thus, we have found that the Jacobian matrix (2.288) of the constitutive relation $\mathbf{B}(\mathbf{H})$ of nonhysteretic magnetic media is symmetric and satisfies inequalities (2.294). It is easy to demonstrate that these restrictions hold for the constitutive equation of isotropic media:

$$\mathbf{B}(\mathbf{H}) = \mu(|\mathbf{H}|)\mathbf{H}. \tag{2.305}$$

Indeed, formula (2.136) suggests that the corresponding Jacobian matrix is symmetric. The validity of inequalities (2.294) can also be verified. To this end, we mention that formulas (2.135)–(2.137) can be written as follows:

$$\frac{\partial B_{x_i}}{\partial H_{x_j}} = \mu(H)\delta_{ij} + \frac{d\mu(H)}{dH} \cdot \frac{H_{x_i} \cdot H_{x_j}}{H}, \tag{2.306}$$

where δ_{ij} is the Kronecker delta.

By using the last formula, we find

$$\begin{aligned} \hat{J}(\mathbf{H})\vec{\xi} \cdot \vec{\xi} &= \sum_i \sum_j \frac{\partial B_{x_i}}{\partial H_{x_j}} \xi_i \xi_j = \mu(H) \sum_i \xi_i^2 \\ &+ \frac{1}{H} \cdot \frac{d\mu(H)}{dH} \sum_i \sum_j H_{x_i}\xi_i H_{x_j}\xi_j = \mu(H)|\xi|^2 \\ &+ \frac{1}{H}\frac{d\mu(H)}{dH}\left(\sum_i H_{x_i}\xi_i\right)^2 = \mu(H)|\vec{\xi}|^2 + \frac{1}{H}\frac{d\mu(H)}{dH}\left(\mathbf{H}\cdot\vec{\xi}\right)^2. \end{aligned} \tag{2.307}$$

Next, we assume that

$$\frac{d\mu(H)}{dH} < 0. \tag{2.308}$$

This assumption is, for instance, valid for the power law approximation (2.25)–(2.27). By using inequality (2.308), from formula (2.307) we derive:

$$\hat{J}(\mathbf{H})\vec{\xi} \cdot \vec{\xi} \leq \mu(H)|\vec{\xi}|^2. \tag{2.309}$$

On the other hand

$$(\mathbf{H} \cdot \vec{\xi})^2 \leq H^2|\vec{\xi}|^2 \tag{2.310}$$

By using the last inequality and (2.308) in formula (2.307), we obtain

$$\hat{J}(\mathbf{H})\vec{\xi} \cdot \vec{\xi} \geq \left(\mu(H) + H\frac{d\mu(H)}{dH}\right)|\vec{\xi}|^2. \tag{2.311}$$

It is clear that

$$\mu_d(H) = \frac{dB(H)}{dH} = \frac{d}{dH}(\mu(H)H) = \mu(H) + H\frac{d\mu(H)}{dH}. \tag{2.312}$$

Thus, we find

$$\hat{J}(\mathbf{H})\vec{\xi} \cdot \vec{\xi} \geq \mu_d(H)|\vec{\xi}|^2. \tag{2.313}$$

By combining inequalities (2.309) and (2.313), we arrive at:

$$\mu_d(H)|\vec{\xi}|^2 \leq \hat{J}(\mathbf{H})\vec{\xi} \cdot \vec{\xi} \leq \mu(H)|\vec{\xi}|^2, \tag{2.314}$$

which means that

$$c = \min_H \mu_d(H), \quad C = \max_H \mu(H), \tag{2.315}$$

where minimum and maximum are taken over the relevant range of variation of H.

It is worthwhile to note that sometimes the situation can be encountered when $c = 0$ and $C = \infty$. This is, for example, the case when we deal with the power law approximation (2.25)–(2.27). In this case, the above situation arises because of the mathematical idealization of actual magnetic properties of media for very small ($H \to 0, \quad \mu(H) \to \infty$) and very large ($H \to \infty, \quad \mu_d(H) \to 0$) values of the magnetic field.

Next, we consider the question of how the constitutive relation (2.305) can be perturbed (modified) in order to account for anisotropic properties of magnetic media without violating the symmetry of Jacobian matrix. The easy and natural way to generate such perturbations is by using the notion of the potential of **B**-field in **H**-space. According to formulas (2.285) and (2.286), this potential is legitimate and can be defined as follows:

$$U(\mathbf{H}) = \int_0^{\mathbf{H}} \mathbf{B}(\mathbf{H}') \cdot d\mathbf{H}', \tag{2.316}$$

where a particular choice of integration path between points 0 and **H** does not matter.

If the potential U is known, then the corresponding constitutive relation can be generated by using the formula:

$$\mathbf{B}(\mathbf{H}) = \mathrm{grad}_{\mathbf{H}} U(\mathbf{H}). \tag{2.317}$$

It is apparent that this constitutive relation will satisfy Eq. (2.286), which is tantamount to the symmetry of the Jacobian matrix. If this matrix is expressed in terms of potential U, it becomes the Hessian matrix:

$$\hat{J}(\mathbf{H}) = \left\{ a_{ij} = \frac{\partial^2 U}{\partial x_i \partial x_j}(\mathbf{H}) \right\}. \tag{2.318}$$

According to formula (2.294), this matrix is positive definite. This implies that the potential U is (strictly) convex [11]. Thus, a **B**-field in **H**-space, which is generated by a constitutive relation $\mathbf{B}(\mathbf{H})$ for nonhysteretic magnetic media, can be described by a convex potential U defined in the same space.

Now let us derive the expression for the potential $U(\mathbf{H})$ in the case of isotropic media. By substituting relation (2.305) into formula (2.316) and integrating along the ray between 0 and **H**, we find

$$U(\mathbf{H}) = \int_0^{\mathbf{H}} \mu(H')H'dH'. \tag{2.319}$$

If we assume the power law approximation (2.27), then from the last equation we obtain

$$U(\mathbf{H}) = \frac{n}{n+1} kH^{\frac{n+1}{n}}, \tag{2.320}$$

which can be rewritten in terms of H_x and H_y as follows:

$$U(H_x, H_y) = \frac{n}{n+1} k \left(H_x^2 + H_y^2\right)^{\frac{n+1}{2n}}. \tag{2.321}$$

It is clear that the isotropicity of media is reflected in the symmetry of the last expression for $U(H_x, H_y)$ with respect to H_x and H_y. To generate the potential for anisotropic media, the above symmetry must be perturbed. The simplest way to do this is to assign different "weight" coefficients for H_x and H_y. By using a perturbation parameter ϵ, this can be accomplished as follows:

$$U_\epsilon(H_x, H_y) = \frac{n}{n+1} k \left[(1+\epsilon)H_x^2 + (1-\epsilon)H_y^2\right]^{\frac{n+1}{2n}}. \tag{2.322}$$

By using formulas (2.322) and (2.317), we generate the following constitutive relations for anisotropic media:

$$B_x(H_x, H_y) = (1+\epsilon)kH_x \left(\sqrt{(1+\epsilon)H_x^2 + (1-\epsilon)H_y^2}\right)^{\frac{1}{n}-1}, \tag{2.323}$$

$$B_y(H_x, H_y) = (1-\epsilon)kH_y \left(\sqrt{(1+\epsilon)H_x^2 + (1-\epsilon)H_y^2}\right)^{\frac{1}{n}-1}, \tag{2.324}$$

which automatically satisfy the symmetry restriction (2.287). In the limiting case of $n \to \infty$, expressions (2.323) and (2.324) lead to the constitutive relations for anisotropic media with abrupt magnetic transitions:

$$B_x(H_x, H_y) = (1+\epsilon)k \frac{H_x}{\sqrt{(1+\epsilon)H_x^2 + (1-\epsilon)H_y^2}}, \tag{2.325}$$

$$B_y(H_x, H_y) = (1-\epsilon)k \frac{H_y}{\sqrt{(1+\epsilon)H_x^2 + (1-\epsilon)H_y^2}}. \tag{2.326}$$

In the cases when $H_y = 0$ or $H_x = 0$, the last relations are reduced to the following ones, respectively:

$$B_x(H_x) = B_{mx}\text{sign}(H_x), \tag{2.327}$$

$$B_y(H_y) = B_{my}\text{sign}(H_y), \tag{2.328}$$

where

$$B_{mx} = k\sqrt{1+\epsilon}, \quad B_{my} = k\sqrt{1-\epsilon}, \quad \frac{B_{mx}}{B_{my}} = \sqrt{\frac{1+\epsilon}{1-\epsilon}}. \tag{2.329}$$

Now we proceed to the analysis of nonlinear diffusion of circularly polarized electromagnetic fields in anisotropic conducting media whose magnetic properties are described by constitutive relations (2.323) and (2.324). Mathematically, this analysis is tantamount to the solution of the following boundary value problem: find the time-periodic solution to the following equations:

$$\frac{\partial^2 H_x(z,t)}{\partial z^2} = \sigma \frac{\partial B_x(H_x, H_y)}{\partial t}, \tag{2.330}$$

$$\frac{\partial^2 H_y(z,t)}{\partial z^2} = \sigma \frac{\partial B_y(H_x, H_y)}{\partial t}, \tag{2.331}$$

subject to the boundary conditions

$$H_x(0,t) = H_m \cos\omega t, \tag{2.332}$$

$$H_y(0,t) = H_m \sin\omega t, \tag{2.333}$$

$$H_x(\infty,t) = H_y(\infty,t) = 0. \tag{2.334}$$

We intend to apply the perturbation technique to the solution of the above boundary value problem. To this end, we shall first find ϵ-expansions for the constitutive relations (2.323) and (2.324). We begin with the ϵ-expansion of the function

$$f(\epsilon) = \left(\sqrt{(1+\epsilon)H_x^2 + (1-\epsilon)H_y^2}\right)^{\frac{1}{n}-1}. \tag{2.335}$$

This expansion has the form

$$f(\epsilon) = f(0) + \epsilon f'(0) + \dots. \tag{2.336}$$

By using simple calculus, we find

$$f(0) = \left(\sqrt{H_x^2 + H_y^2}\right)^{\frac{1}{n}-1} = H^{\frac{1}{n}-1}, \tag{2.337}$$

$$f'(0) = \frac{1-n}{2n} H^{\frac{1}{n}-1} \cdot \frac{H_x^2 - H_y^2}{H^2}. \tag{2.338}$$

By substituting formulas (2.337) and (2.338) into expansion (2.336), we obtain

$$f(\epsilon) = H^{\frac{1}{n}-1}\left(1 + \epsilon\frac{1-n}{2n} \cdot \frac{H_x^2 - H_y^2}{H^2}\right) + \dots . \tag{2.339}$$

From expressions (2.323), (2.324), (2.335), and (2.339), we find

$$B_x(H_x, H_y) = (1+\epsilon)kH^{\frac{1}{n}-1}H_x\left(1 + \epsilon\frac{1-n}{2n} \cdot \frac{H_x^2 - H_y^2}{H^2}\right) + \dots, \tag{2.340}$$

$$B_y(H_x, H_y) = (1-\epsilon)kH^{\frac{1}{n}-1}H_y\left(1 + \epsilon\frac{1-n}{2n} \cdot \frac{H_x^2 - H_y^2}{H^2}\right) + \dots . \tag{2.341}$$

By retaining only zero- and first-order terms in the last two equations, we transform them as follows:

$$B_x\left(H_x, H_y\right) = B_x^0\left(H_x, H_y\right) + \epsilon B_x^0\left(H_x, H_y\right)\left(1 + \frac{1-n}{2n} \cdot \frac{H_x^2 - H_y^2}{H^2}\right) + \dots, \tag{2.342}$$

$$B_y\left(H_x, H_y\right) = B_y^0\left(H_x, H_y\right) - \epsilon B_y^0\left(H_x, H_y\right)\left(1 - \frac{1-n}{2n} \cdot \frac{H_x^2 - H_y^2}{H^2}\right) + \dots, \tag{2.343}$$

where zero-order terms $B_x^0(H_x, H_y)$ and $B_y^0(H_x, H_y)$ coincide with the constitutive relations for isotropic media:

$$B_x^0(H_x, H_y) = kH^{\frac{1}{n}-1}H_x, \quad B_y^0(H_x, H_y) = kH^{\frac{1}{n}-1}H_y. \tag{2.344}$$

We shall look for the solution to the boundary value problem (2.330)–(2.334) in the form

$$H_x(z,t) = H_x^0(z,t) + \epsilon h_x(z,t) + \dots, \tag{2.345}$$

$$H_y(z,t) = H_y^0(z,t) + \epsilon h_y(z,t) + \dots . \tag{2.346}$$

By substituting the last two expressions into (2.342) and (2.343), expanding the results of substitutions with respect to ϵ and retaining only zero- and first-order terms, we obtain

$$B_x(H_x, H_y) = B_x^0(H_x^0, H_y^0) + \epsilon\left[h_x \frac{\partial B_x^0}{\partial H_x}(H_x^0, H_y^0) + h_y \frac{\partial B_x^0}{\partial H_y}(H_x^0, H_y^0)\right] + \epsilon B_x^0(H_x^0, H_y^0)\left[1 + \frac{1-n}{2n} \cdot \frac{(H_x^0)^2 - (H_y^0)^2}{(H^0)^2}\right] + \dots, \tag{2.347}$$

$$B_y(H_x, H_y) = B_y^0(H_x^0, H_y^0) + \epsilon\left[h_x \frac{\partial B_y^0}{\partial H_x}(H_x^0, H_y^0) + h_y \frac{\partial B_y^0}{\partial H_y}(H_x^0, H_y^0)\right] - \epsilon B_y^0(H_x^0, H_y^0)\left[1 - \frac{1-n}{2n} \cdot \frac{(H_x^0)^2 - (H_y^0)^2}{(H^0)^2}\right] + \dots. \tag{2.348}$$

Now, by substituting formulas (2.345)–(2.348) into the boundary value problem (2.330)–(2.334) and by equating the terms of like powers of ϵ, we end up with the following two boundary value problems for zero- and first-order terms, respectively:

$$\frac{\partial^2 H_x^0(z,t)}{\partial z^2} = \sigma \frac{\partial B_x^0(H_x^0, H_y^0)}{\partial t}, \tag{2.349}$$

$$\frac{\partial^2 H_y^0(z,t)}{\partial z^2} = \sigma \frac{\partial B_y^0(H_x^0, H_y^0)}{\partial t}, \tag{2.350}$$

$$H_x^0(0,t) = H_m \cos\omega t, \tag{2.351}$$

$$H_y^0(0,t) = H_m \sin\omega t, \tag{2.352}$$

$$H_x^0(\infty,t) = H_y^0(\infty,t) = 0, \tag{2.353}$$

and

$$\frac{\partial^2 h_x(z,t)}{\partial z^2} - \sigma \frac{\partial}{\partial t}\left[h_x(z,t)\frac{\partial B_x^0}{\partial H_x}(H_x^0, H_y^0) + h_y(z,t)\frac{\partial B_x^0}{\partial H_y}(H_x^0, H_y^0)\right] = \sigma \frac{\partial g_x}{\partial t}, \tag{2.354}$$

$$\frac{\partial^2 h_y(z,t)}{\partial z^2} - \sigma \frac{\partial}{\partial t}\left[h_x(z,t)\frac{\partial B_y^0}{\partial H_x}(H_x^0, H_y^0) + h_y(z,t)\frac{\partial B_y^0}{\partial H_y}(H_x^0, H_y^0)\right] = \sigma \frac{\partial g_y}{\partial t}, \tag{2.355}$$

$$h_x(0,t) = h_y(0,t) = 0, \tag{2.356}$$

$$h_x(\infty,t) = h_y(\infty,t) = 0, \tag{2.357}$$

where the functions $g_x(z,t)$ and $g_y(z,t)$ are given by the following expressions:

$$g_x(z,t) = B_x^0\left(H_x^0, H_y^0\right)\left[1 + \frac{1-n}{2n} \cdot \frac{(H_x^0)^2 - (H_y^0)^2}{(H^0)^2}\right], \tag{2.358}$$

$$g_y(z,t) = -B_y^0\left(H_x^0, H_y^0\right)\left[1 - \frac{1-n}{2n} \cdot \frac{(H_x^0)^2 - (H_y^0)^2}{(H^0)^2}\right]. \tag{2.359}$$

It is obvious that the boundary value problem (2.349)–(2.353) is identical to the boundary value problem (2.122)–(2.126). Consequently, the zero-order terms $H_x^0(z,t)$ and $H_y^0(z,t)$ are given by formulas (2.132) and (2.133). It is also obvious that Eqs. (2.354) and (2.355) are similar to Eqs. (2.127) and (2.128). The only difference is that Eqs. (2.354) and (2.355) are inhomogeneous, that is, they have right-hand sides (i.e., "driving forces"), which are determined by functions $g_x(z,t)$ and $g_y(z,t)$. Thus, it can be concluded that Eqs. (2.354) and (2.355) can be transformed in the same way as we transformed Eqs. (2.127) and (2.128) in Section 2.2. The only new element will be the transformations of the right-hand sides of Eqs. (2.354) and (2.355). These transformations are performed as follows. By substituting expressions (2.132) and (2.133) into formulas (2.358) and (2.359) and by using simple trigonometry, we obtain

$$\begin{aligned} g_x(z,t) &= \mu_m H_m \left(1 - \frac{z}{z_0}\right)^{\frac{2}{n-1}} \cos(\omega t + \theta(z))\left[1 + \frac{1-n}{2n}\cos(2\omega t + 2\theta(z))\right] \\ &= \mu_m H_m \left(1 - \frac{z}{z_0}\right)^{\frac{2}{n-1}} \left[\frac{3n+1}{4n}\cos(\omega t + \theta(z)) + \frac{1-n}{4n}\cos(3\omega t + 3\theta(z))\right]. \end{aligned} \tag{2.360}$$

Similarly, we have

$$\begin{aligned} g_y(z,t) = -\, &\mu_m H_m \left(1 - \frac{z}{z_0}\right)^{\frac{2}{n-1}} \left[\frac{3n+1}{4n}\sin(\omega t + \theta(z)) \right. \\ &\left. - \frac{1-n}{4n}\sin(3\omega t + 3\theta(z))\right]. \end{aligned} \tag{2.361}$$

Next, by introducing new state variables

$$\phi(z,t) = h_x(z,t) + jh_y(z,t), \tag{2.362}$$

$$\psi(z,t) = h_x(z,t) - jh_y(z,t), \tag{2.363}$$

and by using formulas (2.360) and (2.361) and the same line of reasoning as in Section 2.2, we transform the boundary value problem (2.354)–(2.357) to the form:

$$\left(1-\frac{z}{z_0}\right)^2 \frac{\partial^2\phi(z,t)}{\partial z^2} - \sigma\mu_m \frac{1-n}{2n}\frac{\partial}{\partial t}\left[\frac{1+n}{1-n}\phi(z,t) + \left(1-\frac{z}{z_0}\right)^{j2\alpha''}\right.$$

$$\left.\times e^{j2\omega t}\psi(z,t)\right] = \sigma\mu_m H_m \left(1-\frac{z}{z_0}\right)^{\frac{2n}{n-1}} \frac{\partial}{\partial t}\left[\frac{3n+1}{4n}\left(1-\frac{z}{z_0}\right)^{-j\alpha''} e^{-j\omega t}\right.$$

$$\left. + \frac{1-n}{4n}\left(1-\frac{z}{z_0}\right)^{j3\alpha''} e^{j3\omega t}\right], \tag{2.364}$$

$$\left(1-\frac{z}{z_0}\right)^2 \frac{\partial^2\psi(z,t)}{\partial z^2} - \sigma\mu_m \frac{1-n}{2n}\frac{\partial}{\partial t}\left[\frac{1+n}{1-n}\psi(z,t) + \left(1-\frac{z}{z_0}\right)^{-j2\alpha''}\right.$$

$$\left.\times e^{-j2\omega t}\phi(z,t)\right] = \sigma\mu_m H_m \left(1-\frac{z}{z_0}\right)^{\frac{2n}{n-1}} \frac{\partial}{\partial t}\left[\frac{3n+1}{4n}\left(1-\frac{z}{z_0}\right)^{j\alpha''} e^{j\omega t}\right.$$

$$\left. + \frac{1-n}{4n}\left(1-\frac{z}{z_0}\right)^{-j3\alpha''} e^{-j3\omega t}\right], \tag{2.365}$$

$$\phi(0,t) = \psi(0,t) = 0, \tag{2.366}$$

$$\phi(\infty,t) = \psi(\infty,t) = 0. \tag{2.367}$$

We look for the periodic solution to the boundary value problem (2.364)–(2.367) in terms of the Fourier series:

$$\phi(z,t) = \sum_{k=-\infty}^{\infty} \phi_k(z) e^{jk\omega t}, \tag{2.368}$$

$$\psi(z,t) = \sum_{k=-\infty}^{\infty} \psi_k(z) e^{jk\omega t}. \tag{2.369}$$

By substituting these Fourier series into Eqs. (2.364) and (2.365) and equating the terms with the same exponents, we derive the following coupled ordinary differential equations:

$$\left(1-\frac{z}{z_0}\right)^2 \frac{d^2\phi_k(z)}{dz^2} - j\chi_k \left[a\phi_k(z) + \left(1-\frac{z}{z_0}\right)^{j2\alpha''} \cdot \psi_{k-2}(z)\right] =$$
$$-j\nu\left(1-\frac{z}{z_0}\right)^{\frac{2n}{n-1}-j\alpha''} \cdot \delta_{-1,k} + j\zeta\left(1-\frac{z}{z_0}\right)^{\frac{2n}{n-1}+j3\alpha''} \cdot \delta_{3,k}, \tag{2.370}$$

$$\left(1-\frac{z}{z_0}\right)^2 \frac{d^2\psi_{k-2}(z)}{dz^2} - j\chi_{k-2} \left[a\psi_{k-2}(z) + \left(1-\frac{z}{z_0}\right)^{-j2\alpha''} \cdot \phi_k(z)\right] =$$
$$j\nu\left(1-\frac{z}{z_0}\right)^{\frac{2n}{n-1}+j\alpha''} \cdot \delta_{1,k-2} - j\zeta\left(1-\frac{z}{z_0}\right)^{\frac{2n}{n-1}-j3\alpha''} \cdot \delta_{-3,k-2}. \tag{2.371}$$

Here $\delta_{-1,k}$, $\delta_{3,k}$, $\delta_{1,k-2}$, and $\delta_{-3,k-2}$ are the Kronecker deltas, χ_k and a are specified by formulas (2.165) and (2.166), respectively, while ν and ζ are given by the following expressions:

$$\nu = \omega\sigma\mu_m H_m \frac{3n+1}{4n}, \tag{2.372}$$

$$\zeta = 3\omega\sigma\mu_m H_m \frac{1-n}{4n}. \tag{2.373}$$

As before, we find that simultaneous differential Eqs. (2.370) and (2.371) are coupled in separate pairs. The solution to these differential equations is subject to the following boundary conditions:

$$\phi_k(0) = \psi_{k-2}(0) = 0, \tag{2.374}$$

$$\phi_k(\infty) = \psi_{k-2}(\infty) = 0. \tag{2.375}$$

It is obvious that for all $k \geq 0$ and $k \neq 3$, Eqs. (2.370) and (2.371) and the corresponding boundary conditions are homogeneous. This implies that

$$\phi_k(z) = 0, \qquad \text{if } k \geq 0 \text{ and } k \neq 3, \tag{2.376}$$

$$\psi_k(z) = 0, \qquad \text{if } k \geq 0 \text{ and } k \neq 1. \tag{2.377}$$

From these formulas, we conclude that within the first-order perturbation theory all harmonics of the electromagnetic field, except the first and third harmonics, are equal to zero. To find the first and third harmonics of magnetic field perturbations, we need to solve the following boundary value problem:

$$\left(1-\frac{z}{z_0}\right)^2 \frac{d^2\phi_3(z)}{dz^2} - j\chi_3\left[a\phi_3(z) + \left(1-\frac{z}{z_0}\right)^{j2\alpha''} \cdot \psi_1(z)\right] = j\zeta\left(1-\frac{z}{z_0}\right)^{\frac{2n}{n-1}+j3\alpha''}, \tag{2.378}$$

$$\left(1-\frac{z}{z_0}\right)^2 \frac{d^2\psi_1(z)}{dz^2} - j\chi_1\left[a\psi_1(z) + \left(1-\frac{z}{z_0}\right)^{-j2\alpha''} \cdot \phi_3(z)\right] = j\nu\left(1-\frac{z}{z_0}\right)^{\frac{2n}{n-1}+j\alpha''}, \tag{2.379}$$

$$\phi_3(0) = \psi_1(0) = 0, \tag{2.380}$$

$$\phi_3(\infty) = \psi_1(\infty) = 0. \tag{2.381}$$

We look for the particular solution to Eqs. (2.378) and (2.379) in the form

$$\phi_3^{(p)}(z) = C_3\left(1-\frac{z}{z_0}\right)^{\lambda_3}, \tag{2.382}$$

$$\psi_1^{(p)}(z) = C_1\left(1-\frac{z}{z_0}\right)^{\lambda_1}, \tag{2.383}$$

where

$$\lambda_3 = \frac{2n}{n+1} + j3\alpha'', \quad \lambda_1 = \alpha = \frac{2n}{n-1} + j\alpha''. \tag{2.384}$$

By substituting expressions (2.382) and (2.383) into Eqs. (2.378) and (2.379), we end up with the following simultaneous equations for C_3 and C_1:

$$\left[\lambda_3(\lambda_3 - 1) - j\chi_3 a z_0^2\right] C_3 - j\chi_3 z_0^2 C_1 = j\zeta z_0^2, \tag{2.385}$$

$$\left[\lambda_1(\lambda_1 - 1) - j\chi_1 a z_0^2\right] C_1 - j\chi_1 z_0^2 C_3 = j\nu z_0^2. \tag{2.386}$$

By taking into account Eq. (2.271) and the fact that $\lambda_1 = \alpha$, the last two equations can be easily solved. As a result, we arrive at the following expressions for C_3 and C_1:

$$C_3 = j\frac{\zeta z_0^2 - 3\nu z_0^2}{\lambda_3(\lambda_3 - 1) - j\chi_3(a-1)z_0^2}, \tag{2.387}$$

$$C_1 = -\frac{3\nu[\lambda_3(\lambda_3 - 1) - j\chi_3 a z_0^2] + j\chi_3 z_0^2 \zeta}{\chi_3[\lambda_3(\lambda_3 - 1) - j\chi_3(a-1)z_0^2]}. \tag{2.388}$$

A general solution to the homogeneous differential equations corresponding to Eqs. (2.378) and (2.379) can be found in the same form (2.236)–(2.237) as in the previous section. Thus, the complete solution to the differential Eqs. (2.378) and (2.379) can be written in the form

$$\phi_3(z) = A'\left(1 - \frac{z}{z_0}\right)^{\beta'} + A''\left(1 - \frac{z}{z_0}\right)^{\beta''} + C_3\left(1 - \frac{z}{z_0}\right)^{\lambda_3}, \tag{2.389}$$

$$\psi_1(z) = B'\left(1 - \frac{z}{z_0}\right)^{\beta' - j2\alpha''} + B''\left(1 - \frac{z}{z_0}\right)^{\beta'' - j2\alpha''} + C_1\left(1 - \frac{z}{z_0}\right)^{\lambda_1}. \tag{2.390}$$

In formulas (2.389) and (2.390), β' and β'' are the roots of the characteristic Eq. (2.238) with positive real parts, while coefficients A', A'', B', and B'' can be found from the boundary conditions (2.380)–(2.381) and equations similar to Eqs. (2.188)–(2.189). This leads to the following set of simultaneous equations with respect to the above coefficients:

$$A' + A'' = -C_3, \tag{2.391}$$

$$B' + B'' = -C_1, \tag{2.392}$$

$$\left[(\beta')^2 - \beta' - j\chi_3 a z_0^2\right] A' - j\chi_3 z_0^2 B' = 0, \tag{2.393}$$

$$\left[(\beta'')^2 - \beta'' - j\chi_3 a z_0^2\right] A'' - j\chi_3 z_0^2 B'' = 0. \tag{2.394}$$

By solving the above equations, coefficients A', A'', B', and B'' can be determined. By using these coefficients and the same line of reasoning as before, the following formulas for the phasors of the first harmonics of the magnetic field components can be derived:

$$\hat{H}_{x,1}(z) = (H_m + \epsilon C_1)\left(1 - \frac{z}{z_0}\right)^{\alpha} + \epsilon\left[B'\left(1 - \frac{z}{z_0}\right)^{\beta' - j2\alpha''} + B''\left(1 - \frac{z}{z_0}\right)^{\beta'' - j2\alpha''}\right], \tag{2.395}$$

$$\hat{H}_{y,1}(z) = -j(H_m - \epsilon C_1)\left(1 - \frac{z}{z_0}\right)^{\alpha} + j\epsilon\left[B'\left(1 - \frac{z}{z_0}\right)^{\beta' - j2\alpha''} + B''\left(1 - \frac{z}{z_0}\right)^{\beta'' - j2\alpha''}\right]. \tag{2.396}$$

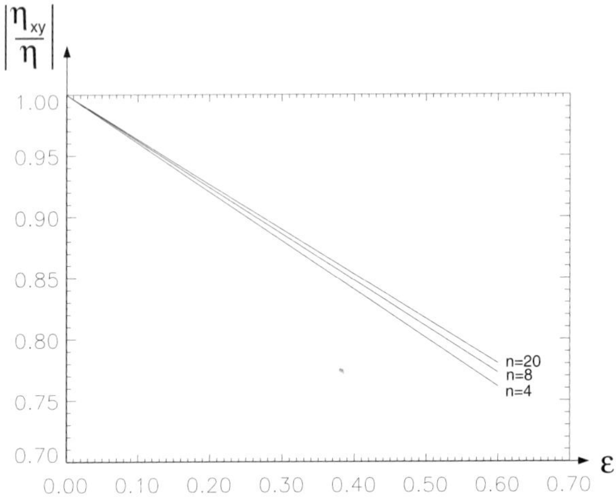

Fig. 2.17

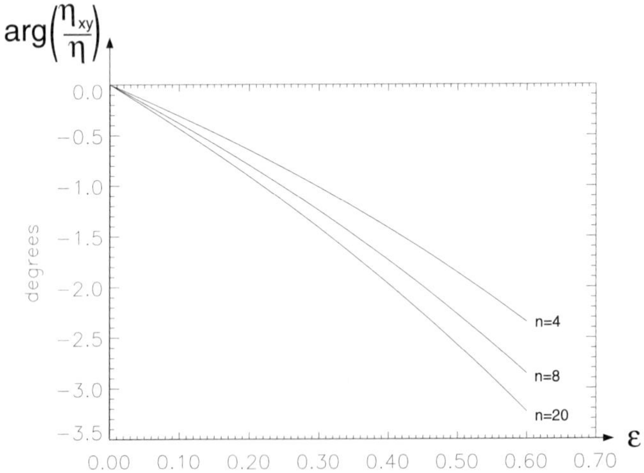

Fig. 2.18

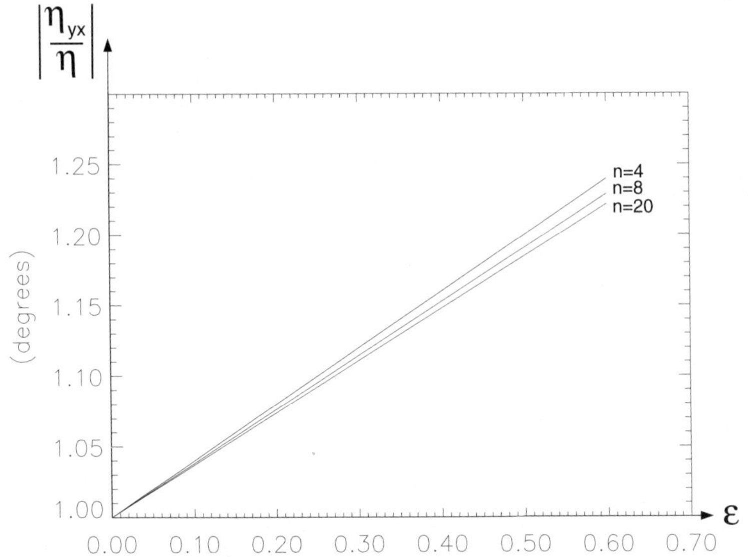

Fig. 2.19

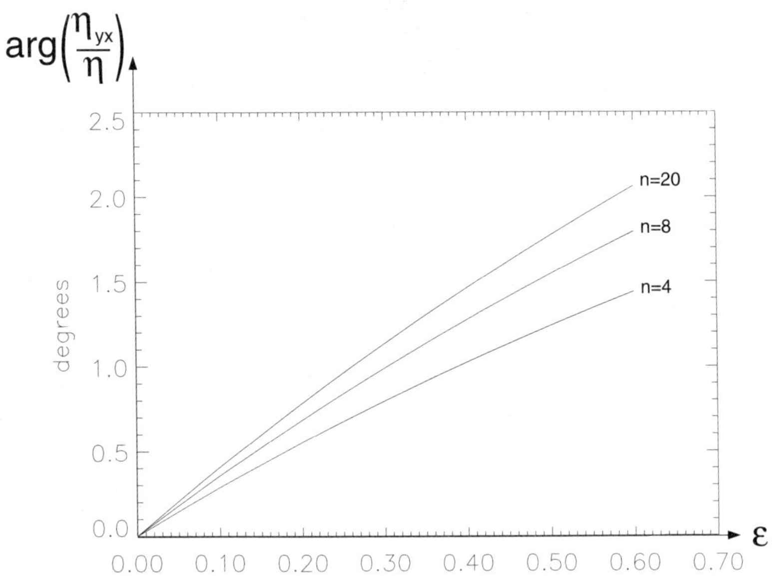

Fig. 2.20

By using the last two equations and by invoking the formulas (2.63) and

(2.65), we obtain the following expressions for surface impedances:

$$\frac{\eta_{xy}}{\eta} = \frac{\alpha(H_m - \epsilon C_1) - \epsilon[B'(\beta' - j2\alpha'') + B''(\beta'' - j2\alpha'')]}{\alpha H_m}, \tag{2.397}$$

$$\frac{\eta_{yx}}{\eta} = \frac{\alpha(H_m + \epsilon C_1) + \epsilon[B'(\beta' - j2\alpha'') + B''(\beta'' - j2\alpha'')]}{\alpha H_m}, \tag{2.398}$$

where η is the surface impedance in the case of isotropic media.

Formulas (2.397) and (2.398) allow one to evaluate the affect of magnetic anisotropy on the surface impedance. It is important to stress that the right-hand sides of formulas (2.397) and (2.398) do not depend on H_m. This is because, according to formulas (2.372), (2.373), (2.387), (2.388), and (2.391)–(2.394), coefficients C_1, B', and B'' are directly proportional to H_m.

Thus, ratios η_{xy}/η and η_{yx}/η are functions only of n and ϵ. These functions have been computed for various values of n and ϵ and the results of computations are shown in Figs. 2.17, 2.18, 2.19, and 2.20.

2.5 NONLINEAR DIFFUSION OF ELLIPTICALLY POLARIZED ELECTROMAGNETIC FIELDS IN ANISOTROPIC MEDIA

In this section, we shall further generalize the perturbation technique to analyze nonlinear diffusion of elliptically polarized electromagnetic fields in anisotropic media. To this end, both elliptical polarizations and constitutive relations for anisotropic media will be treated as perturbations of circular polarizations and constitutive relations for isotropic media, respectively. This will require the introduction of two perturbation parameters: one for boundary conditions and another for constitutive relations.

As far as the constitutive relations are concerned, expressions (2.323) and (2.324) will be used in the sequel for the description of anisotropic and nonlinear magnetic properties of media. These expressions imply that axes x and y are chosen as anisotropy axes. For anisotropic media (contrary to the case of isotropic media), various elliptical polarizations (which differ by the orientation of the ellipse traced by the endpoint of the magnetic field $\mathbf{H}(0,t)$) are not equivalent to one another. This requires a certain modification of boundary conditions (2.213) and (2.214). To achieve this modification, consider major and minor axes of the polarization ellipse, which will be denoted as x' and y' respectively. Let θ_0 be the angle between the major axis x' and the anisotropy axis x. In the x' and y' coordinate, the elliptical polarization can be characterized by the equations:

$$H_{x'}(0,t) = H_{mx'} \cos \omega t, \tag{2.399}$$

$$H_{y'}(0,t) = H_{my'} \sin \omega t. \tag{2.400}$$

By introducing the perturbation parameter $\tilde{\epsilon}$:

$$\tilde{\epsilon} = \frac{H_{mx'} - H_{my'}}{H_{mx'} + H_{my'}}, \tag{2.401}$$

the last two formulas can be rewritten as follows:

$$H_{x'}(0,t) = H_m \cos \omega t + \tilde{\epsilon} H_m \cos \omega t, \tag{2.402}$$

$$H_{y'}(0,t) = H_m \sin \omega t - \tilde{\epsilon} H_m \sin \omega t, \tag{2.403}$$

where, as before,

$$H_m = \frac{H_{mx'} + H_{my'}}{2}. \tag{2.404}$$

The x- and y-component of the magnetic field $\mathbf{H}(0,t)$ can be expressed in terms of x'- and y'-components of the same field and the orientation angle θ_0 as follows:

$$H_x(0,t) = H_{x'}(0,t) \cos \theta_0 - H_{y'}(0,t) \sin \theta_0, \tag{2.405}$$

$$H_y(0,t) = H_{x'}(0,t) \sin \theta_0 + H_{y'}(0,t) \cos \theta_0. \tag{2.406}$$

By substituting formulas (2.402) and (2.403) in the last two equations and by using simple trigonometry, we end up with the following relations for x- and y-components of the magnetic field:

$$H_x(0,t) = H_m \cos(\omega t + \theta_0) + \tilde{\epsilon} H_m \cos(\omega t - \theta_0), \tag{2.407}$$

$$H_y(0,t) = H_m \sin(\omega t + \theta_0) - \tilde{\epsilon} H_m \sin(\omega t - \theta_0). \tag{2.408}$$

The last two formulas are transparent from the physical point of view, and they reflect the well-known fact that any elliptical polarization can be represented as a superposition of two oppositely rotating circular polarizations. One of these circular polarizations is treated as a perturbation.

Now, we proceed to the analysis of nonlinear diffusion of elliptically polarized electromagnetic fields in anisotropic media with constitutive relations (2.323) and (2.324). Mathematically, this analysis is tantamount to the solution of the following equations:

$$\frac{\partial^2 H_x(z,t)}{\partial z^2} = \sigma \frac{\partial B_x(H_x, H_y)}{\partial t}, \tag{2.409}$$

$$\frac{\partial^2 H_y(z,t)}{\partial z^2} = \sigma \frac{\partial B_y(H_x, H_y)}{\partial t}, \tag{2.410}$$

subject to the boundary conditions (2.407)–(2.408) and zero conditions at infinity.

By using the general idea of the perturbation technique, we look for the solution in the form

$$H_x(z,t) = H_x^0(z,t) + \epsilon h_x(z,t) + \tilde{\epsilon}\tilde{h}_x(z,t) + \ldots, \tag{2.411}$$

$$H_y(z,t) = H_y^0(z,t) + \epsilon H_y(z,t) + \tilde{\epsilon}\tilde{h}_y(z,t) + \ldots. \tag{2.412}$$

In other words, we use expansions with respect to two perturbation parameters (ϵ and $\tilde{\epsilon}$), which are present in the constitutive relations and the boundary conditions, respectively.

By substituting the last two expressions into formulas (2.323) and (2.324) and by using the same line of reasoning as in Sections 2.2 and 2.4, we arrive at the following expansions for $B_x(H_x, H_y)$ and $B_y(H_x, H_y)$:

$$\begin{aligned} B_x\left(H_x, H_y\right) &= B_x^0\left(H_x^0, H_y^0\right) + \epsilon \left[h_x \frac{\partial B_x^0}{\partial H_x}\left(H_x^0, H_y^0\right) \right. \\ &\left. + h_y \frac{\partial B_x^0}{\partial H_y}\left(H_x^0, H_y^0\right) + B_x^0\left(H_x^0, H_y^0\right)\left(1 + \frac{1-n}{2n} \cdot \frac{(H_x^0)^2 - (H_y^0)^2}{(H^0)^2}\right)\right] \\ &+ \tilde{\epsilon}\left[\tilde{h}_x \frac{\partial B_x^0}{\partial H_x}\left(H_x^0, H_y^0\right) + \tilde{h}_y \frac{\partial B_x^0}{\partial H_y}\left(H_x^0, H_y^0\right)\right] + \ldots, \end{aligned} \tag{2.413}$$

$$\begin{aligned} B_y\left(H_x, H_y\right) &= B_y^0\left(H_x^0, H_y^0\right) + \epsilon \left[h_x \frac{\partial B_y^0}{\partial H_x}\left(H_x^0, H_y^0\right) \right. \\ &\left. + h_y \frac{\partial B_y^0}{\partial H_y}\left(H_x^0, H_y^0\right) - B_y^0\left(H_x^0, H_y^0\right)\left(1 - \frac{1-n}{2n} \cdot \frac{(H_x^0)^2 - (H_y^0)^2}{(H^0)^2}\right)\right] \\ &+ \tilde{\epsilon}\left[\tilde{h}_x \frac{\partial B_y^0}{\partial H_x}\left(H_x^0, H_y^0\right) + \tilde{h}_y \frac{\partial B_y^0}{\partial H_y}\left(H_x^0, H_y^0\right)\right] + \ldots, \end{aligned} \tag{2.414}$$

where, as before, $B_x^0(H_x^0, H_y^0)$ and $B_y^0(H_x^0, H_y^0)$ are constitutive relations for isotropic media defined by Eq. (2.344). It is easy to see that expansions (2.413) and (2.414) are simple combinations of expansions (2.347), (2.120) and (2.348), (2.121), respectively, as they must be.

By substituting expansions (2.411), (2.412), (2.413), and (2.414) into Eqs. (2.409)–(2.408) and by equating similar terms, we end up with the following three boundary value problems:

$$\frac{\partial^2 H_x^0(z,t)}{\partial z^2} = \sigma \frac{\partial B_x^0(H_x^0, H_y^0)}{\partial t}, \tag{2.415}$$

$$\frac{\partial^2 H_y^0(z,t)}{\partial z^2} = \sigma \frac{\partial B_y^0(H_x^0, H_y^0)}{\partial t}, \tag{2.416}$$

$$H_x^0(0,t) = H_m \cos(\omega t + \theta_0), \tag{2.417}$$

$$H_y^0(0,t) = H_m \sin(\omega t + \theta_0), \tag{2.418}$$

$$H_x^0(\infty,t) = H_y^0(\infty,t) = 0. \tag{2.419}$$

Next,

$$\begin{aligned} &\frac{\partial^2 h_x(z,t)}{\partial z^2} - \sigma \frac{\partial}{\partial t}\left[h_x(z,t) \frac{\partial B_x^0}{\partial H_x}\left(H_x^0, H_y^0\right) \right. \\ &\left. + h_y(z,t) \frac{\partial B_x^0}{\partial H_y}\left(H_x^0, H_y^0\right) \right] = \sigma \frac{\partial g_x(z,t)}{\partial t}, \end{aligned} \tag{2.420}$$

$$\begin{aligned} &\frac{\partial^2 h_y(z,t)}{\partial z^2} - \sigma \frac{\partial}{\partial t}\left[h_x(z,t) \frac{\partial B_y^0}{\partial H_x}\left(H_x^0, H_y^0\right) \right. \\ &\left. + h_y(z,t) \frac{\partial B_y^0}{\partial H_y}(H_x^0, H_y^0) \right] = \sigma \frac{\partial g_y(z,t)}{\partial t}, \end{aligned} \tag{2.421}$$

$$h_x(0,t) = h_y(0,t) = 0, \tag{2.422}$$

$$h_x(\infty,t) = h_y(\infty,t) = 0, \tag{2.423}$$

where functions $g_x(z,t)$ and $g_y(z,t)$ are given by formulas (2.358) and (2.359), respectively; and, finally:

$$\begin{aligned} &\frac{\partial^2 \tilde{h}_x(z,t)}{\partial z^2} - \sigma \frac{\partial}{\partial t}\left[\tilde{h}_x(z,t) \frac{\partial B_x^0}{\partial H_x}\left(H_x^0, H_y^0\right) \right. \\ &\left. + \tilde{h}_y(z,t) \frac{\partial B_x^0}{\partial H_y}\left(H_x^0, H_y^0\right) \right] = 0, \end{aligned} \tag{2.424}$$

$$\begin{aligned} &\frac{\partial^2 \tilde{h}_y(z,t)}{\partial z^2} - \sigma \frac{\partial}{\partial t}\left[\tilde{h}_x(z,t) \frac{\partial B_y^0}{\partial H_x}(H_x^0, H_y^0) \right. \\ &\left. + \tilde{h}_y(z,t) \frac{\partial B_x^0}{\partial H_y}(H_x^0, H_y^0) \right] = 0, \end{aligned} \tag{2.425}$$

$$\tilde{h}_x(0,t) = H_m \cos(\omega t - \theta_0), \tag{2.426}$$

$$\tilde{h}_y(0,t) = -H_m \sin(\omega t - \theta_0), \tag{2.427}$$

$$\tilde{h}_x(\infty, t) = \tilde{h}_y(\infty, t) = 0. \tag{2.428}$$

It is clear that the boundary value problems (2.415)–(2.419), (2.420)–(2.423), and (2.424)–(2.418) are almost identical to the boundary value problems (2.349)–(2.353), (2.354)–(2.357), and (2.127)–(2.131), respectively. The only difference is the presence of the "orientation" angle θ_0 in formulas (2.417), (2.418), (2.426), and (2.427). For this reason, $H_x^0(z,t)$ and $H_y^0(z,t)$ will be given by formulas (2.57)–(2.58) instead of formulas (2.132)–(2.133). This will result in the replacement of relation (2.149) for $\theta(z)$ by the following formula:

$$\theta(z) = \theta_0 + \alpha'' \ln\left(1 - \frac{z}{z_0}\right). \tag{2.429}$$

By using literally the same line of reasoning as in the previous section, it can be shown that the solution to the boundary value problem (2.420)–(2.423) contain only the first and third harmonics, while all other harmonics are equal to zero. To find these first and third harmonics, the boundary value problem (2.420)–(2.423) can be reduced to the following boundary value problem for the ordinary differential equations:

$$\left(1 - \frac{z}{z_0}\right)^2 \frac{d^2\phi_3(z)}{dz^2} - j\chi_3\left[a\phi_3(z) + e^{j2\theta_0}\left(1 - \frac{z}{z_0}\right)^{j2\alpha''} \cdot \psi_1(z)\right]$$
$$= j\zeta e^{j3\theta_0}\left(1 - \frac{z}{z_0}\right)^{\frac{2n}{n-1} + j3\alpha''}, \tag{2.430}$$

$$\left(1 - \frac{z}{z_0}\right)^2 \frac{d^2\psi_1}{dz^2} - j\chi_1\left[a\psi_1(z) + e^{-j2\theta_0}\left(1 - \frac{z}{z_0}\right)^{-j2\alpha''} \cdot \phi_3(z)\right]$$
$$= j\nu e^{j\theta_0}\left(1 - \frac{z}{z_0}\right)^{\frac{2n}{n-1} + j\alpha''}, \tag{2.431}$$

$$\phi_3(0) = \psi_1(0) = 0, \tag{2.432}$$

$$\phi_3(\infty) = \psi_1(\infty) = 0, \tag{2.433}$$

where $\phi_3(z)$ and $\psi_1(z)$ are the third and first harmonics of the functions $\phi(z)$ and $\psi(z)$ defined by formulas (2.362) and (2.363), respectively.

Equations (2.430) and (2.431) are very similar to Eqs. (2.378) and (2.379). The only difference is the presence of exponential factors $e^{j2\theta_0}$ and $e^{j3\theta_o}$ in Eq. (2.430) and factors $e^{-j2\theta_0}$ and $e^{j\theta_0}$ in Eq. (2.431). However, this difference can be accommodated by looking for the particular solution to Eqs. (2.430) and (2.431) in the form

$$\phi_3^{(p)}(z) = e^{j3\theta_0} C_3 \left(1 - \frac{z}{z_0}\right)^{\lambda_3}, \tag{2.434}$$

$$\psi_1^{(p)}(z) = e^{j\theta_0} C_1 \left(1 - \frac{z}{z_0}\right)^{\lambda_1}, \tag{2.435}$$

where λ_3 and λ_1 are given by formulas (2.384).

By substituting functions (2.434) and (2.435) into Eqs. (2.430) and (2.431), we end up with the simultaneous equations for coefficients C_3 and C_1 which are identical to Eqs. (2.385) and (2.386). This means that these coefficients can be computed by using formulas (2.387) and (2.388).

The complete solution to differential Eqs. (2.430) and (2.431) can be sought in the form

$$\phi_3(z) = e^{j3\theta_0} \left[A' \left(1 - \frac{z}{z_0}\right)^{\beta'} + A'' \left(1 - \frac{z}{z_0}\right)^{\beta''} + C_3 \left(1 - \frac{z}{z_0}\right)^{\lambda_3}\right], \tag{2.436}$$

$$\begin{aligned} \psi_1(z) =& e^{j\theta_0} \left[B' \left(1 - \frac{z}{z_0}\right)^{\beta' - j2\alpha''} + B'' \left(1 - \frac{z}{z_0}\right)^{\beta'' - j2\alpha''}\right. \\ & \left. + C_1 \left(1 - \frac{z}{z_0}\right)^{\lambda_1}\right]. \end{aligned} \tag{2.437}$$

By using the same line of reasoning as before, it can be shown that β' and β'' are the roots of the characteristic Eq. (2.238) with positive real parts, while coefficients A', A'', B', and B'' can be found by solving simultaneous Eqs. (2.391)–(2.394). Thus, we can see that the algorithm of calculation of coefficients C_3, C_1, A', A'', B', and B'' is exactly the same as in the previous section, and that the peculiarity of Eqs. (2.430) and (2.431) is fully accounted for by factors $e^{j3\theta_0}$ and $e^{j\theta_0}$ in formulas (2.436) and (2.437), respectively.

Next, we proceed to the solution of the boundary value problem (2.424)–(2.428). By introducing the functions

$$\tilde{\phi}(z,t) = \tilde{h}_x(z,t) + j\tilde{h}_y(z,t), \tag{2.438}$$

$$\tilde{\psi}(z,t) = \tilde{h}_x(z,t) - j\tilde{h}_y(z,t), \tag{2.439}$$

and by literally repeating the same line of reasoning as in Sections 2.2 and 2.3, it can be established that only the first and third harmonics of $\tilde{\psi}(z,t)$ and $\tilde{\phi}(z,t)$ respectively, are not equal to zero. To find these harmonics, the boundary value problem (2.424)–(2.428) can be reduced to the following boundary value problem for the ordinary differential equations:

$$\left(1 - \frac{z}{z_0}\right)^2 \frac{d^2\tilde{\phi}_3(z)}{dz^2} - j\chi_3 \left[a\tilde{\phi}_3(z) + e^{j2\theta_0}\left(1 - \frac{z}{z_0}\right)^{j2\alpha''} \cdot \tilde{\psi}_1(z)\right] = 0, \tag{2.440}$$

$$\left(1-\frac{z}{z_0}\right)^2 \frac{d^2\tilde{\psi}_1(z)}{dz^2} - j\chi_1 \left[a\tilde{\psi}_1(z) + e^{-j2\theta_0}\left(1-\frac{z}{z_0}\right)^{-j2\alpha''} \cdot \tilde{\phi}_3(z)\right] = 0, \tag{2.441}$$

$$\tilde{\phi}_3(0) = 0, \quad \tilde{\psi}_1(0) = H_m e^{-j\theta_0}, \tag{2.442}$$

$$\tilde{\phi}_3(\infty) = \tilde{\psi}_1(\infty) = 0. \tag{2.443}$$

Again, we find that the boundary value problem (2.440)–(2.443) is very similar to the boundary value problem (2.225)–(2.228). The only difference is the presence of exponential factors $e^{j2\theta_0}$ in Eq. (2.440), $e^{-j2\theta_0}$ in Eq. (2.441), and $e^{-j\theta_0}$ in the second boundary condition (2.442). This difference can be accommodated by looking for the solution to the boundary value problem (2.440)–(2.443) in the form

$$\tilde{\phi}_3(z) = e^{j\theta_0}\left[\tilde{A}'\left(1-\frac{z}{z_0}\right)^{\beta'} + \tilde{A}''\left(1-\frac{z}{z_0}\right)^{\beta''}\right], \tag{2.444}$$

$$\tilde{\psi}_1(z) = e^{-j\theta_0}\left[\tilde{B}'\left(1-\frac{z}{z_0}\right)^{\beta'-j2\alpha''} + \tilde{B}''\left(1-\frac{z}{z_0}\right)^{\beta''-j2\alpha''}\right]. \tag{2.445}$$

By using the same line of reasoning as before, we find that β' and β'' are the roots of the characteristic Eq. (2.238) with positive real parts, while coefficients $\tilde{A},'$, $\tilde{A}''$, $\tilde{B}'$ and $\tilde{B}''$ can be computed by using formulas (2.242)–(2.244). Thus, the algorithm of calculations of $\beta', \beta'', \tilde{A}', \tilde{A}'', \tilde{B}'$, and $\tilde{B}''$ is exactly the same as in Section 2.3, and the peculiarity of Eqs. (2.440)–(2.441) and the boundary conditions (2.442) is fully accounted for by factors $e^{j\theta_0}$ and $e^{-j\theta_0}$ in formulas (2.444) and (2.445), respectively.

Having determined all coefficients and exponents in formulas (2.436)–(2.437) and (2.444)–(2.445), the phasors of the first harmonics of the magnetic field components can be found as follows:

$$\begin{aligned}\hat{H}_{x_1}(z) = {} & (H_m + \epsilon C_1)\, e^{j\theta_0}\left(1-\frac{z}{z_0}\right)^{\alpha} \\ & + \epsilon e^{j\theta_0}\left[B'\left(1-\frac{z}{z_0}\right)^{\beta'-j2\alpha''} + B''\left(1-\frac{z}{z_0}\right)^{\beta''-j2\alpha''}\right] \\ & + \tilde{\epsilon} e^{-j\theta_0}\left[\tilde{B}'\left(1-\frac{z}{z_0}\right)^{\beta'-j2\alpha''} + \tilde{B}''\left(1-\frac{z}{z_0}\right)^{\beta''-j2\alpha''}\right],\end{aligned} \tag{2.446}$$

$$
\begin{aligned}
\hat{H}_{y_1}(z) = & -j(H_m - \epsilon C_1)e^{j\theta_0}\left(1 - \frac{z}{z_0}\right)^{\alpha} \\
& + j\epsilon e^{j\theta_0}\left[B'\left(1 - \frac{z}{z_0}\right)^{\beta' - j2\alpha''} + B''\left(1 - \frac{z}{z_0}\right)^{\beta'' - j2\alpha''}\right] \\
& + j\tilde{\epsilon} e^{-j\theta_0}\left[\tilde{B}'\left(1 - \frac{z}{z_0}\right)^{\beta' - j2\alpha''} + \tilde{B}''\left(1 - \frac{z}{z_0}\right)^{\beta'' - j2\alpha''}\right].
\end{aligned}
\tag{2.447}
$$

By using the last two formulas and by invoking the relations (2.63) and (2.65), we obtain the following equations for surface impedances:

$$
\begin{aligned}
\frac{\eta_{xy}}{\eta} = & \frac{1}{\alpha(e^{j\theta_0} - \tilde{\epsilon} e^{-j\theta_0})H_m}\Big\{\alpha\left(H_m - \epsilon C_1\right)e^{j\theta_0} \\
& - \epsilon e^{j\theta_0}\left[B'\left(\beta' - j2\alpha''\right) + B''\left(\beta'' - j2\alpha''\right)\right] \\
& - \tilde{\epsilon} e^{-j\theta_0}\left[\tilde{B}'\left(\beta' - j2\alpha''\right) + \tilde{B}''\left(\beta'' - j2\alpha''\right)\right]\Big\},
\end{aligned}
\tag{2.448}
$$

$$
\begin{aligned}
\frac{\eta_{yx}}{\eta} = & \frac{1}{\alpha(e^{j\theta_0} + \tilde{\epsilon} e^{-j\theta_0})H_m}\Big\{\alpha\left(H_m + \epsilon C_1\right)e^{j\theta_0} \\
& + \epsilon e^{j\theta_0}\left[B'\left(\beta' - j2\alpha''\right) + B''\left(\beta'' - j2\alpha''\right)\right] \\
& + \tilde{\epsilon} e^{-j\theta_0}\left[\tilde{B}'\left(\beta' - j2\alpha''\right) + \tilde{B}''\left(\beta'' - j2\alpha''\right)\right]\Big\},
\end{aligned}
\tag{2.449}
$$

where η is the surface impedance in the case of isotropic media.

As before, it is easy to see that ratios η_{xy}/η and η_{yx}/η do not depend on H_m. This is because coefficients $C_1, B', B'', \tilde{B}'$, and $\tilde{B}''$ are directly proportional to H_m. It is also clear from the above discussion that these coefficients do not depend on the "orientation" angle θ_0 and are only functions of exponent n used in the power law approximation. In this sense, formulas (2.448) and (2.449) give explicit dependence of surface impedance ratios in terms of the perturbation parameters ϵ and $\tilde{\epsilon}$ and the "orientation" angle θ_0. Some sample results of calculation of the ratios η_{xy}/η and η_{yx}/η are presented in Figs. 2.21 through 2.28.

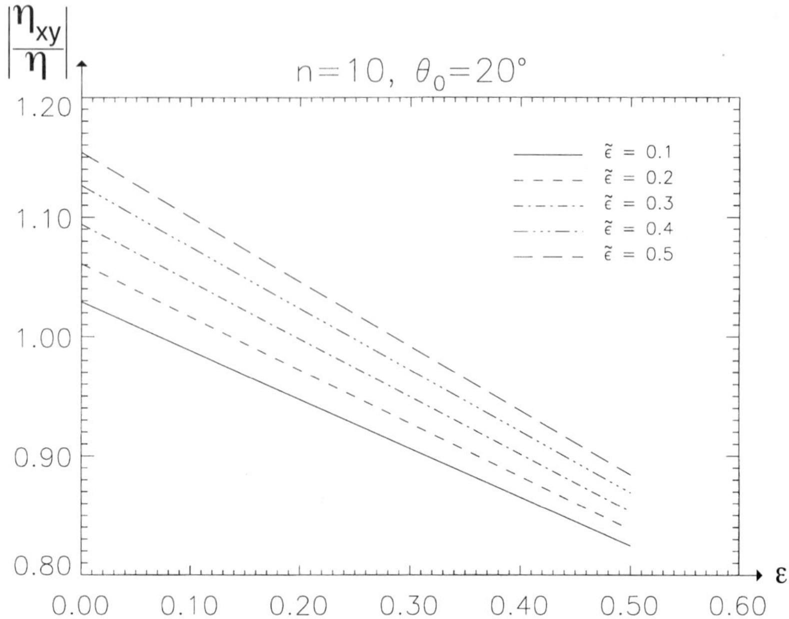

Fig. 2.21

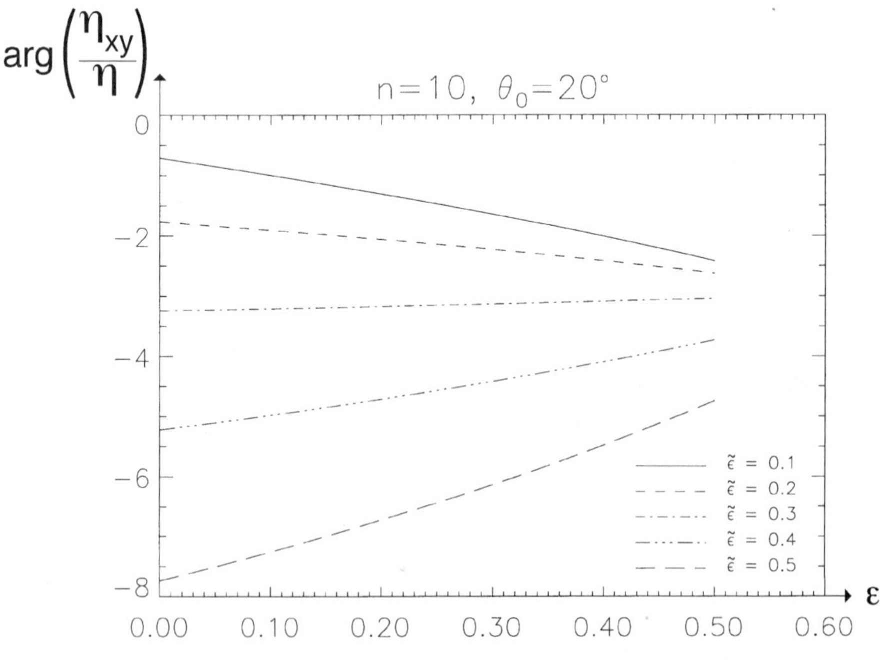

Fig. 2.22

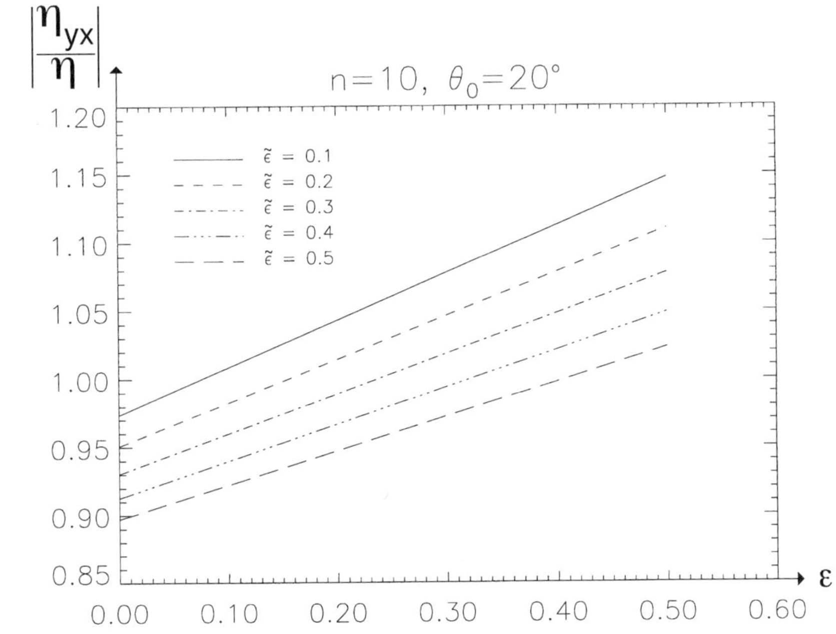

Fig. 2.23

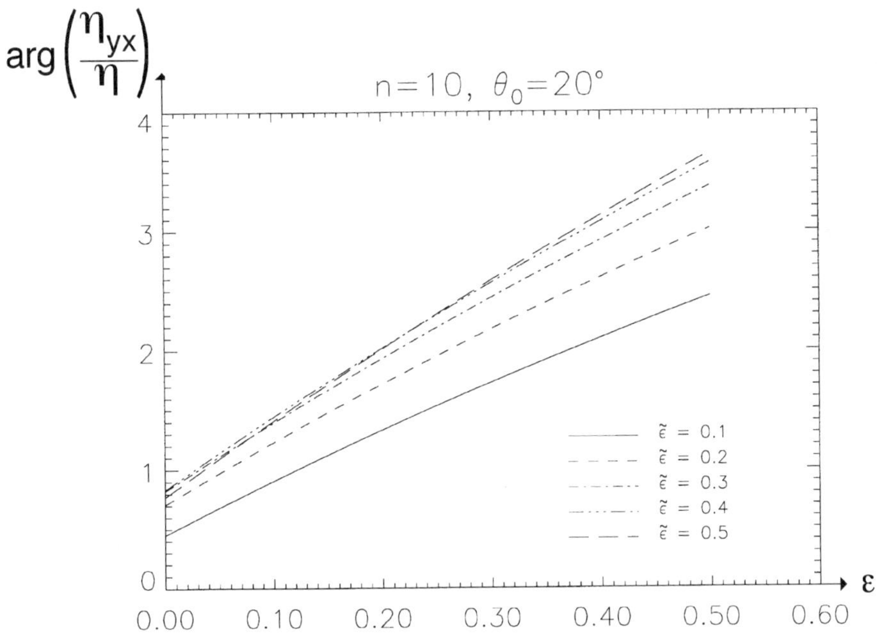

Fig. 2.24

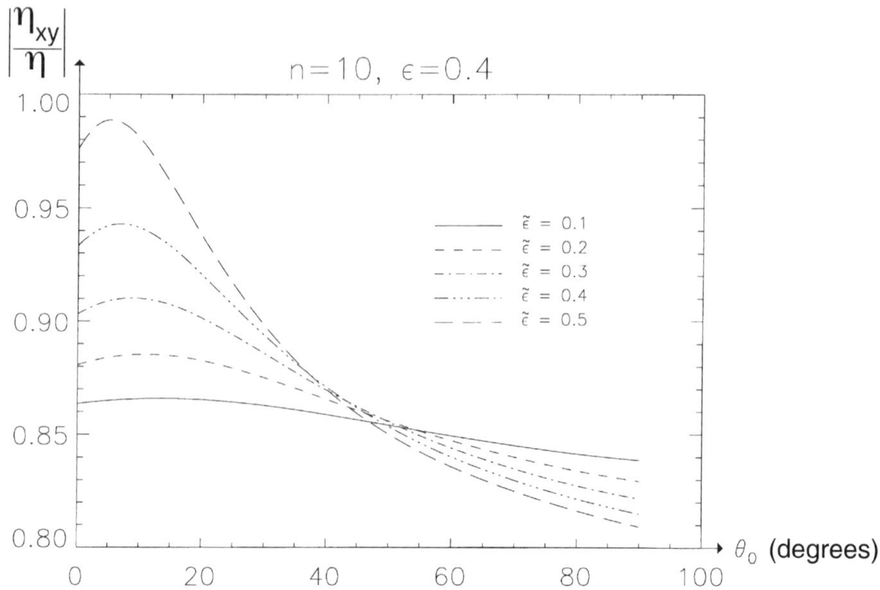

Fig. 2.25

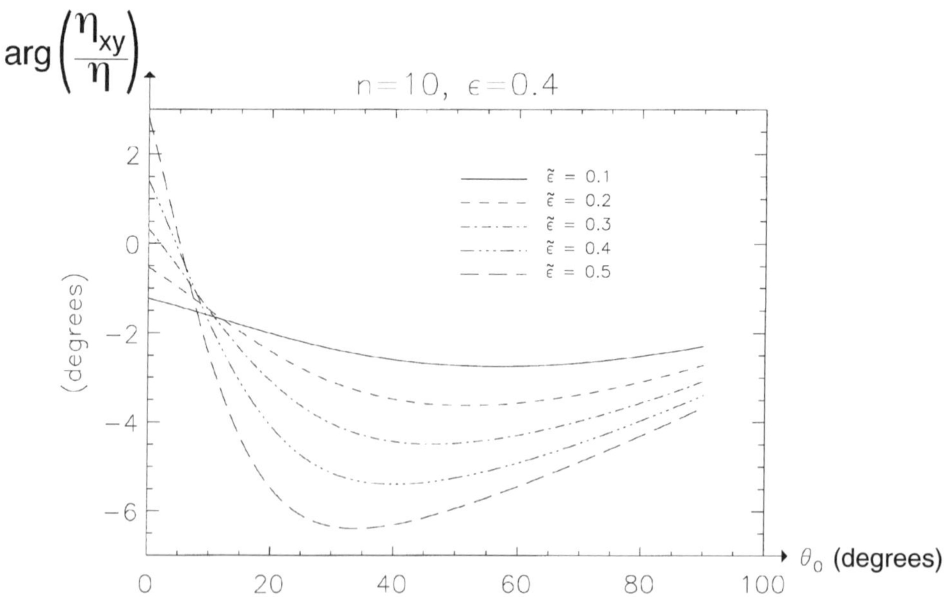

Fig. 2.26

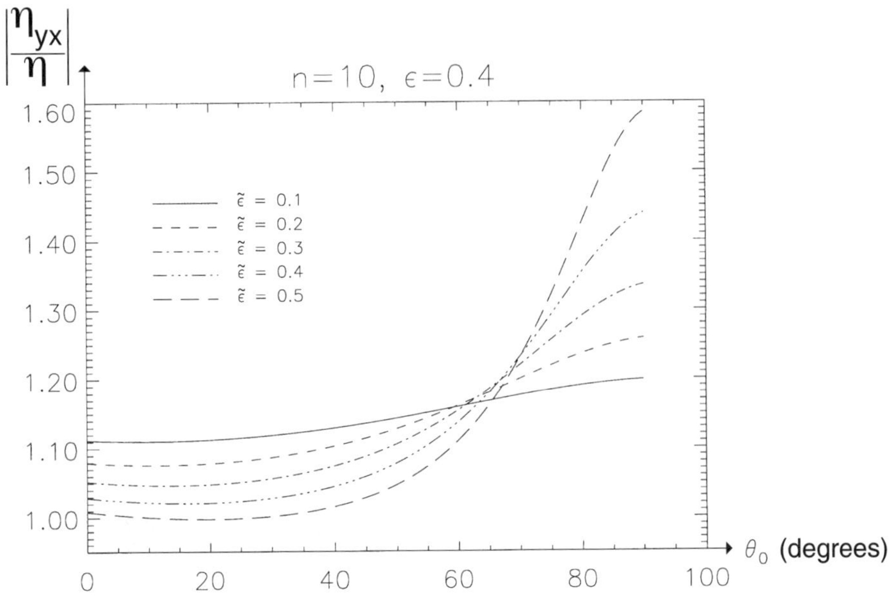

Fig. 2.27

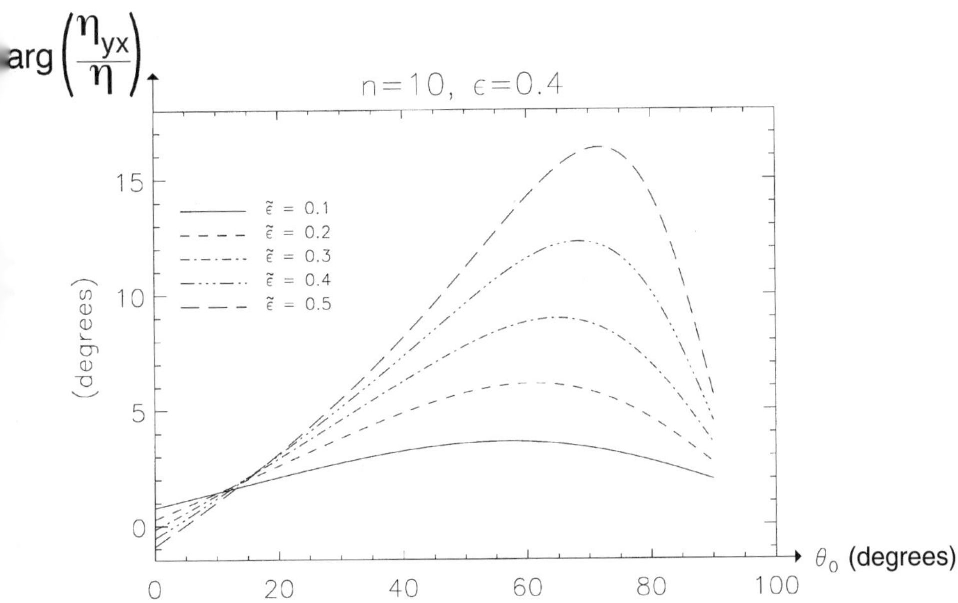

Fig. 2.28

2.6 EDDY CURRENT LOSSES IN THIN LAMINATIONS

It is well known that rotating magnetic fields occur in various types of electric machinery, actuators and other devices. It was realized that eddy current losses in steel laminations caused by these fields are appreciably higher than losses associated with unidirectional alternating magnetic fields of comparable magnitude. For this reason, "rotational" eddy current losses have been a focus of active research for many years. These losses were first investigated experimentally for some specific field values and flux patterns (see [12], [13], [8]). Then some efforts were made to study these losses theoretically [3] by using numerical techniques for the solution of nonlinear diffusion equations. In this section, we shall revisit the issue of "rotational" eddy current losses. By using the results obtained in the previous sections, we shall derive analytical expressions for these losses and clarify certain aspects and questions related to this matter.

We begin with the discussion of eddy current losses under the assumption that a distribution of magnetic flux density over a lamination cross-section is uniform. We start with the case of unidirectional alternating magnetic fields. This is a classical problem that has been extensively treated in the literature. Then the discussion of this classical problem will be generalized to the case of rotating magnetic fields.

Consider a magnetic conducting lamination with height h, width w, and thickness Δ (see Fig. 2.29). It is assumed that the magnetic flux density is uniform over the lamination cross-section and has only the x-component

$$\mathbf{B}(t) = \vec{e}_x B_m \cos \omega t. \tag{2.450}$$

This time-varying magnetic flux density induces eddy currents whose closed lines lie in planes normal to the x-axis. Let L_x be one of these eddy current lines. By applying Faraday's law of electromagnetic induction to path L_x, we find

$$\oint_{L_x} \vec{E} \cdot d\vec{l} = -\frac{d\Phi_x(z,t)}{dt}, \tag{2.451}$$

where $\Phi_x(z,t)$ is the flux that links L_x.

By taking into account that the lamination is thin ($\Delta \ll h$), the left-hand and right-hand sides of formula (2.451) can be approximated as follows:

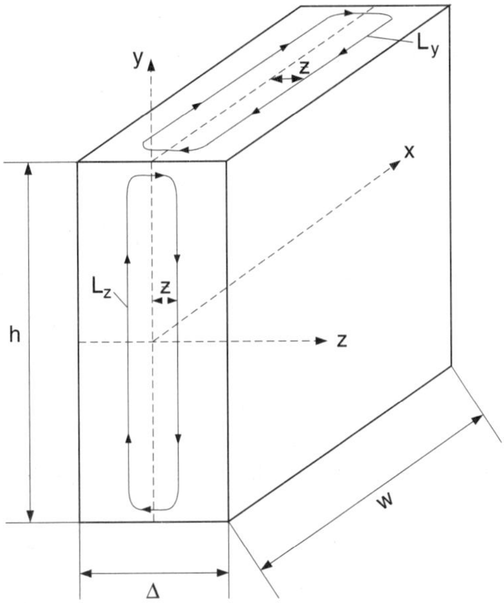

Fig. 2.29

$$\oint_L \vec{E} \cdot d\vec{l} \approx E_y(z,t) \cdot 2h, \tag{2.452}$$

$$\Phi_x(z,t) \approx 2hzB_m \cos\omega t. \tag{2.453}$$

By substituting (2.452) and (2.453) into (2.451), we arrive at

$$E_y(z,t) = \omega z B_m \sin\omega t, \tag{2.454}$$

which suggests that

$$E_{my}(z) = \omega z B_m. \tag{2.455}$$

By using expression (2.454), we can compute local power loss density $p(z,t)$:

$$p(z,t) = \sigma E_y^2(z,t) = \sigma\omega^2 z^2 B_m^2 \sin^2\omega t. \tag{2.456}$$

This power loss density varies with time. For this reason, it is customary to characterize eddy current losses by the average power loss density $\overline{p}(z)$, which can be computed as follows:

$$\overline{p}(z) = \frac{1}{2}\sigma E_{my}^2. \tag{2.457}$$

By taking into account formula (2.455), the last equation can be written in the form

$$\overline{p}(z) = \frac{\omega^2\sigma}{2} B_m^2 z^2. \tag{2.458}$$

The eddy current power losses per unit surface area of lamination can be obtained by integration of $\overline{p}(z)$ with respect to z from $-\frac{\Delta}{2}$ to $\frac{\Delta}{2}$:

$$\overline{p}^{\text{lin}} = 2\int_0^{\frac{\Delta}{2}} \overline{p}(z)dz = \omega^2\sigma B_m^2 \frac{\Delta^3}{24}, \tag{2.459}$$

where superscript "lin" indicates that the losses are computed for the case of linear polarization of the magnetic field.

By using the obvious relation

$$\Phi_m = B_m\Delta, \tag{2.460}$$

formula (2.459) can be reduced to the form

$$\overline{p}^{\text{lin}} = \Phi_m^2 \frac{\omega^2\sigma\Delta}{24}. \tag{2.461}$$

Now suppose that the magnetic flux density is uniform over the lamination cross-section and circularly polarized:

$$\mathbf{B} = \mathbf{e}_x B_m \cos\omega t + \mathbf{e}_y B_m \sin\omega t. \tag{2.462}$$

This magnetic flux density induces the electric field, which has x- and y-components. The y-component of this field can be computed in the same way as before. In other words, formula (2.454) is valid for this component. To compute the x-component of the electric field, we consider a path L_y in a plane normal to y-axis (see Fig. 2.25) and apply Faraday's law to this path:

$$\oint_{L_y} \vec{E}\cdot d\vec{l} = -\frac{d\Phi_y(z,t)}{dt}, \tag{2.463}$$

where $\Phi_y(z,t)$ stands for the magnetic flux that links L_y.

By taking into account that the lamination is thin ($\Delta \ll w$) we find

$$\oint_{L_y} \vec{E}\cdot d\vec{l} \approx E_x(z,t)\cdot 2w, \tag{2.464}$$

$$\Phi_y(z,t) \approx 2wzB_m \sin\omega t. \tag{2.465}$$

By substituting two last equations into formula (2.463), we obtain

$$E_x(z,t) = -\omega z B_m \cos\omega t. \tag{2.466}$$

By comparing formulas (2.454) and (2.466) we conclude that the induced electric field is circularly polarized. This is expected because the magnetic flux density is circularly polarized. By using formulas (2.454) and (2.466), we can compute the instantaneous power loss density

$$p(z,t) = \sigma\left[E_x^2(z,t) + E_y^2(z,t)\right] = \sigma\omega^2 z^2 B_m^2. \tag{2.467}$$

It is apparent from the last equation that this power loss density is constant in time. In other words, in the case of circular polarization of the magnetic flux density the "eddy current" energy dissipation occurs at a constant rate in time. This clearly explains why the rotational eddy current losses are higher than those for unidirectional magnetic fields.

From formula (2.467) we conclude that

$$\overline{p}(z) = p(z,t) = \sigma\omega^2 z^2 B_m^2. \tag{2.468}$$

By integrating the last expression with respect to z, we obtain losses per unit surface area of lamination:

$$\overline{p}^{\text{cir}} = 2\int_0^{\frac{\Delta}{2}} \overline{p}(z)dz = \omega^2\sigma B_m^2 \frac{\Delta^3}{12}, \tag{2.469}$$

where superscript "cir" indicates that the losses are computed for the case of circular polarization of magnetic field.

By using formula (2.460), the last equation can be transformed as follows:

$$\overline{p}^{\text{cir}} = \Phi_m^2 \frac{\omega^2\sigma\Delta}{12}. \tag{2.470}$$

By comparing expressions (2.461) and (2.470), we conclude that

$$\overline{p}^{\text{cir}} = 2\overline{p}^{\text{lin}}. \tag{2.471}$$

The above discussion can be easily generalized to the case of elliptical polarization of magnetic flux density

$$\mathbf{B}(t) = \mathbf{e}_x B_{mx}\cos\omega t + \mathbf{e}_y B_m \sin\omega t. \tag{2.472}$$

For this case, formulas (2.466) and (2.454) can be respectively written as follows:

$$E_x(z,t) = -\omega z B_{my}\cos\omega t, \tag{2.473}$$

$$E_y(z,t) = \omega z B_{mx}\sin\omega t. \tag{2.474}$$

This leads to the following expressions for the instantaneous power loss density

$$p(z,t) = \sigma(E_x^2 + E_y^2) = \sigma\omega^2 z^2 (B_{mx}^2 \sin^2 \omega t + B_{my}^2 \cos^2 \omega t), \tag{2.475}$$

and average power loss density:

$$\overline{p}(z) = \frac{1}{2}\sigma \left(E_{mx}^2 + E_{my}^2\right) = \frac{\sigma\omega^2 z^2}{2} \left(B_{mx}^2 + B_{my}^2\right). \tag{2.476}$$

Now, by literally repeating the same line of reasoning as before, we obtain the following expression for the power losses per unit surface area of lamination:

$$\overline{p}^{el} = \left(\Phi_{mx}^2 + \Phi_{my}^2\right) \frac{\omega^2 \sigma \Delta}{24}. \tag{2.477}$$

By comparing formulas (2.461) and (2.477) we conclude that the last equation can be written in the form

$$\boxed{\overline{p}^{el} = \overline{p}_x^{\text{lin}} + \overline{p}_y^{\text{lin}}.} \tag{2.478}$$

The last expression clearly suggests that eddy current losses in the case of elliptical polarizations of the magnetic flux density are equal to the sum of eddy current losses associated with two unidirectional and orthogonal components of magnetic flux density acting separately. This fact was first observed experimentally (see [12], [13]) and later was confirmed by numerical computations [3]. It is important to point out that we have analytically derived this fact without invoking any assumptions concerning magnetic properties of laminations. For this reason, this fact as well as formulas (2.470) and (2.477) hold for magnetically isotropic and anisotropic laminations with (and without) hysteresis. The main limitation of our derivation is the assumption that the magnetic flux density is uniform over a lamination cross-section. If this assumption does not hold, the above fact and formulas (2.470) and (2.477) are not valid. In other words, in the case of nonuniform distributions of magnetic flux density, eddy current power losses are affected by the magnetic properties of laminations.

To treat the case of nonuniform magnetic flux density, we shall use the results obtained in the previous sections of this chapter. These results have been derived for nonlinear diffusion of electromagnetic fields in magnetically nonlinear conducting half-space, and the existence of finite penetration depth z_0 has been established. Therefore, it is obvious that if the thickness of lamination exceeds $2z_0$, then nonlinear diffusion of electromagnetic fields at each side of the lamination will occur in the same way as in the case of the semi-infinite conducting half-space. Consequently, we can use the previously derived results for the case of conducting lamination. Namely, when the magnetic field is circularly polarized and conducting media are magnetically isotropic, we can write the following formulas for the phasors of the magnetic field:

$$\hat{H}_x(z) = \begin{cases} H_m \left(1 - \frac{\frac{\Delta}{2}+z}{z_0}\right)^{\alpha}, & \text{if } -\frac{\Delta}{2} \le z \le -\frac{\Delta}{2} + z_0, \\ 0, & \text{if } -\frac{\Delta}{2} + z_0 \le z \le \frac{\Delta}{2} - z_0, \\ H_m \left(1 - \frac{\frac{\Delta}{2}-z}{z_0}\right)^{\alpha}, & \text{if } \frac{\Delta}{2} - z_0 \le z \le \frac{\Delta}{2}, \end{cases} \tag{2.479}$$

$$\hat{H}_y(z) = \begin{cases} -jH_m \left(1 - \frac{\frac{\Delta}{2}+z}{z_0}\right)^{\alpha}, & \text{if } -\frac{\Delta}{2} \le z \le -\frac{\Delta}{2} + z_0, \\ 0, & \text{if } -\frac{\Delta}{2} + z_0 \le z \le \frac{\Delta}{2} - z_0, \\ -jH_m \left(1 - \frac{\frac{\Delta}{2}-z}{z_0}\right)^{\alpha}, & \text{if } \frac{\Delta}{2} - z_0 \le z \le \frac{\Delta}{2}, \end{cases} \tag{2.480}$$

where z_0 and α are given by expressions (2.53) and (2,40), (2.47), (2.51), respectively.

By using these formulas, we can compute surface impedances on each side of the lamination. It is obvious that these impedances will be the same and given by formulas (2.66)–(2.68). Now, by invoking the notion of the Poynting vector $\vec{S}$, we can compute eddy current losses in the lamination as follows:

$$\begin{aligned} \overline{p}^{\text{cir}} =& 2\text{Re}\left(\mathbf{S}\cdot\mathbf{a}_z\right) = \text{Re}\left[\left(\hat{\mathbf{E}}\left(\frac{\Delta}{2}\right)\times\hat{\mathbf{H}}^*\left(\frac{\Delta}{2}\right)\right)\cdot\mathbf{a}_z\right] = \\ &= \text{Re}\left[\hat{E}_x\left(\frac{\Delta}{2}\right)\hat{H}_y^*\left(\frac{\Delta}{2}\right) - \hat{E}_y\left(\frac{\Delta}{2}\right)\hat{H}_x^*\left(\frac{\Delta}{2}\right)\right], \end{aligned} \tag{2.481}$$

where the symbol $*$ is used for the notation of complex conjugate quantity.

By employing the relations

$$\hat{E}_x\left(\frac{\Delta}{2}\right) = \eta\hat{H}_y\left(\frac{\Delta}{2}\right), \quad \hat{E}_y\left(\frac{\Delta}{2}\right) = -\eta\hat{H}_x\left(\frac{\Delta}{2}\right), \tag{2.482}$$

and by substituting them into formula (2.481), we derive

$$\overline{p}^{\text{cir}} = \text{Re}(\eta)\left[\left|\hat{H}_x\left(\frac{\Delta}{2}\right)\right|^2 + \left|\hat{H}_y\left(\frac{\Delta}{2}\right)\right|^2\right] = 2H_m^2\text{Re}(\eta), \tag{2.483}$$

where we used the fact that in the case of circular polarizations $\left|\hat{H}_x\left(\frac{\Delta}{2}\right)\right| = \left|\hat{H}_y\left(\frac{\Delta}{2}\right)\right| = H_m$.

Now, by recalling expressions (2.66) and (2.67), we arrive at

$$\overline{p}^{\mathrm{cir}} = H_m^2 \sqrt{\frac{\omega \mu_m}{\sigma}} \frac{4n}{[2n(n+1)(3n+1)^2]^{\frac{1}{4}}}. \tag{2.484}$$

In this formula, "rotational" eddy current losses are expressed in terms of the magnetic field magnitude at the lamination boundary. In many applications, the total flux through the lamination is given. For this reason, it is desirable to express rotational eddy current losses in terms of this flux. To this end, we recall the equation

$$\frac{\partial \hat{E}_x}{\partial z} = -j\omega \hat{B}_y. \tag{2.485}$$

By integrating this equation with respect to z, we obtain

$$\hat{\Phi}_y = 2\int_{\frac{\Delta}{2}-z_0}^{\frac{\Delta}{2}} \hat{B}_y(z)dz = j\frac{2}{\omega}\int_{\frac{\Delta}{2}-z_0}^{\frac{\Delta}{2}} \frac{\partial \hat{E}_x}{\partial z}(z)dz = j\frac{2}{\omega}\hat{E}_x\left(\frac{\Delta}{2}\right), \tag{2.486}$$

where we used the fact that $\hat{E}_x\left(\frac{\Delta}{2} - z_0\right) = 0$, and Φ is a flux per unit width (or height).

By utilizing the impedance relation (2.482) in the last formula, we derive

$$\hat{\Phi}_y = j\frac{2\eta}{\omega}\hat{H}_y\left(\frac{\Delta}{2}\right). \tag{2.487}$$

From the last equation, we find

$$\Phi_m = \frac{2|\eta|}{\omega} H_m$$

or by invoking expression (2.65), we have

$$\Phi_m = \frac{2|\alpha|}{\omega \sigma z_0} H_m. \tag{2.488}$$

Since z_0 depends on μ_m (see (2.53)), which in turn, is a function of H_m, the last formula can be construed as a nonlinear equation for H_m. By solving this equation, we can find H_m and μ_m for the given Φ_m. By plugging this value of μ_m in the expression for z_0 and by using formula (2.488), we find the following expression for H_m in terms of Φ_m:

$$H_m = \Phi_m \frac{\omega \sigma z_0}{2|\alpha|}. \tag{2.489}$$

From formulas (2.40), (2.47), and (2.51), we easily derive

$$|\alpha|^2 = \frac{2n(3n+1)}{(n-1)^2}. \tag{2.490}$$

By substituting formula (2.489) into expression (2.484) and by using Eqs. (2.53) and (2.490), after simple transformations we arrive at the following result:

$$\overline{p}^{\text{cir}} = \Phi_m^2 \omega \sqrt{\frac{\omega\sigma}{\mu_m}} \cdot \frac{[2n(n+1)(3n+1)^2]^{\frac{1}{4}}}{2(3n+1)}. \tag{2.491}$$

By comparing the last result with formula (2.484), we can observe that the above rotational eddy current losses have different frequency dependencies. These losses increase as $\sim \omega^{\frac{1}{2}}$ in the case of the fixed magnetic field magnitude at the lamination boundary, and they grow as $\sim \omega^{\frac{3}{2}}$ for the given magnetic flux through the lamination. It is also clear that these losses depend differently on σ and μ_m. According to formulas (2.484) and (2.491), the rotational eddy current losses are nonlinear functions of H_m and Φ_m, respectively.

So far, we have discussed the case when $2z_0 < \Delta$. Consider the limiting case when

$$z_0 = \frac{\Delta}{2}, \tag{2.492}$$

and let us express $\overline{p}^{\text{cir}}$ explicitly in terms of z_0. According to formulas (2.483), (2.65), and (2.40), we have

$$\overline{p}^{\text{cir}} = 2H_m^2 \frac{\alpha'}{\sigma z_0}. \tag{2.493}$$

Now, by substituting relation (2.489) into the last equation, we obtain

$$\overline{p}^{\text{cir}} = \Phi_m^2 \frac{\omega^2 \sigma z_0 \alpha'}{2|\alpha|^2}. \tag{2.494}$$

By taking into account condition (2.492) in the last formula, we derive

$$\overline{p}^{\text{cir}} = \Phi_m^2 \frac{\omega^2 \sigma \Delta \alpha'}{4|\alpha|^2}. \tag{2.495}$$

From expression (2.47) and (2.490), we find

$$\frac{\alpha'}{4|\alpha|^2} = \frac{n-1}{4(3n+1)}, \tag{2.496}$$

and formula (2.495) can now be rewritten as follows:

$$\overline{p}^{\text{cir}} = \Phi_m^2 \omega^2 \sigma \Delta \frac{n-1}{4(3n+1)}. \tag{2.497}$$

For sufficiently large n, we have

$$\frac{n-1}{4(3n+1)} \approx \frac{1}{12}, \tag{2.498}$$

and formula (2.497) assumes the form

$$\overline{p}^{\text{cir}} = \Phi_m^2 \frac{\omega^2 \sigma \Delta}{12}. \tag{2.499}$$

The last equation coincides with formula (2.470) derived by using a different line of reasoning under the assumption that the magnetic flux density within the lamination is uniform. Although encouraging, this coincidence is not very surprising. This is because under condition (2.492) the distribution of magnetic flux density is almost uniform (see Fig. 2.30). Thus, we can conclude that for $z_0 \leq \frac{\Delta}{2}$ we can use formula (2.491) for the calculation of rotational eddy current losses, while for $z_0 > \frac{\Delta}{2}$ formula (2.470) is appropriate, and there is a more or less smooth transition from the values of eddy current losses predicted by formula (2.491) to the values predicted by formula (2.470). It is also important to point out that in the cases $z_0 \leq \frac{\Delta}{2}$ and $z_0 > \frac{\Delta}{2}$ there are different frequency dependencies for the rotational eddy current losses. In the first case, these losses grow as $\sim \omega^{\frac{3}{2}}$, whereas in the second case these losses increase as $\sim \omega^2$. Another important remark is that the above two cases are determined not only by the frequency of the magnetic field but by medium saturation as well. This is because the penetration depth z_0 is field dependent. Thus, the level of saturation may affect the frequency dependence of the rotational eddy current losses.

We shall next consider the case of eddy current losses for elliptical polarizations of the magnetic field at the boundary. For this case, formula (2.481) can be modified as follows:

$$\overline{p}^{\text{el}} = \text{Re}\left[\hat{E}_{x,1}\left(\frac{\Delta}{2}\right)\hat{H}_y^*\left(\frac{\Delta}{2}\right) - \hat{E}_{y,1}\left(\frac{\Delta}{2}\right)\hat{H}_x^*\left(\frac{\Delta}{2}\right)\right], \tag{2.500}$$

where $\hat{E}_{x,1}\left(\frac{\Delta}{2}\right)$ and $\hat{E}_{y,1}\left(\frac{\Delta}{2}\right)$ are the phasors of the first harmonics of x- and y-components of electric field at the boundary.

These phasors are related to the phasors of the x- and y-components of the magnetic field by the equations

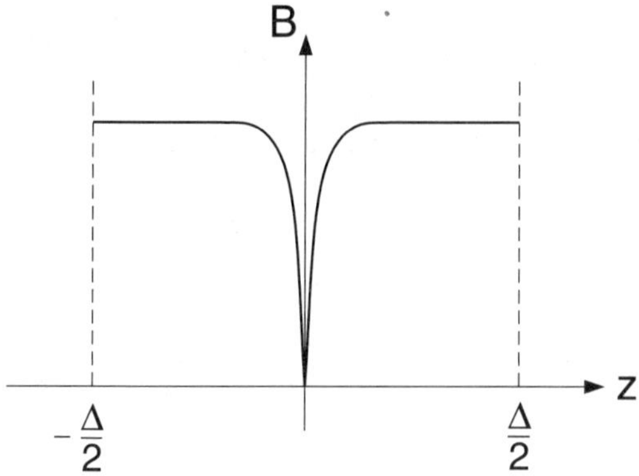

Fig. 2.30

$$\hat{E}_{x,1}\left(\frac{\Delta}{2}\right) = \eta_{xy}\hat{H}_y\left(\frac{\Delta}{2}\right), \quad \hat{E}_{y,1}\left(\frac{\Delta}{2}\right) = -\eta_{yx}\hat{H}_x\left(\frac{\Delta}{2}\right). \tag{2.501}$$

By using the last two relations in formula (2.500), we arrive at

$$\overline{p}^{\rm el} = |\hat{H}_x\left(\frac{\Delta}{2}\right)|^2 \mathrm{Re}\left(\eta_{xy}\right) + |\hat{H}_y\left(\frac{\Delta}{2}\right)|^2 \mathrm{Re}\left(\eta_{yx}\right). \tag{2.502}$$

In the case of elliptical polarizations specified by formulas (2.213) and (2.214), we have

$$|\hat{H}_x\left(\frac{\Delta}{2}\right)|^2 = (1+\epsilon)^2 H_m^2, \quad |\hat{H}_y\left(\frac{\Delta}{2}\right)|^2 = (1-\epsilon)^2 H_m^2. \tag{2.503}$$

By substituting the last two expressions into formula (2.502), we find

$$\overline{p}^{\rm el} = H_m^2\left[(1+\epsilon)^2 \mathrm{Re}\left(\eta_{xy}\right) + (1-\epsilon)^2 \mathrm{Re}\left(\eta_{yx}\right)\right]. \tag{2.504}$$

Finally, by employing formula (2.483), we derive

$$\frac{\overline{p}^{\rm el}}{\overline{p}^{\rm cir}} = \frac{(1+\epsilon)^2 \mathrm{Re}(\eta_{xy}) + (1-\epsilon)^2 \mathrm{Re}(\eta_{yx})}{2\mathrm{Re}(\eta)}. \tag{2.505}$$

In Section 2.3, we have discussed in detail the technique for calculations of impedances η_{yx} and η_{xy}. By using this technique and formula (2.505), the dependence of ration $\overline{p}^{\rm el}/\overline{p}^{\rm cir}$ on ϵ for different values of n has been calculated. The sample results of these calculations are shown in Fig. 2.31.

By using the results of Sections 2.4 and 2.5, similar calculations can be performed for anisotropic (oriented steel) laminations in the case of circular and elliptical polarizations of the magnetic field at the boundary.

However, in applications, the total magnetic flux through the lamination is usually given. This requires some modifications of the techniques presented in Sections 2.3, 2.4, and 2.5. Next, we shall proceed to outline these modifications.

Suppose that the total magnetic flux through the lamination is elliptically polarized:

$$\Phi_x(t) = \Phi_{mx} \sin \omega t, \quad \Phi_y(t) = -\Phi_{my} \cos \omega t. \tag{2.506}$$

We shall find the boundary condition corresponding to this situation. We start with the nonlinear diffusion equation

$$\frac{\partial^2 H_x}{\partial z^2} = \sigma \frac{\partial B_x}{\partial t}, \tag{2.507}$$

and integrate this equation with respect to z, which leads to

$$\int_0^\infty \frac{\partial^2 H_x}{\partial z^2} dz = \sigma \frac{d\Phi_x}{dt}. \tag{2.508}$$

By performing integration in the left-hand side of the last formula and by taking into account that $\frac{\partial H_x}{\partial z}(\infty, t) = 0$, we find

$$\frac{\partial H_x}{\partial z}(0,t) = -\sigma \frac{d\Phi_x}{dt}. \tag{2.509}$$

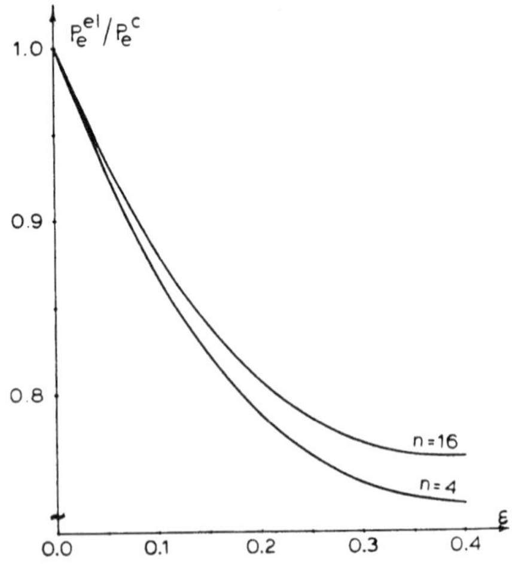

Fig. 1.31

Similarly, we derive

$$\frac{\partial H_y}{\partial z}(0,t) = -\sigma \frac{d\Phi_y}{dt}. \tag{2.510}$$

By substituting expression (2.506) into the relations (2.509) and (2.510), we obtain the boundary conditions

$$\frac{\partial H_x}{\partial z}(0,t) = -\sigma\omega\Phi_{mx}\cos\omega t, \tag{2.511}$$

$$\frac{\partial H_y}{\partial z}(0,t) = -\sigma\omega\Phi_{my}\sin\omega t. \tag{2.512}$$

The last boundary conditions can be written in the form that is appropriate for the development of the perturbation technique:

$$\frac{\partial H_x}{\partial z}(0,t) = -\sigma\omega\Phi_m(\cos\omega t + \epsilon\cos\omega t), \tag{2.513}$$

$$\frac{\partial H_y}{\partial z}(0,t) = -\sigma\omega\Phi_m(\sin\omega t - \epsilon\sin\omega t), \tag{2.514}$$

where

$$\Phi_m = \frac{\Phi_{mx} + \Phi_{my}}{2}, \qquad \epsilon = \frac{\Phi_{mx} - \Phi_{my}}{\Phi_{mx} + \Phi_{my}}. \tag{2.515}$$

The boundary conditions (2.513)–(2.514) suggest that the elliptical polarization of the total flux can be mathematically treated as a perturbation of the circular polarization of the total flux. Thus, the previously developed machinery of the perturbation technique can be fully employed. The solution for the circular polarization of the flux is the same as (2.479) and (2.480) with the only difference that H_m in these formulas should be replaced by $-j\omega\Phi_m/2\eta$. This fact can be inferred from relation (2.487). All the differential equations for perturbations will be identical to those derived in Sections 2.2 through 2.5. However, instead of the previous Dirichlet boundary conditions, now we have to deal with the Neumann boundary conditions (2.513)–(2.514). This is a relatively minor difference and, for this reason, all relevant algebraic transformations are omitted.

Finally, for the sake of completeness (and comparison), we consider eddy current losses for the case of linear polarizations of the magnetic field or the magnetic flux. We limit our discussion to the situation when the penetration depth is less than $\frac{\Delta}{2}$. To make our analysis simple, we assume that we deal with abrupt magnetic transitions. First, consider the case when the lamination is subject to the sinusoidal magnetic field at the boundary:

$$H_0(t) = H_m \sin\omega t. \tag{2.516}$$

In this case, by using the Poynting vector, we can easily derive that the eddy current losses are given by the expression

$$\overline{p}^{\text{lin}} = H_m^2 \text{Re}(\eta), \tag{2.517}$$

where the surface impedance η is defined by formula (see (1.57)):

$$\eta = 1.34\sqrt{\frac{\omega\mu_m}{\sigma}}\, e^{j\frac{\pi}{6.77}}. \tag{2.518}$$

By using the last equation in expression (2.517), we obtain

$$\overline{p}^{\text{lin}} = 1.2H_m^2\sqrt{\frac{\omega\mu_m}{\sigma}}. \tag{2.519}$$

Now suppose that the lamination is subject to the sinusoidal flux:

$$\Phi_x(t) = \Phi_m \cos\omega t. \tag{2.520}$$

By using the same line of reasoning as in the derivation of formula (2.486), we find

$$E_y(0,t) = -\frac{d\Phi_x(t)}{dt}. \tag{2.521}$$

By substituting (2.520) into (2.521), we obtain

$$E_y(0,t) = \omega\Phi_m \sin\omega t. \tag{2.522}$$

By using the Poynting theorem and the surface impedance, we can express eddy current losses in terms of electric field at the boundary:

$$\overline{p}^{\text{lin}} = \frac{E_m^2}{\text{Re}(\eta)}. \tag{2.523}$$

In the last formula, the surface impedance η should be determined for the case of sinusoidal electric field at the boundary. This impedance is given by formula (see (1.72)):

$$\eta = 1.47\sqrt{\frac{\omega\mu_m}{\sigma}}\, e^{\frac{j\pi}{7.83}}. \tag{2.524}$$

By substituting (2.524) into (2.523) and taking into account (2.522), we derive

$$\overline{p}^{\text{lin}} = 0.74\Phi_m^2\omega\sqrt{\frac{\omega\sigma}{\mu_m}}. \tag{2.525}$$

Thus, as before we can observe that eddy current losses increase as $\sim \omega^{\frac{1}{2}}$ in the case of the fixed magnetic field at the boundary, and these losses grow as $\sim \omega^{\frac{3}{2}}$ in the case of the given flux through the lamination.

Consider the case when the penetration depth is exactly equal to $\frac{\Delta}{2}$. By using formula (1.73), this can be mathematically expressed as follows:

$$\sqrt{\frac{1.57}{\omega\sigma\mu_m}} = \frac{\Delta}{2}. \tag{2.526}$$

To use the last equation in formula (2.525), we transform this formula in the following way:

$$\overline{p}^{\text{lin}} = 0.74\Phi_m^2 \frac{\omega^2\sigma}{\sqrt{1.57}} \cdot \sqrt{\frac{1.57}{\omega\sigma\mu_m}}. \tag{2.527}$$

Now, by using (2.526) in (2.527), we arrive at

$$\overline{p}^{\text{lin}} = \Phi_m^2 \frac{\omega^2\sigma\Delta}{3.4}. \tag{2.528}$$

By comparing the last equation with expression (2.461), we observe that, although they have similar mathematical forms, their meanings are quite different. First, relation (2.528) does not give the frequency dependence of eddy current losses, whereas formula (2.461) does. This is because relation (2.528) is valid only for one frequency, namely, the frequency at which formula (2.526) holds. Second, even for this frequency, relation (2.528) predicts substantially higher eddy current losses than those predicted by formula (2.461). This discrepancy can be explained as follows. Formula (2.461) has been derived under the assumption of uniform magnetic flux density within the lamination at all instants of time, whereas the last equation has been derived without invoking this assumption. Furthermore, it is quite clear that the distribution of magnetic flux density is quite nonuniform at the instants of time when zero front $z_0(t)$ is appreciably smaller than $\frac{\Delta}{2}$. Nonuniform magnetic flux density distributions usually result in larger eddy currents and hence in higher losses.

It is interesting to point out that, in the case of abrupt magnetic transitions and in the situation when the penetration depth is larger than one half of the lamination thickness, the eddy currents are not continuous in time but are rather intermittent. Indeed, after two rectangular fronts of magnetic flux density meet at the middle of the lamination, the magnetic flux through the lamination becomes constant in time and it remains this way until new rectangular fronts of opposite polarity are formed. This results in the intermittency of eddy currents. This intermittency is not natural but rather it is due to idealizations introduced by abrupt magnetic transitions. In the case of gradual magnetic transitions and rectangular profile approximation (see Section 1.5), the merger of rectangular profiles of the magnetic flux density at the middle of the lamination does not result in the constancy in time of the magnetic flux through the lamination. Instead, this merger results in the uniform distribution of the magnetic flux density over the lamination cross-section, but this uniform magnetic flux density still varies with time. In other words, the merger of rectangular profiles results in the conditions under which formula (2.461) has been derived. Thus, it can be concluded that the sooner the rectangular profiles of the magnetic flux density meet at the middle of the lamination, the more accurate the formula (2.461) for eddy current losses. In general, eddy current losses may have two distinct components. The first component represents eddy current losses occurred prior to the merger of the rectangular profiles of the magnetic flux density. This component increases with frequency as $\sim \omega^{3/2}$. The second component represents the eddy current losses that occurred after the merger of the rectangular profiles of the magnetic flux density. According to formula (2.461), this component of the eddy current losses grows with frequency as $\sim \omega^2$. These two components together may produce the frequency dependence of eddy current losses of the type $\sim \omega^{\frac{3}{2}} + \omega^2$ that has been observed in experiments. Another explanation for this experimentally observed frequency dependence of eddy current losses has been given by G. Bertotti ([1], [2]) and it is based on a different conceptual foundation.

REFERENCES

[1] G. Bertotti, IEEE Transactions on Magnetics, **24**, No. 1, pp. 621–630, (1988).

[2] G. Bertotti, A. Boglietti, M. Chiampi, D. Chiarabaglio, F. Fiorillo, M. Lazzari, IEEE Transactions on Magnetics, **27**, No. 6, (1991).

[3] R.M. DelVecchio, IEEE Transactions on Magnetics, **18**, No. 6, pp. 1707–1709, (1982).

[4] I.D. Mayergoyz, Archiv für Electrotechnik, **64**, No. 314, pp. 153–162, (1981).

[5] I.D. Mayergoyz, F.M. Abdel-Kader, and F.P. Emad, Journal of Applied Physics, **55**, No. 3, pp. 618–629, (1984).

[6] I.D. Mayergoyz and F.M. Abdel-Kader, IEEE Transactions on Magnetics, **20**, No. 5, pp. 2007–2009, (1984).

[7] I.D. Mayergoyz, "Iterative Methods for the Calculation of Static Fields in Inhomogeneous Anisotropic and Nonlinear Media," Naukova Dumka, Kiev, (1979).

[8] A.J. Moses and B. Thomas, IEEE Transactions on Magnetics, **9**, pp. 651–654, (1973).

[9] L.R. Neumann, "Skin Effect in Ferromagnetics," Gosenergoisdat, Moscow, (1949).
[10] H. Poincaré, "Science and Hypothesis," Dover Publications, New York, (1952).
[11] G. Strang, "Introduction to Applied Mathematics," Wellesley-Cambridge Press, (1986).
[12] R.D. Strattan and J.F. Young, Journal of Applied Physics, **33**, No. 3, pp. 1285–1286, (1962).
[13] F.J. Young and H.L. Schenk, Journal of Applied Physics, **37**, No. 3, pp. 1210–1211, (1966).

CHAPTER 3

Nonlinear Diffusion of Weak Magnetic Fields

3.1 NONLINEAR DIFFUSION OF LINEARLY POLARIZED ELECTROMAGNETIC FIELDS

In the previous chapters, we discussed nonlinear diffusion of electromagnetic fields in magnetically nonlinear conducting media in the case when differential magnetic permeability $\mu_d(H)$ of media is decreased as the magnetic field is increased. This type of variation of the differential magnetic permeability typically occurs for sufficiently strong magnetic fields and it reflects magnetic saturation of media. In this chapter, we shall discuss nonlinear diffusion of magnetic fields in another case when the differential magnetic permeability of media is increased with the increase in the magnetic field. This case is realized for relatively weak magnetic fields. Typical B vs. H and μ_d vs. H relations for this case are shown in Figs. 3.1 a and 3.1 b, respectively. In the sequel, we shall use the following power law approximation for this type of relations

$$B = (kH)^{\frac{1}{n}}, \qquad (n < 1). \tag{3.1}$$

This approximation is formally similar to ones used in the previous chapters. However, the important difference is that approximation (3.1) is valid for $n < 1$, whereas in the previously used power law approximations we had $n > 1$. The formal similarity of these two power law approximations will allow us to use almost identical mathematical machinery for the analysis of nonlinear diffusion of electromagnetic fields. However, the physical

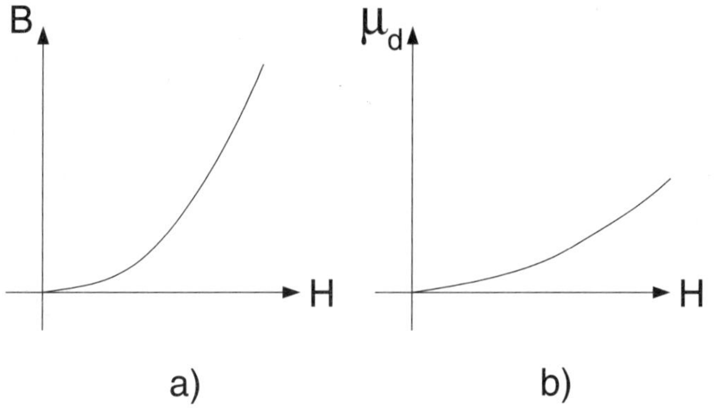

Fig. 3.1

features of nonlinear diffusion itself will be quite different. This is because for $n > 1$ the nonlinearity of medium results in the increase in μ_d and, consequently, in the increase in the attenuation rate of electromagnetic fields as they are diffused deeper and deeper inside the conducting medium. On the other hand, for $n < 1$ the nonlinearity of medium results in the decrease in μ_d and, consequently, in the decrease in the attenuation rate of electromagnetic fields as they are further diffused inside the medium.

Next, we point out that variations of differential magnetic permeability similar to that shown in Fig. 3.1 b are observed in other situations as well. Examples of these situations include symmetric and nonsymmetric minor hysteresis loops (see Figs. 3.2 a and 3.2 b) formed for sufficiently small magnetic field variations, and initial parts of ascending and descending branches of symmetric hysteresis loops (Fig. 3.2 c) formed for sufficiently large field variations. In all these situations, the power law approximation of the form

$$b = (kh)^{\frac{1}{n}}, \qquad (n < 1) \tag{3.2}$$

is appropriate. In formula (3.2), b and h are a "shifted" magnetic flux density and magnetic field, respectively. They are related to the actual magnetic flux density and magnetic field by the expressions

$$b = B_m \pm B, \qquad h = H_m \pm H \tag{3.3}$$

in the case of ascending and descending branches of hysteresis loops shown in Figs. 3.2 a and 3.2 c, and by the formula

$$b = \Delta B, \qquad h = \Delta H \tag{3.4}$$

in the case of the hysteresis loop shown in Fig. 3.2 b. In all these cases, the differential magnetic permeability is increased as h is increased (Fig. 3.2 d).

It is worthwhile to point out that approximations (3.1) and (3.2) idealize actual magnetic properties of media for very small values of H and h.

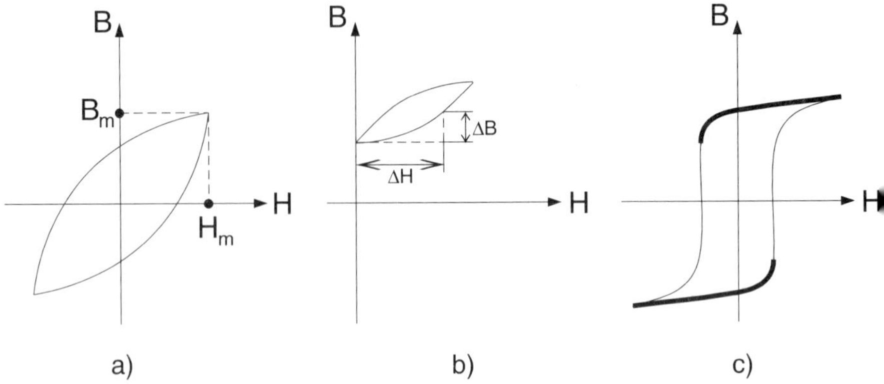

Fig. 3.2

This is because these approximations predict vanishing values of differential magnetic permeability, whereas actual values should be finite. Because this discrepancy occurs for very small values of magnetic fields, it will affect magnetic field distributions for very large depths, which is hardly of paramount importance.

In this section, we shall discuss nonlinear diffusion of electromagnetic fields in the case when these fields are linearly polarized. This leads to the scalar nonlinear diffusion equation

$$\frac{\partial^2 H}{\partial z^2} = \sigma \frac{\partial B(H)}{\partial t}. \tag{3.5}$$

By using constitutive relations (3.1) and (3.2), this equation can be written in the following forms, respectively:

$$\frac{\partial^2 B^n}{\partial z^2} = k\sigma \frac{\partial B}{\partial t}, \tag{3.6}$$

$$\frac{\partial^2 b^n}{\partial z^2} = k\sigma \frac{\partial b}{\partial t}. \tag{3.7}$$

It is clear that Eqs. (3.6) and (3.7) are mathematically identical. Consequently, all the results obtained for constitutive relation (3.2) and nonlinear diffusion Eqs. (3.7) can be easily reformulated for constitutive relation (3.1) and Eq. (3.6). For this reason, we shall only discuss the case of constitutive relation (3.2) and nonlinear diffusion Eq. (3.7). To start the discussion, we consider the "model" problem specified by the following boundary and initial conditions:

$$b_0(t) = b(0,t) = ct^p, \qquad (p > 0), \tag{3.8}$$

$$b(\infty, t) = 0, \tag{3.9}$$

$$b(z, 0) = 0. \tag{3.10}$$

This model problem is similar to one posed in Section 1.3, and it corresponds to nonlinear diffusion of electromagnetic fields in a conducting half-space.

The initial-boundary value problem (3.7)–(3.10) can be reduced to the boundary value problem for a certain ordinary differential equation. As in Chapter 1, this can be achieved through the dimensional analysis of Eq. (3.7) and boundary condition (3.8). This analysis leads to the following dimensional relations:

$$\frac{[b]^n}{[z]^2} = [k][\sigma]\frac{[b]}{[t]}, \tag{3.11}$$

$$[b] = [c][t]^p, \tag{3.12}$$

which suggest that the following variable is dimensionless:

$$\xi = \frac{z}{(k^{-1}\sigma^{-1}c^{n-1})^{\frac{1}{2}}\, t^m}, \tag{3.13}$$

where

$$m = \frac{p(n-1)+1}{2}. \tag{3.14}$$

By using the above dimensionless variable, we shall look for the solution of the initial-boundary value problem (3.7)–(3.10) in the form

$$b(z, t) = ct^p f(\xi), \tag{3.15}$$

where $f(\xi)$ is a dimensionless function of dimensionless variable ξ.

By literally repeating the same line of reasoning as in Section 1.4, we find that $b(z, t)$ given by formula (3.15) will be the solution to the initial-boundary value problem (3.7)–(3.10) if function $f(\xi)$ satisfies the following equation and boundary conditions:

$$\frac{d^2 f^n}{d\xi^2} + m\xi\frac{df}{d\xi} - pf = 0, \tag{3.16}$$

$$f(0) = 1, \tag{3.17}$$

$$f(\infty) = 0. \tag{3.18}$$

Function $f(\xi)$ can be construed as the normalized profile, $\tilde{b}(z, t)$, of magnetic flux density:

$$\tilde{b}(z, t) = \frac{b(z, t)}{b_0(t)} = f(\xi). \tag{3.19}$$

This profile completely characterizes the nonuniformity of magnetic flux density distribution at any instant of time. The analysis of the normalized profile reveals that there is a very important difference between the self-similar solutions discussed in Section 1.4 and the self-similar solutions (3.15). This difference stems from the fact that in Chapter 1 we had the situation when $n > 1$ and, for this reason, exponent m was positive for all values of $p \geq 0$. In the situation being discussed, we have the inequality $n < 1$ (see formulas (3.10) and (3.2)), which suggests that exponent m in formula (3.13) may change its sign. Indeed, according to formula (3.14), we have three distinct cases:

case 1, when

$$0 < p < \frac{1}{1-n} \quad \text{and} \quad m > 0; \tag{3.20}$$

case 2, when

$$p = \frac{1}{1-n} \quad \text{and} \quad m = 0; \tag{3.21}$$

and case 3, when

$$p > \frac{1}{1-n} \quad \text{and} \quad m < 0. \tag{3.22}$$

In the first case, from formulas (3.13) and (3.19) we conclude that the normalized profiles of the magnetic flux density are dilated (stretched) along the z-axis as time t is increased (see Fig.3.3 a). In other words, the nonuniformity of the magnetic flux density distribution is decreased with time.

In the second case, from formulas (3.13) and (3.19) we find that the normalized profile of the magnetic flux density does not change as time t is increased (see Fig. 3.3 b). This means that the nonuniformity of the magnetic flux density distribution remains the same with time. This property can be interpreted as "standing" diffusion of electromagnetic fields.

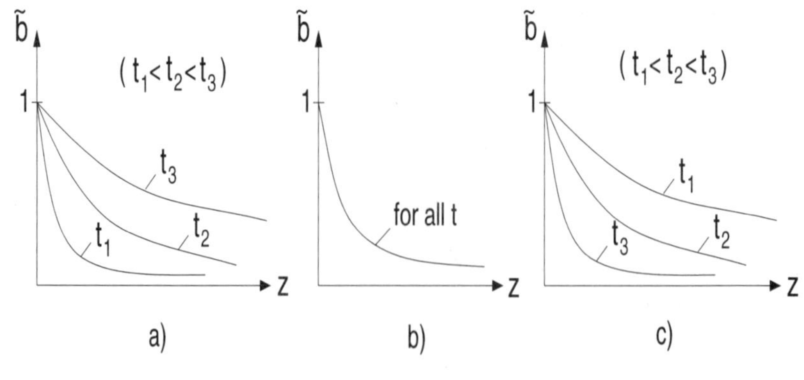

Fig. 3.3

Finally, it is apparent from formulas (3.13) and (3.19) that in the third case the normalized profiles of the magnetic flux density are contracted along the z-axis as time t is increased (see Fig. 3.3 c). In other words, the nonuniformity of the magnetic flux density distribution is increased with time. This property of self-similar solutions is quite unusual, and it can be construed as "backward" diffusion. "Backward" diffusion is peculiar for nonlinearity (3.2) and it does not exist for linear media ($n = 1$) or media with saturation ($n > 1$). Indeed, according to formula (3.22), the set of self-similar solutions that exhibit "backward" diffusion becomes smaller and smaller as n tends to 1, and it disappears in the limit when $n = 1$.

It turns out that the simple analytical solution to the boundary value problem (3.16)–(3.18) can be found in the case of "standing" diffusion. Since in this case $m = 0$ and $p = \frac{1}{1-n}$, Eq. (3.16) can be written as follows:

$$\frac{d^2 f^n}{d\xi^2} + \frac{1}{n-1} f = 0. \tag{3.23}$$

To integrate this equation, we introduce auxiliary functions:

$$f^n(\xi) = \phi(\xi), \tag{3.24}$$

$$\frac{d\phi}{d\xi} = \mathcal{F}(\xi). \tag{3.25}$$

By using these functions, we find

$$\frac{d^2 f^n}{d\xi^2} = \frac{d\mathcal{F}}{d\xi} = \frac{d\mathcal{F}}{d\phi} \cdot \frac{d\phi}{d\xi} = \mathcal{F}\frac{d\mathcal{F}}{d\phi} = \frac{1}{2}\frac{d(\mathcal{F}^2)}{d\phi}. \tag{3.26}$$

By using formulas (3.24) and (3.26), the second-order differential Eq. (3.23) can be reduced to the following first-order equation:

$$\frac{d(\mathcal{F}^2)}{d\phi} = \frac{2}{1-n}\phi^{\frac{1}{n}}. \tag{3.27}$$

By integrating the last equation, we obtain

$$\mathcal{F}^2 = \frac{2n}{1-n^2}\phi^{\frac{n+1}{n}} + A_1, \tag{3.28}$$

where A_1 is an integration constant. By recalling formulas (3.24) and (3.25), the last equation can be transformed as follows:

$$\frac{df^n}{d\xi} = \left(\frac{2n}{1-n^2} f^{n+1} + A_1\right)^{\frac{1}{2}}. \tag{3.29}$$

To find the constant A_1, we demonstrate that

$$\frac{df^n}{d\xi}(\infty) = 0. \tag{3.30}$$

Indeed, we can easily establish the following expression for the electric field:

$$E(z,t) = -\frac{1}{\sigma}\frac{\partial H}{\partial z} = -\frac{1}{\sigma}\frac{\partial h}{\partial z} = -\frac{1}{\sigma k}\frac{\partial b^n}{\partial z}. \tag{3.31}$$

Now, by using formulas (3.15) and (3.13), we find

$$E(z,t) = -\frac{c^n t^{pn}}{\sigma k}\frac{df^n}{dz} = -\frac{c^n t^{pn}}{\sigma k (k^{-1}\sigma^{-1}c^{n-1})^{\frac{1}{2}} t^m} \cdot \frac{df^n}{d\xi}. \tag{3.32}$$

From formula (3.32) and the fact that the electric field vanishes at infinity, we establish the validity of condition (3.30). From this condition and formulas (3.18) and (3.29), we find that

$$A_1 = 0. \tag{3.33}$$

As a result, differential Eq. (3.29) is simplified as follows:

$$\frac{df^n}{d\xi} = \sqrt{\frac{2n}{1-n^2}} f^{\frac{n+1}{2}}. \tag{3.34}$$

By separating the variables in the last equation, we obtain

$$f^{\frac{n-3}{2}} df = \sqrt{\frac{2}{n(1-n^2)}} d\xi. \tag{3.35}$$

By integrating the last equation, we derive

$$f^{\frac{n-1}{2}} = -\left(\sqrt{\frac{1-n}{2n(1+n)}}\xi + A_2\right), \tag{3.36}$$

where A_2 is a constant of integration. The last formula can be transformed as follows:

$$f^{n-1}(\xi) = \left(\sqrt{\frac{1-n}{2n(1+n)}}\xi + A_2\right)^2, \tag{3.37}$$

which, in turn, yields the expression

$$f(\xi) = \frac{1}{\left(\sqrt{\frac{1-n}{2n(1+n)}}\xi + A_2\right)^{\frac{2}{1-n}}}. \tag{3.38}$$

Now, by invoking the boundary condition (3.17), from formula (3.38) we derive

$$A_2 = 0. \tag{3.39}$$

This leads to the following final expression for $f(\xi)$:

$$f(\xi) = \frac{1}{\left(1 + \sqrt{\frac{1-n}{2n(1+n)}}\xi\right)^{\frac{2}{1-n}}}. \tag{3.40}$$

By using this expression and formulas (3.13), (3.15), and (3.21), we derive the following equation for the magnetic flux density:

$$b(z,t) = \frac{ct^{\frac{1}{1-n}}}{\left(1 + z\sqrt{\frac{(1-n)k\sigma c^{1-n}}{2n(1+n)}}\right)^{\frac{2}{1-n}}}, \tag{3.41}$$

which yields the following expression for the magnetic field:

$$h(z,t) = \frac{c^n t^{\frac{n}{1-n}}}{k\left(1 + z\sqrt{\frac{(1-n)k\sigma c^{1-n}}{2n(1+n)}}\right)^{\frac{2n}{1-n}}}. \tag{3.42}$$

A few remarks are in order concerning the "standing" mode solution (3.41). This solution is similar (in some ways) to the standing mode solution (1.285) obtained in Section 1.6. In both cases, the solutions are represented as the product of two functions, which depend only on t and z, respectively. It is this mathematical property of the above solutions that leads to their physical interpretations as "standing" modes. This property also suggests that the solution (3.41) can be obtained by using the method of separation of variables, that is, in the same way as was done for the solution (1.285). There are, however, substantial differences between the above two solutions. The solution (1.285) is localized in time and space, whereas the solution (3.41) is "unlimited" in both time and space. In other words, the latter solution exists for the semi-infinite time interval $0 \leq t \leq \infty$ and at any instant of time $t > 0$ this solution extends to infinity in the direction of the z-axis. The above solution also suggests the asymptotic behavior of the magnetic flux density at infinity. Namely, formula (3.41) shows that $b(z,t)$ tends to zero as $z^{\frac{2}{n-1}}$ when z approaches infinity. Next, we demonstrate that this is a general property that is valid for all self-similar solutions (3.15). To this end, we shall find the asymptotics of function $f(\xi)$ at infinity. We shall use the notation

$$f(\xi) \sim A(1 + a\xi)^{\alpha}, \qquad (\alpha < 0), \tag{3.43}$$

if

$$\lim_{\xi\to\infty} \frac{f(\xi)}{A(1+a\xi)^{\alpha}} = 1. \tag{3.44}$$

It is clear that formula (3.43) is an asymptotic equality.

To find A, a, and α in formula (3.43), we turn to differential Eq. (3.16) and rewrite this equation in the form

$$\frac{d^2 f^n}{d\xi^2} = -\frac{m}{a}(1+a\xi)\frac{df}{d\xi} + \frac{m}{a}\frac{df}{d\xi} + pf. \tag{3.45}$$

By assuming asymptotics (3.43) for function $f(\xi)$, we can transform Eq. (3.45) into the following asymptotic equality:

$$A^n \alpha n(\alpha n - 1)a^2(1+a\xi)^{\alpha n-2} \sim -Am\alpha(1+a\xi)^{\alpha}$$
$$+Am\alpha(1+a\xi)^{\alpha-1} + pA(1+a\xi)^{\alpha}. \tag{3.46}$$

Because $\alpha < 0$, the second term on the right-hand side of asymptotic equality (3.46) decays faster than the two other terms. For this reason, this term can be omitted. With this fact in mind, we conclude that asymptotic equality (3.46) can hold only if

$$\alpha n - 2 = \alpha, \tag{3.47}$$

and

$$A^n a^2 \alpha n(\alpha n - 1) = A(p - m\alpha). \tag{3.48}$$

From formula (3.47), we find

$$\alpha = \frac{2}{n-1}. \tag{3.49}$$

By substituting the last expression into Eq. (3.48) and invoking formula (3.14) for m, we obtain

$$A^{n-1}a^2 = \frac{1-n}{2n(n+1)}. \tag{3.50}$$

Thus, A and a cannot be chosen independently; any choice of one of these two constants determines the value of another. For instance, if we make the choice

$$A = 1, \tag{3.51}$$

then, according to formula (3.50), we have

$$a = \sqrt{\frac{1-n}{2n(n+1)}}. \tag{3.52}$$

Now, by recalling the expression (3.40) for $f(\xi)$, we observe that it has the form of (3.43) with the values of α, A, and a given by formulas (3.49), (3.51), and (3.52), respectively.

Thus, we have established that

$$f(\xi) \sim \frac{A}{(1+a\xi)^{\frac{2}{1-n}}}. \tag{3.53}$$

By using the last asymptotic equality in formula (3.15) and taking into account expression (3.13) for ξ, we easily conclude that the following asymptotics is valid for the magnetic flux density

$$b(z,t) \sim z^{-\frac{2}{1-n}}. \tag{3.54}$$

This is the fact we intended to prove. It is established for boundary conditions(3.8), that is, for any $p > 0$. These boundary conditions describe a very broad class of monotonically increasing functions as p varies from 0 to infinity. This suggests that asymptotics (3.54) is valid for any monotonically increasing boundary condition. This fact can be mathematically proven by using the "maximum principle" for Eq. (3.7).

The fact that in asymptotics (3.53) A and a are not independent but related by the expression (3.50) is not accidental. It can be traced to the property of Eq. (3.16) being invariant with respect to the transformations:

$$\mathcal{F}(\xi) = \lambda^{-\frac{2}{n-1}} f(\lambda\xi). \tag{3.55}$$

In other words, if $f(\xi)$ is a solution to Eq. (3.16) then $\mathcal{F}(\xi)$ is a solution to the same equation. This can be established by literally repeating the reasoning given in Section 1.4. This property can be utilized in the following way. Suppose we can find a solution to Eq. (3.16) that satisfies the boundary condition (3.18) but does not satisfy the boundary condition (3.17):

$$f(0) = q \neq 1. \tag{3.56}$$

Then, by taking

$$\lambda = q^{\frac{n-1}{2}}, \tag{3.57}$$

we find that the function

$$\mathcal{F}(\xi) = \frac{1}{q} f\left(q^{\frac{n-1}{2}}\xi\right) \tag{3.58}$$

will be the solution to Eq.(3.16) that satisfies both boundary conditions (3.17) and (3.18). Thus, if we know any solution to Eq. (3.16) satisfying the boundary condition (3.18), then, by using transformation (3.55), we can map this solution into the solution that satisfies the boundary condition (3.17) as well.

Transformation (3.55) along with asymptotics (3.53) can be utilized for the solution of Eq. (3.16). Namely, we can look for the solution in the form

$$f(\xi) = A(1+\xi)^{\frac{2}{n-1}}\left[1 + a_1(1+\xi)^{-1} + a_2(1+\xi)^{-2} + \ldots\right]. \tag{3.59}$$

By substituting the last expression into Eq. (3.16) and by equating terms of the same order of $(1+\xi)$, we can sequentially determine coefficients $a_1, a_2, \ldots$. Then, by using transformation (3.58), we can map solution (3.59) into the solution that will satisfy the boundary condition (3.17). This is the strategy that was utilized and worked well in Section 1.4. This was the case because exponent n in power approximation (1.92) (or (1.93)) was appreciably larger than 1. This is not true for exponent n in power approximation (3.2). For this reason, we pursue another approach that exploits smallness of m. Indeed, for $m = 0$ we have found the exact analytical solution (3.40) to Eq. (3.16). Now, by assuming that m is sufficiently small, we shall use the perturbation technique to find solutions to Eq. (3.16). To this end, we express p in terms of m:

$$p = \frac{2m-1}{n-1}, \tag{3.60}$$

and transform Eq. (3.16) as follows:

$$\frac{d^2 f^n}{d\xi^2} + m\xi\frac{df}{d\xi} - \frac{2m}{n-1}f + \frac{1}{n-1}f = 0. \tag{3.61}$$

In this way, we absorbed p into m and reduced differential Eq. (3.16) to the form (3.61) that contains only parameter m. Next, we shall look for the solution of Eq. (3.61) in the form

$$f(\xi) = f_0(\xi) + mf_1(\xi) + m^2 f_2(\xi) + \ldots + m^k f_k(\xi) + \ldots. \tag{3.62}$$

To proceed further, we have to find the power series expansion with respect to m for f^n:

$$f^n = \left(f_0 + mf_1 + m^2 f_2 + \ldots m^k f_k + \ldots\right)^n. \tag{3.63}$$

To do this, we use the formula

$$f^n = f^n\Big|_{m=0} + m\frac{df^n}{dm}\Big|_{m=0} + \frac{m^2}{2!}\frac{d^2 f^n}{dm^2}\Big|_{m=0} + \frac{m^3}{3!}\frac{d^3 f^n}{dm^3}\Big|_{m=0} + \ldots. \tag{3.64}$$

It is apparent from (3.62) that

$$f^n\Big|_{m=0} = f_0^n. \tag{3.65}$$

Then, from formula (3.63), we find

$$\frac{df^n}{dm} = n\left(f_0 + mf_1 + m^2 f_2 + \ldots\right)^{n-1} \cdot \left(f_1 + 2mf_2 + 3m^2 f_3 + \ldots\right). \tag{3.66}$$

Consequently,

$$\frac{df^n}{dm}\Big|_{m=0} = nf_0^{n-1} f_1. \tag{3.67}$$

Similarly, from (3.63), we derive

$$\begin{aligned}\frac{d^2 f^n}{dm^2} =& n(n-1)\left(f_0 + mf_1 + m^2 f_2 + \ldots\right)^{n-2} \cdot \left(f_1 + 2mf_2 + 3m^2 f_3 + \ldots\right)^2 \\ &+ n\left(f_0 + mf_1 + m^2 f_2 + \ldots\right)^{n-1} \cdot \left(2f_2 + 6mf_3 + \ldots\right).\end{aligned} \tag{3.68}$$

As a result,

$$\frac{d^2 f^n}{dm^2}\Big|_{m=0} = 2!nf_0^{n-1} f_2 + n(n-1) f_0^{n-2} f_1^2. \tag{3.69}$$

By using the same line of reasoning, we obtain

$$\frac{d^3 f^n}{dm^3}\Big|_{m=0} = 3!nf_0^{n-1} f_3 + 6n(n-1) f_0^{n-2} f_1 f_2 + n(n-1)(n-2) f_0^{n-3} f_1^3. \tag{3.70}$$

By substituting formulas (3.65), (3.67), (3.69), and (3.70) into expansion (3.64), we find

$$\begin{aligned}f^n =& f_0^n + mnf_0^{n-1} f_1 + m^2\left[nf_0^{n-1} f_2 + \frac{n(n-1)}{2!} f_0^{n-2} f_1^2\right] \\ &+ m^3\left[nf_0^{n-1} f_3 + \frac{n(n-1)(n-2)}{3!} f_0^{n-3} f_1^3 + n(n-1) f_0^{n-2} f_1 f_2\right] + \ldots.\end{aligned} \tag{3.71}$$

By plugging expressions (3.62) and (3.71) into differential Eq. (3.61), we arrive at

$$\frac{d^2 f_0^n}{d\xi^2} + mn\frac{d^2 f_0^{n-1} f_1}{d\xi^2} + m^2\left[n\frac{d^2 f_0^{n-1} f_2}{d\xi^2} + \frac{n(n-1)}{2!}\frac{d^2 f_0^{n-2} f_1^2}{d\xi^2}\right]$$

$$+m^3\left[n\frac{d^2f_0^{n-1}f_3}{d\xi^2}+\frac{n(n-1)(n-2)}{3!}\frac{d^2f_0^{n-3}f_1^3}{d\xi^2}+n(n-1)\frac{d^2f_0^{n-2}f_1f_2}{d\xi^2}\right]$$
$$+\ldots+m\xi\left(\frac{df_0}{d\xi}+m\frac{df_1}{d\xi}+m^2\frac{df_2}{d\xi}+m^3\frac{df_3}{d\xi}+\ldots\right)$$
$$-\frac{2m}{n-1}\left(f_0+mf_1+m^2f_2+m^3f_3+\ldots\right)$$
$$+\frac{1}{n-1}\left(f_0+mf_1+m^2f_2+m^3f_3+\ldots\right)=0. \tag{3.72}$$

By collecting the terms of the same order of m and by equating them to zero, we obtain the following differential equations:

$$\frac{d^2f_0^n}{d\xi^2}+\frac{1}{n-1}f_0=0, \tag{3.73}$$

$$n\frac{d^2f_0^{n-1}f_1}{d\xi^2}+\frac{1}{n-1}f_1=\frac{2}{n-1}f_0-\xi\frac{df_0}{d\xi}, \tag{3.74}$$

$$n\frac{d^2f_0^{n-1}f_2}{d\xi^2}+\frac{1}{n-1}f_2=\frac{2}{n-1}f_1-\xi\frac{df_1}{d\xi}-\frac{n(n-1)}{2}\frac{d^2f_0^{n-2}f_1^2}{d\xi^2}, \tag{3.75}$$

$$\begin{aligned}n\frac{d^2f_0^{n-1}f_3}{d\xi^2}+\frac{1}{n-1}f_3&=\frac{2}{n-1}f_2-\xi\frac{df_2}{d\xi}\\ &-\frac{n(n-1)(n-2)}{3!}\frac{d^2f_0^{n-3}f_1^3}{d\xi^2}-n(n-1)\frac{d^2f_0^{n-2}f_1f_2}{d\xi^2}.\end{aligned} \tag{3.76}$$

Differential Eq. (3.73) is identical to differential Eq. (3.23), which we have already solved (see formula (3.40)). Thus, a solution to Eq. (3.73) that goes to zero at infinity can be written as follows:

$$f_0(\xi)=(1+a\xi)^{\frac{2}{n-1}}, \tag{3.77}$$

where a is given by formula (3.52).

Differential Eqs.(3.74), (3.75), and (3.76) have similar mathematical structures. All these equations can be written in the following generic form:

$$\frac{d^2f_0^{n-1}f_k}{d\xi^2}+\frac{1}{n(n-1)}f_k=Q_k(\xi), \quad (k=1,2,\ldots), \tag{3.78}$$

where the right-hand sides $Q_k(\xi)$ are determined from the solution of Eq. (3.78) for all $k'<k$.

By using expression (3.77), we find

$$\frac{d^2f_0^{n-1}f_k}{d\xi^2}=\frac{d}{d\xi^2}\left[(1+a\xi)^2f_k\right]=(1+a\xi)^2\frac{d^2f_k}{d\xi^2}+4a(1+a\xi)\frac{df_k}{d\xi}+2a^2f_k. \tag{3.79}$$

By substituting formula (3.79) into Eqs. (3.78), we arrive at

$$(1+a\xi)^2\frac{d^2 f_k}{d\xi^2}+4a(1+a\xi)\frac{df_k}{d\xi}+\left(2a^2+\frac{1}{n(n-1)}\right)f_k=Q_k(\xi). \quad (3.80)$$

Thus, for all k's we need to solve similar Euler's type differential equations. These are linear equations with variable coefficients. Since Euler's type equations are well studied, in principle, Eq. (3.80) can be solved up to any value of k for a given n. However, if we want to perform these calculations in general (symbolic) form, formulas become very complicated and convoluted as k is increased. For this reason, we shall consider only the calculation of the first order term f_1 and then shall draw some conclusions. For $k=1$, from formulas (3.74) and (3.77) we find

$$Q_1(\xi)=\frac{1}{n}\left[\frac{2}{n-1}f_0-\xi\frac{df_0}{d\xi}\right]=\frac{1}{n}\left[\frac{2}{n-1}(1+a\xi)^{\frac{2}{n-1}}-\frac{2a\xi}{n-1}(1+a\xi)^{\frac{3-n}{n-1}}\right]=$$
$$\frac{2}{n(n-1)}(1+a\xi)^{\frac{3-n}{n-1}}(1+a\xi-a\xi)=\frac{2}{n(n-1)}(1+a\xi)^{\frac{3-n}{n-1}}. \quad (3.81)$$

Consequently, for $k=1$ Eq. (3.80) can be written as follows:

$$(1+a\xi)^2\frac{d^2 f_1}{d\xi^2}+4a(1+a\xi)\frac{df_1}{d\xi}+(2a^2+\frac{1}{n(n-1)})f_1=\frac{2}{n(n-1)}(1+a\xi)^{\frac{3-n}{n-1}}. \quad (3.82)$$

The corresponding homogeneous equation

$$(1+a\xi)^2\frac{d^2 f_1^h}{d\xi^2}+4a(1+a\xi)\frac{df_1^h}{d\xi}+\left(2a^2+\frac{1}{n(n-1)}\right)f_1^h=0 \quad (3.83)$$

has solutions of the form

$$f_1^h(\xi)=A(1+a\xi)^\alpha. \quad (3.84)$$

Here, α is a root of characteristic equation

$$a^2\alpha(\alpha-1)+4a^2\alpha+2a^2+\frac{1}{n(n-1)}=0. \quad (3.85)$$

By taking into account formula (3.52), the last equation can be simplified and written as follows:

$$\alpha^2+3\alpha+2\frac{n^2-3n}{(n-1)^2}=0. \quad (3.86)$$

By solving the last equation, we find

$$\alpha_1 = \frac{3-n}{n-1}, \qquad \alpha_2 = \frac{2n}{1-n}. \tag{3.87}$$

Since α_1 coincides with the exponent of the right-hand side of Eq. (3.82), we look for the solution of this equation in the form

$$f_1(\xi) = A(1+a\xi)^{\alpha_1} \ln(1+a\xi). \tag{3.88}$$

By substituting the last formula into Eq. (3.82) and performing all necessary transformations, we end up with

$$2Aa^2\alpha_1 + 3a^2A = \frac{2}{n(n-1)}. \tag{3.89}$$

By taking into account formulas (3.52) and (3.87), we derive

$$A = \frac{4(n+1)}{(3+n)(1-n)}. \tag{3.90}$$

Thus,

$$f_1(\xi) = \frac{4(n+1)}{(3+n)(1-n)}(1+a\xi)^{\frac{3-n}{n-1}} \ln(1+a\xi). \tag{3.91}$$

Consequently,

$$\begin{aligned} f(\xi) \simeq f_0(\xi) + mf_1(\xi) &= (1+a\xi)^{\frac{2}{n-1}} \\ &+ m\frac{4(n+1)}{(3+n)(1-n)}(1+a\xi)^{\frac{3-n}{n-1}} \ln(1+a\xi). \end{aligned} \tag{3.92}$$

It is easy to see that function $f(\xi)$ satisfies both boundary conditions (3.17) and (3.18). If this were not the case, we would use transformation (3.55) to impose the boundary condition (3.17). It is clear from formula (3.91) that

$$f_1(0) = 0 \qquad \text{and} \qquad f_1(\infty) = 0. \tag{3.93}$$

It is also apparent that function $f(\xi)$ is positive for $0 < \xi < \infty$. This means that this function assumes at least one maximum value. To find this value, we consider the equation

$$f_1'(\xi_0) = 0. \tag{3.94}$$

According to formula (3.88), the last equation is equivalent to

$$\ln(1+a\xi_0) = -\frac{1}{\alpha_1}, \tag{3.95}$$

which, in turn, is tantamount to

$$(1 + a\xi_0) = e^{-\frac{1}{\alpha_1}}. \tag{3.96}$$

Thus, function $f_1(\xi)$ has only one maximum. This function is schematically shown in Fig. 3.4. By substituting formulas (3.95) and (3.96) into expression (3.91), we find

$$\max f_1(\xi) = f_1(\xi_0) = \frac{4(n+1)}{e(3+n)(3-n)}. \tag{3.97}$$

By using the last formula, we derive

$$\max |mf_1(\xi)| \leq |m| \frac{4(n+1)}{e(3+n)(3-n)}. \tag{3.98}$$

Consider the case when

$$n = 0.5 \qquad \text{and} \qquad |m| < 0.5. \tag{3.99}$$

For this case, according to formula (3.60), we have

$$0 \leq p \leq 4. \tag{3.100}$$

For these values of p, the boundary conditions (3.8) describe a sufficiently broad class of monotonically increasing functions of time (see Fig. 3.5). However, for all these monotonically increasing boundary conditions, according to formulas (3.98) and (3.99) we have

$$\max |mf_1(\xi)| < 0.13. \tag{3.101}$$

Thus, as a first approximation, we can neglect the term $mf_1(\xi)$ and use the following expression for $f(\xi)$:

$$f(\xi) \approx (1 + a\xi)^{\frac{2}{n-1}}. \tag{3.102}$$

By substituting the last equation into formula (3.15) and taking into account (3.13), we end up with the following approximate expression for the magnetic flux density:

$$b(z,t) \approx \frac{b_0(t)}{(1 + \frac{z}{z_0(t)})^{\frac{2}{1-n}}}. \tag{3.103}$$

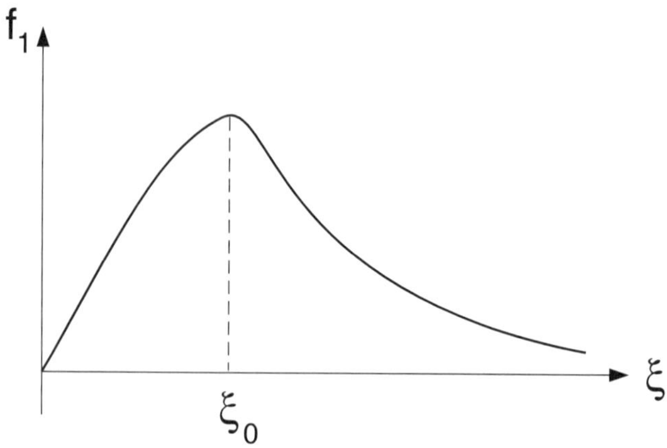

Fig. 3.4

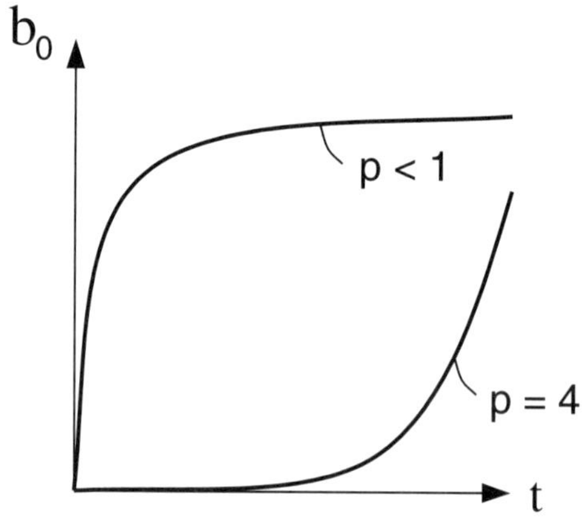

Fig. 3.5

We can make the next step in our approximations and assume that formula (3.103) is fairly accurate for all monotonically increasing boundary conditions. Unknown function $z_0(t)$ in this formula then can be determined from the first moment equation for nonlinear diffusion Eq. (3.7)

$$\int_0^\infty z \frac{\partial^2 h}{\partial z^2} dz = \sigma \int_0^\infty z \frac{\partial b}{\partial t} dz. \tag{3.104}$$

By literally repeating the same line of reasoning as in Section 1.5, we can transform the last formula as follows:

$$\frac{1}{\sigma} \int_0^t h_0(\tau) d\tau = \int_0^\infty z b(z,t) dz, \tag{3.105}$$

where $h_0(t)$ is the boundary value of the magnetic field.

By substituting expression (3.103) into (3.105) and performing integration, we derive

$$z_0(t) = \left[\frac{2n(n+1) \int_0^t h_0(\tau) d\tau}{(1-n)^2 \sigma b_0(t)} \right]^{\frac{1}{2}} . \tag{3.106}$$

Formulas (3.103) and (3.106) provide a complete (albeit quite approximate) description for nonlinear diffusion of electromagnetic fields. This description is not as accurate (and universal) as the one we obtained in Chapter 1 for the case of strong magnetic fields. Moreover, this description cannot be extended to treat time-periodic problems of nonlinear diffusion. It turns out that the time-periodic problems of nonlinear diffusion are more or less manageable in the case of circular and elliptical polarizations of electromagnetic fields. A detailed analysis of these problems is presented in the following sections.

3.2 NONLINEAR DIFFUSION OF CIRCULARLY POLARIZED ELECTROMAGNETIC FIELDS IN ISOTROPIC MEDIA

From the mathematical point of view, this problem is tantamount to the finding of time-periodic solutions of the following coupled nonlinear partial differential equations:

$$\frac{\partial^2 H_x}{\partial z^2} = \sigma \frac{\partial}{\partial t} \left[\mu \left(\sqrt{H_x^2 + H_y^2} \right) H_x \right], \tag{3.107}$$

$$\frac{\partial^2 H_y}{\partial z^2} = \sigma \frac{\partial}{\partial t} \left[\mu \left(\sqrt{H_x^2 + H_y^2} \right) H_y \right], \tag{3.108}$$

subject to the following boundary conditions:

$$H_x(0,t) = H_m \cos(\omega t + \theta_0), \tag{3.109}$$

$$H_y(0,t) = H_m \sin(\omega t + \theta_0), \tag{3.110}$$

$$H_x(\infty, t) = 0, \tag{3.111}$$

$$H_y(\infty, t) = 0. \tag{3.112}$$

As before, we observe that the mathematical structure of nonlinear differential Eqs. (3.107)–(3.108) and boundary conditions (3.109)–(3.112) is invariant with respect to rotations of the x- and y-axes. In other words, the mathematical form of these equations and boundary conditions will remain the same for any choice of x- and y-axes in the plane $z = 0$. This implies that the solution of the boundary value problem (3.107)–(3.112) should also be rotationally invariant. This means that the magnetic field is circularly polarized anywhere inside the conducting media:

$$H_x(z) = H(z)\cos(\omega t + \theta(z)), \tag{3.113}$$

$$H_y(z) = H(z)\sin(\omega t + \theta(z)). \tag{3.114}$$

Next, we shall formally demonstrate that the "circularly polarized" solution (3.113)–(3.114) is consistent with the mathematical structure of the boundary value problem (3.107)–(3.112).

First, we note that according to formulas (3.113) and (3.114) we have

$$|\mathbf{H}(z)| = \sqrt{H_x^2(z,t) + H_y^2(z,t)} = H(z). \tag{3.115}$$

Consequently, the magnitude of the magnetic field and the magnetic permeability, $\mu(|\mathbf{H}|)$, remain constant with time at any point within the conducting media.

Now we represent formulas (3.113) and (3.114) in the phasor form:

$$\hat{H}_x(z) = H(z)e^{j\theta(z)} \tag{3.116}$$

$$\hat{H}_y(z) = -jH(z)e^{j\theta(z)}. \tag{3.117}$$

It is clear from the last three expressions that

$$|\mathbf{H}(z,t)| = H(z) = \left|\hat{H}_x(z)\right| = \left|\hat{H}_y(z)\right|. \tag{3.118}$$

Consequently,

$$\mu(|\mathbf{H}(z,t)|) = \mu(\left|\hat{H}_x(z)\right|) = \mu(\left|\hat{H}_y(z)\right|). \tag{3.119}$$

By using formula (3.119) as well as anticipated forms (3.113) and (3.114) for the solution, we can transform the boundary value problem (3.107)–(3.112) into the following two boundary value problems:

$$\frac{d^2\hat{H}_x}{dz^2} = j\omega\sigma\mu\left(\left|\hat{H}_x\right|\right)\hat{H}_x. \tag{3.120}$$

$$\hat{H}_x(0) = \hat{H}_m, \tag{3.121}$$

$$\hat{H}_x(\infty) = 0, \tag{3.122}$$

and

$$\frac{d^2\hat{H}_y}{dz^2} = j\omega\sigma\mu\left(\left|\hat{H}_y\right|\right)\hat{H}_y, \tag{3.123}$$

$$\hat{H}_y(0) = -j\hat{H}_m, \tag{3.124}$$

$$\hat{H}_y(\infty) = 0, \tag{3.125}$$

where

$$\hat{H}_m = H_m e^{j\theta_0}. \tag{3.126}$$

The very fact that the boundary value problem (3.107)–(3.112) is **exactly** transformed into the boundary value problems (3.120)–(3.122) and (3.123)–(3.125) for magnetic field phasors indicates that the "circularly polarized" solution is consistent with the mathematical structure of the boundary value problem (3.107)–(3.112). This fact also proves that there are no higher-order time-harmonics of the magnetic field anywhere within the conducting media despite its nonlinear magnetic properties.

The achieved simplification is quite remarkable from the purely mathematical point of view. First, partial differential equations (3.107)–(3.108) are exactly reduced to ordinary differential Eqs. (3.120) and (3.123). Second, the boundary value problems (3.120)–(3.122), and (3.123)–(3.125) are entirely decoupled. Third, the decoupled boundary value problems have identical mathematical structures. For this reason, the same solution technique can be used to handle these problems. It is instructive to stress that all these simplifications stem from the rotational symmetry of the original boundary value problem (3.107)–(3.112).

The above simplifications are valid for any isotropic nonlinear media. To proceed further, we shall use the following constitutive equation for isotropic media:

$$B = kH^n, \qquad (n > 1). \tag{3.127}$$

This constitutive equation describes B vs. H relations exemplified by Fig. 3.1 a. In other words, this constitutive equation is similar to the one given by formula (3.1), however, the meanings of n and k are different. We have made this change because it leads to simpler expressions.

From formula (3.127), we find

$$\mu(H) = \frac{B}{H} = kH^{n-1}, \tag{3.128}$$

which leads to the following expression for the magnetic permeability at the boundary of media:

$$\mu_m = kH_m^{n-1} = k\left|\hat{H}_m\right|^{n-1}. \tag{3.129}$$

From the last two formulas, we obtain

$$\mu(H) = \mu_m \left(\frac{H}{H_m} \right)^{n-1}. \tag{3.130}$$

Consequently,

$$\mu(|\hat{H}_x|) = \mu_m \left| \frac{\hat{H}_x}{\hat{H}_m} \right|^{n-1}, \quad \mu(|\hat{H}_y|) = \mu_m \left| \frac{\hat{H}_y}{\hat{H}_m} \right|^{n-1}. \tag{3.131}$$

By using relations (3.131), Eqs. (3.120) and (3.123) can be written as follows:

$$\frac{d^2 \hat{H}_x}{dz^2} = j\omega\sigma\mu_m \left| \frac{\hat{H}_x}{\hat{H}_m} \right|^{n-1} \hat{H}_x, \tag{3.132}$$

$$\frac{d^2 \hat{H}_y}{dz^2} = j\omega\sigma\mu_m \left| \frac{\hat{H}_y}{\hat{H}_m} \right|^{n-1} \hat{H}_y. \tag{3.133}$$

We shall look for the solution of Eq. (3.132) in the form

$$\boxed{\hat{H}_x(z) = \hat{H}_m \left(1 + \frac{z}{z_0} \right)^{-\alpha},} \tag{3.134}$$

where

$$\boxed{\alpha = \alpha' + j\alpha''.} \tag{3.135}$$

It is clear that (under the condition $\alpha' > 0$) the function (3.134) satisfies the boundary conditions (3.121) and (3.122). From formula (3.134), we find

$$\left| \frac{\hat{H}_x(z)}{H_m} \right| = \left(1 + \frac{z}{z_0} \right)^{-\alpha'}, \tag{3.136}$$

which yields

$$\left| \frac{\hat{H}_x(z)}{H_m} \right|^{n-1} = \left(1 + \frac{z}{z_0} \right)^{-\alpha'(n-1)}. \tag{3.137}$$

By substituting expressions (3.134) and (3.137) into Eq. (3.132), we obtain

$$\alpha(\alpha+1)\hat{H}_m \left(1 + \frac{z}{z_0} \right)^{-\alpha-2} = j\omega\sigma\mu_m \hat{H}_m z_0^2 \left(1 + \frac{z}{z_0} \right)^{-\alpha-\alpha'(n-1)}. \tag{3.138}$$

It is clear that the last equality can be valid only if

$$\alpha(\alpha+1) = j\omega\sigma\mu_m z_o^2, \tag{3.139}$$

and

$$2 = \alpha'(n-1). \tag{3.140}$$

The last equation yields

$$\boxed{\alpha' = \frac{2}{n-1}.} \tag{3.141}$$

Next, the characteristic Eq. (3.139) will be used in order to find α'' and z_0. To this end, we shall write complex characteristic Eq. (3.139) as the following two real equations:

$$\alpha'(\alpha'+1) = (\alpha'')^2, \tag{3.142}$$

$$(2\alpha'+1)\alpha'' = \omega\sigma\mu_m z_0^2. \tag{3.143}$$

By substituting formula (3.141) into Eq. (3.142), we find

$$\boxed{\alpha'' = \frac{\sqrt{2(n+1)}}{n-1}.} \tag{3.144}$$

Next, by substituting formulas (3.141) and (3.144) into Eq. (3.143), we obtain

$$\boxed{z_0 = \frac{[2(n+1)(n+3)^2]^{\frac{1}{4}}}{(n-1)\sqrt{\omega\sigma\mu_m}}.} \tag{3.145}$$

Thus, formulas (3.134), (3.135), (3.141), (3.144), and (3.145) completely define the solution of the boundary value problem (3.120)–(3.122). The boundary value problem (3.123)–(3.125) is identical (up to notations) to the boundary value problem (3.120)–(3.122). As a result, the solution to the former boundary value problem can be written as follows:

$$\boxed{\hat{H}_y(z) = -j\hat{H}_m\left(1+\frac{z}{z_0}\right)^{-\alpha}.} \tag{3.146}$$

By converting formulas (3.134) and (3.146) from the phasor form into the time-domain form and taking into account relations (3.141) and (3.144), we obtain the following solution of the boundary value problem (3.107)–(3.112):

$$H_x(z,t) = H_m\left(1+\frac{z}{z_0}\right)^{-\frac{2}{n-1}}\cos\left(\omega t+\theta_0-\frac{\sqrt{2(n+1)}}{n-1}\ln\left(1+\frac{z}{z_0}\right)\right), \tag{3.147}$$

$$H_y(z,t) = H_m\left(1+\frac{z}{z_0}\right)^{-\frac{2}{n-1}}\sin\left(\omega t+\theta_0-\frac{\sqrt{2(n+1)}}{n-1}\ln\left(1+\frac{z}{z_0}\right)\right), \tag{3.148}$$

where z_0 is given by Eq. (3.145).

It is instructive to show that in the limit of $n \to 1$ formulas (3.147) and (3.148) give classical expressions for linear media. First, it is apparent from formula (3.145) for z_0 that

$$\left(1+\frac{z}{z_0}\right)^{-\frac{2}{n-1}} = \left\{[1+(n-1)\gamma]^{\frac{1}{(n-1)\gamma}}\right\}^{-2\gamma}, \tag{3.149}$$

where

$$\gamma = \frac{z\sqrt{\omega\sigma\mu_m}}{[2(n+1)(n+3)^2]^{\frac{1}{4}}}. \tag{3.150}$$

Consequently,

$$\lim_{n\to 1}\left(1+\frac{z}{z_0}\right)^{-\frac{2}{n-1}} = e^{-\beta z}, \tag{3.151}$$

where

$$\beta = \sqrt{\frac{\omega\sigma\mu_m}{2}}. \tag{3.152}$$

Similarly, it can be established that

$$\lim_{n\to 1}\frac{\sqrt{2(n+1)}}{n-1}\ln\left(1+\frac{z}{z_0}\right) = \beta z. \tag{3.153}$$

By using relations (3.151) and (3.153) in formulas (3.147) and (3.148), we arrive at the classical expressions:

$$H_x(z,t) = H_m e^{-\beta z}\cos(\omega t+\theta_0-\beta z). \tag{3.154}$$

$$H_y(z,t) = H_m e^{-\beta z}\sin(\omega t+\theta_0-\beta z). \tag{3.155}$$

By comparing these classical expressions with formulas (3.147) and (3.148), we conclude that nonlinearity (3.127) results in slower (than exponential) decay of the magnetic field magnitude and in logarithmic variations of initial phase $\theta(z)$ with respect to z rather than in linear variations of this phase. It is also interesting to compare formulas (3.147) and (3.148) with solution (2.57)–(2.58) from the previous chapter. We observe that in the case of strong magnetic fields the process of nonlinear diffusion is localized near the boundary of conducting media, whereas in the case of weak magnetic fields nonlinear diffusion extends up to infinity. In formulas (2.57) and (2.58) parameter z_0 has the meaning of penetration depth, that is, there is no electromagnetic field beyond z_0. In contrast, parameter z_0 in formulas (3.147) and (3.148) does not have a transparent relation to the penetration depth, although it clearly affects the rate of decay of the magnetic field, that is, the smaller z_0 the faster the magnetic field decays.

From formulas (3.147) and (3.148), we find

$$|\mathbf{H}(z,t)| = H(z) = H_m \left(1 + \frac{z}{z_0}\right)^{-\frac{2}{n-1}}. \tag{3.156}$$

By using the last equation in formula (3.130), we obtain

$$\mu(H(z)) = \mu_m \left(1 + \frac{z}{z_0}\right)^{-2}. \tag{3.157}$$

By combining expressions (3.156) and (3.157), we derive

$$B(z) = B_m \left(1 + \frac{z}{z_0}\right)^{-\frac{2n}{n-1}}, \tag{3.158}$$

where

$$B_m = \mu_m H_m \tag{3.159}$$

is the magnetic flux density at the boundary.

A typical plot of $B(z)$ vs. z is shown in Fig. 3.6. Again, we can observe that the rate of decay of $B(z)$ is lower than exponential. Physically, this can be explained by the fact that the magnetic permeability of media is decreased as the electromagnetic field diffuses deeper in the conducting media.

Next, let us proceed to the calculation of surface impedance in the case of circular polarization. By using formulas (3.134) and (3.146) along with the following expressions for the phasors of the electric field

$$\hat{E}_x(z) = -\frac{1}{\sigma}\frac{d\hat{H}_y(z)}{dz}, \quad \hat{E}_y(z) = \frac{1}{\sigma}\frac{d\hat{H}_x(z)}{dz}, \tag{3.160}$$

we find that the surface impedance is given by

$$\eta = \frac{\hat{E}_x(0)}{\hat{H}_y(0)} = -\frac{\hat{E}_y(0)}{\hat{H}_x(0)} = \frac{\alpha}{\sigma z_0}. \tag{3.161}$$

By invoking formulas (3.135), (3.141), (3.144), and (3.145), from the last equation we derive

$$\eta = \sqrt{\frac{\omega\mu_m}{\sigma}} \cdot \frac{2 + j\sqrt{2(n+1)}}{[2(n+1)(n+3)^2]^{\frac{1}{4}}}. \tag{3.162}$$

Thus, the surface impedance is expressed in terms of frequency and physical properties of conducting media. This impedance is field dependent because μ_m is determined by the value of the magnetic field at the medium boundary. By setting $n = 1$, we recover from the last equation the well-known formula for the surface impedance of linear conducting media:

$$\eta = (1 + j)\sqrt{\frac{\omega\mu}{2\sigma}}. \tag{3.163}$$

Of course, in this case the magnetic permeability does not depend on the magnetic field and is the same everywhere within the media. For this reason, subscript m for μ in the last formula is omitted.

It is often desired to represent the surface impedance in the polar form

$$\eta = |\eta| e^{j\varphi}. \tag{3.164}$$

By using formula (3.162), we find

$$|\eta| = \left(\frac{2}{n+1}\right)^{\frac{1}{4}} \sqrt{\frac{\omega\mu_m}{\sigma}}, \tag{3.165}$$

and

$$\tan\varphi = \sqrt{\frac{n+1}{2}}. \tag{3.166}$$

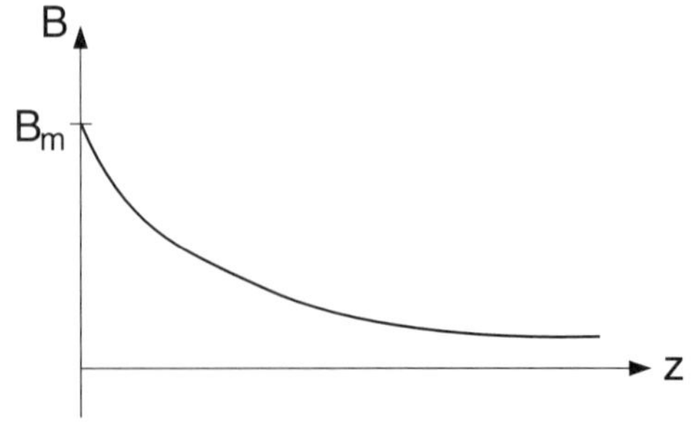

Fig. 3.6

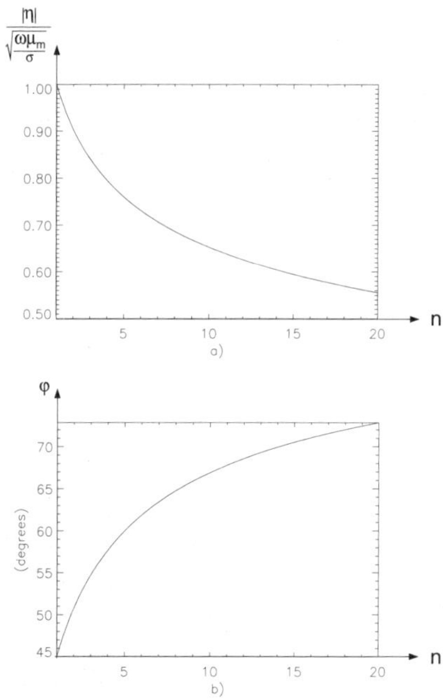

Fig. 3.7

These remarkably simple formulas reveal that the magnitude of the surface impedance tends to zero and its polar angle tends to $\frac{\pi}{2}$ as n is increased. This is in contrast with the results obtained in Section 2.1. There, we found that the magnetic nonlinearity of conducting media results in the increase in the magnitude of the surface impedance and in the decrease in its polar angle φ. The plots of $|\eta|$ vs. n and φ vs. n are shown in Figs. 3.7 a and 3.7 b, respectively.

Previously, it has been assumed that the magnetic field components at the boundary are specified (see formulas (3.109) and (3.110)). However, the found solutions can be easily generalized to the case when total fluxes are given. Indeed, suppose that

$$\Phi_x(t) = \Phi_m \sin(\omega t + \theta_0), \quad \Phi_y(t) = \Phi_m \cos(\omega t + \theta_0), \tag{3.167}$$

where $\Phi_x(t)$ and $\Phi_y(t)$ are total fluxes per unit length in y and x directions, respectively.

Using formulas

$$E_x(0,t) = -\frac{d\Phi_y(t)}{dt}, \quad E_y(0,t) = -\frac{d\Phi_x(t)}{dt}, \tag{3.168}$$

we derive

$$\hat{E}_x(0) = -j\omega\hat{\Phi}_m, \qquad \hat{E}_y(0) = -\omega\hat{\Phi}_m, \tag{3.169}$$

where

$$\hat{\Phi}_m = \Phi_m e^{j\theta_0}. \tag{3.170}$$

Now, by invoking formula (3.161), we find

$$\hat{H}_x(0) = \hat{H}_m = \frac{\omega\sigma z_0}{\alpha}\hat{\Phi}_m, \tag{3.171}$$

$$\hat{H}_y(0) = -j\hat{H}_m = -j\frac{\omega\sigma z_0}{\alpha}\hat{\Phi}_m. \tag{3.172}$$

By substituting the last two equations into (3.134) and (3.146), respectively, we end up with

$$\hat{H}_x(z) = \frac{\omega\sigma z_0}{\alpha}\hat{\Phi}_m\left(1+\frac{z}{z_0}\right)^{-\alpha}, \tag{3.173}$$

$$\hat{H}_y(z) = -j\frac{\omega\sigma z_0}{\alpha}\hat{\Phi}_m\left(1+\frac{z}{z_0}\right)^{-\alpha}, \tag{3.174}$$

which are the formulas we intended to establish.

It is important to note that the last two formulas do not give fully explicit expressions for magnetic field components in terms of $\hat{\Phi}$. This is because z_0 depends on μ_m that, in turn, depends on H_m. This difficulty can be circumvented if we write relation (3.171) in the form

$$\frac{|\alpha|}{\omega\sigma z_0}H_m = \Phi_m \tag{3.175}$$

and consider it as a nonlinear equation for H_m. By solving this equation, we can find H_m, which then can be utilized for the calculation of z_0. Having done that, we can use formulas (3.173) and (3.174).

3.3 NONLINEAR DIFFUSION OF ELLIPTICALLY POLARIZED MAGNETIC FIELDS IN ISOTROPIC MEDIA

Nonlinear diffusion of elliptically polarized magnetic fields in isotropic media can be analytically treated by using the perturbation technique. The development of this technique closely parallels the discussion presented in Sections 2.2 and 2.3. For this reason, in our exposition only the major steps of this technique will be outlined, while minor details will be omitted. The same style of exposition will be adopted in the next section as well.

To start the discussion, let us suppose that the following boundary conditions are realized at the interface of conducting half-space:

$$H_x(0,t) = H_m\cos\omega t + \epsilon H_m f_x(t), \tag{3.176}$$

$$H_y(0,t) = H_m \sin \omega t + \epsilon H_m f_y(t). \tag{3.177}$$

Here ϵ is a small parameter, while $f_x(t)$ and $f_y(t)$ are known periodic functions of time with period $T = \frac{2\pi}{\omega}$.

By using the general idea of perturbation techniques, we look for the solution of Eqs. (3.107)–(3.108) subject to boundary conditions (3.176)–(3.177) and (3.111)–(3.112) in the following form:

$$H_x(z,t) = H_x^0(z,t) + \epsilon h_x(z,t), \tag{3.178}$$

$$H_y(z,t) = H_y^0(z,t) + \epsilon h_y(z,t). \tag{3.179}$$

By substituting the last two formulas into the equations and boundary conditions mentioned above and by equating the terms of like powers of ϵ, we end up with the following boundary value problems:

$$\frac{\partial^2 H_x^0}{\partial z^2} = \sigma \frac{\partial}{\partial t} \left[\mu \left(\sqrt{(H_x^0)^2 + (H_y^0)^2} \right) H_x^0 \right], \tag{3.180}$$

$$\frac{\partial^2 H_y^0}{\partial z^2} = \sigma \frac{\partial}{\partial t} \left[\mu \left(\sqrt{\left(H_x^0\right)^2 + \left(H_y^0\right)^2} \right) H_y^0 \right], \tag{3.181}$$

$$H_x^0(0,t) = H_m \cos \omega t, \tag{3.182}$$

$$H_y^0(0,t) = H_m \sin \omega t, \tag{3.183}$$

$$H_x^0(\infty,t) = 0, \tag{3.184}$$

$$H_y^0(\infty,t) = 0, \tag{3.185}$$

and

$$\frac{\partial^2 h_x}{\partial z^2} = \sigma \frac{\partial}{\partial t} \left[h_x \frac{\partial B_x}{\partial H_x} \left(H_x^0, H_y^0\right) + h_y \frac{\partial B_x}{\partial H_y} \left(H_x^0, H_y^0\right) \right], \tag{3.186}$$

$$\frac{\partial^2 h_y}{\partial z^2} = \sigma \frac{\partial}{\partial t} \left[h_x \frac{\partial B_y}{\partial H_x} \left(H_x^0, H_y^0\right) + h_y \frac{\partial B_y}{\partial H_y} \left(H_x^0, H_y^0\right) \right], \tag{3.187}$$

$$h_x(0,t) = H_m f_x(t), \tag{3.188}$$

$$h_y(0,t) = H_m f_y(t). \tag{3.189}$$

$$h_x(\infty,t) = 0, \tag{3.190}$$

$$h_y(\infty,t) = 0. \tag{3.191}$$

Here, as before, we have used the following notations:

$$B_x\left(H_x, H_y\right) = \mu \left(\sqrt{\left(H_x^2 + H_y^2\right)} \right) H_x, \tag{3.192}$$

$$B_y\left(H_x, H_y\right) = \mu\left(\sqrt{\left(H_x^2 + H_y^2\right)}\right) H_y. \tag{3.193}$$

The boundary value problem (3.180)–(3,.185) is identical to the boundary value problem (3.107)–(3.112), which describes nonlinear diffusion of circularly polarized electromagnetic fields. The solution to this problem was found in the previous section and it can be written in the following form:

$$H_x^0(z,t) = H_m\left(1+\frac{z}{z_0}\right)^{-\frac{2}{n-1}} \cos(\omega t + \theta(z)), \tag{3.194}$$

$$H_y^0(z,t) = H_m\left(1+\frac{z}{z_0}\right)^{-\frac{2}{n-1}} \sin(\omega t + \theta(z)), \tag{3.195}$$

where

$$\theta(z) = -\frac{\sqrt{2(n+1)}}{n-1}\ln\left(1+\frac{z}{z_0}\right) = -\alpha'' \ln\left(1+\frac{z}{z_0}\right). \tag{3.196}$$

By substituting formulas (3.194) and (3.195) into Eqs. (3.186)–(3.187) and by performing the same transformations as in Section 2.2, we arrive at the following equations:

$$\begin{aligned}\frac{\partial^2 h_x(z,t)}{\partial z^2} &= \sigma\mu_m\left(1+\frac{z}{z_0}\right)^{-2}\frac{\partial}{\partial t}\left\{h_x(z,t)\left[\frac{n+1}{2}\right.\right.\\ &\left.\left.+\frac{n-1}{2}\cos(2\omega t + 2\theta(z))\right] + h_y(z,t)\frac{n-1}{2}\sin(2\omega t + 2\theta(z))\right\},\end{aligned} \tag{3.197}$$

$$\begin{aligned}\frac{\partial^2 h_y(z,t)}{\partial z^2} =&\sigma\mu_m\left(1+\frac{z}{z_0}\right)^{-2}\frac{\partial}{\partial t}\left\{h_x(z,t)\frac{n-1}{2}\sin(2\omega t + 2\theta(z))\right.\\ &\left.+h_y(z,t)\left[\frac{n+1}{2}-\frac{n-1}{2}\cos(2\omega t + 2\theta(z))\right]\right\}.\end{aligned} \tag{3.198}$$

To further simplify the above equations, we introduce the auxiliary function:

$$\phi(z,t) = h_x(z,t) + jh_y(z,t), \tag{3.199}$$

$$\psi(z,t) = h_x(z,t) - jh_y(z,t). \tag{3.200}$$

By using these functions and literally repeating the transformations described in Section 2.2, we end up with the equations:

$$\frac{\partial^2 \phi(z,t)}{\partial z^2} = \sigma\mu_m\left(1+\frac{z}{z_0}\right)^{-2}\frac{\partial}{\partial t}\left[\frac{n+1}{2}\phi(z,t)\right.$$

$$+\frac{n-1}{2}\left(1+\frac{z}{z_0}\right)^{-j2\alpha''} e^{j2\omega t}\psi(z,t)\Bigg], \tag{3.201}$$

$$\begin{aligned}\frac{\partial^2\psi(z,t)}{\partial z^2} &= \sigma\mu_m\left(1+\frac{z}{z_0}\right)^{-2}\frac{\partial}{\partial t}\left[\frac{n+1}{2}\psi(z,t)\right.\\ &\left.+\frac{n-1}{2}\left(1+\frac{z}{z_0}\right)^{j2\alpha''} e^{-j2\omega t}\phi(z,t)\right].\end{aligned} \tag{3.202}$$

Next, we shall look for the solution of Eqs. (3.201) and (3.202) in the form of Fourier series:

$$\phi(z,t) = \sum_{k=-\infty}^{\infty} \phi_k(z)e^{jk\omega t}, \tag{3.203}$$

$$\psi(z,t) = \sum_{k=-\infty}^{\infty} \psi_k(z)e^{jk\omega t}. \tag{3.204}$$

By substituting the above Fourier series into Eqs. (3.201)–(3.202) and by equating similar terms, we obtain the following infinite set of coupled ordinary differential equations:

$$\left(1+\frac{z}{z_0}\right)^2\frac{d^2\phi_k(z)}{dz^2} = j\chi_k\left[a\phi_k(z)+\left(1+\frac{z}{z_0}\right)^{-j2\alpha''}\cdot\psi_{k-2}(z)\right], \tag{3.205}$$

$$\left(1+\frac{z}{z_0}\right)^2\frac{d^2\psi_{k-2}(z)}{dz^2} = j\chi_{k-2}\left[a\psi_{k-2}(z)+\left(1+\frac{z}{z_0}\right)^{j2\alpha''}\cdot\phi_k(z)\right],$$
$$(k=0,\pm1,\pm2,\ldots), \tag{3.206}$$

where

$$a=\frac{n+1}{n-1}, \quad \chi_k = k\frac{n-1}{2}\omega\sigma\mu_m. \tag{3.207}$$

The remarkable feature of these simultaneous ordinary differential equations is that they are coupled in separate pairs. For this reason, each pair of coupled differential Eqs. (3.205)–(3.206) can be solved separately. The solution of these differential equations must be subject to the boundary conditions

$$\phi_k(0) = H_m\left(f_{x,k}+jf_{y,k}\right), \tag{3.208}$$

$$\psi_{k-2}(0) = H_m\left(f_{x,k-2}-jf_{y,k-2}\right), \tag{3.209}$$

$$\phi_k(\infty) = \psi_{k-2}(\infty) = 0. \tag{3.210}$$

Boundary conditions (3.208)–(3.209) follow from formulas (3.199)–(3.200) and boundary conditions (3.188)–(3.189), whereas notations $f_{x,k}$ and $f_{y,k}$ stand for Fourier coefficients of functions $f_x(t)$ and $f_y(t)$, respectively.

Having solved Eqs. (3.205)–(3.206), we can find Fourier coefficients of perturbations:

$$h_{x,k}(z) = \frac{1}{2}\left[\phi_k(z) + \psi_k(z)\right], \tag{3.211}$$

$$h_{y,k}(z) = \frac{1}{2j}\left[\phi_k(z) - \psi_k(z)\right], \tag{3.212}$$

and then perturbations themselves:

$$h_x(z,t) = \sum_{k=-\infty}^{\infty} h_{x,k}(z)e^{jk\omega t}, \tag{3.213}$$

$$h_y(z,t) = \sum_{k=-\infty}^{\infty} h_{y,k}(z)e^{jk\omega t}. \tag{3.214}$$

Now we shall apply the perturbation technique just outlined to the case when the magnetic field on the boundary is elliptically polarized. Since we deal with isotropic media, we can choose axes x and y to be coincident with major and minor axes of the "polarization" ellipse, respectively. For this choice of coordinate axes, we have the following boundary conditions:

$$H_x(0,t) = H_{mx}\cos\omega t, \tag{3.215}$$

$$H_y(0,t) = H_{my}\sin\omega t. \tag{3.216}$$

By introducing notations

$$H_m = \frac{H_{mx} + H_{my}}{2}, \qquad \epsilon = \frac{H_{mx} - H_{my}}{H_{mx} + H_{my}}, \tag{3.217}$$

the above boundary conditions can be written in the perturbation form:

$$H_x(0,t) = H_m\cos\omega t + \epsilon H_m\cos\omega t, \tag{3.218}$$

$$H_y(0,t) = H_m\sin\omega t - \epsilon H_m\sin\omega t. \tag{3.219}$$

By comparing boundary conditions (3.218)–(3.219) with boundary conditions (3.176)–(3.177), we can identify $f_x(t)$ and $f_y(t)$ as follows:

$$f_x(t) = \cos\omega t, \quad f_y(t) = -\sin\omega t. \tag{3.220}$$

From formula (3.220), we find

$$f_{x,1} = \frac{1}{2}, \quad f_{x,-1} = \frac{1}{2}, \quad f_{x,k} = 0 \quad \text{if } |k| \neq 1, \tag{3.221}$$

$$f_{y,1} = -\frac{1}{2j}, \quad f_{y,-1} = \frac{1}{2j}, \quad f_{y,k} = 0 \quad \text{if } |k| \neq 1. \tag{3.222}$$

By using formulas (3.221) and (3.222) in boundary conditions (3.208) and (3.209), we conclude that

$$\phi_k(0) = 0, \quad \psi_k(0) = 0 \quad \text{if } |k| \neq 1, \tag{3.223}$$

$$\phi_1(0) = 0, \quad \psi_1(0) = H_m. \tag{3.224}$$

Consequently, for $k \geq 0$ and $k \neq 3$ we have the boundary value problems for pairs of coupled Eqs. (3.205)–(3.206) subject to zero boundary conditions. Equations (3.205)–(3.206) are homogeneous, so we conclude that for $k \geq 0$ and $k \neq 3$ these equations have zero solutions:

$$\phi_k(z) = 0, \quad \psi_{k-2}(z) = 0 \quad \text{if } k \geq 0 \text{ and } k \neq 3. \tag{3.225}$$

According to formulas (3.211) and (3.212), this implies that

$$h_{x,k}(z) = 0, \quad h_{y,k}(z) = 0 \quad \text{if } k \geq 0, k \neq 1, k \neq 3. \tag{3.226}$$

Thus, only the first and third harmonics are not equal to zero. To find these harmonics, we shall write Eqs. (3.205)–(3.206) for $k = 3$ and supplement them with boundary conditions (3.224) and (3.210). In this way, we arrive at the following boundary value problem:

$$\left(1 + \frac{z}{z_0}\right)^2 \frac{d^2\phi_3}{dz^2} - j\chi_3 \left[a\phi_3 + \left(1 + \frac{z}{z_0}\right)^{-j2\alpha''} \cdot \psi_1\right] = 0, \tag{3.227}$$

$$\left(1 + \frac{z}{z_0}\right)^2 \frac{d^2\psi_1}{dz^2} - j\chi_1 \left[a\psi_1 + \left(1 + \frac{z}{z_0}\right)^{j2\alpha''} \cdot \phi_3\right] = 0, \tag{3.228}$$

$$\phi_3(0) = 0, \quad \psi_1(0) = H_m, \tag{3.229}$$

$$\phi_3(\infty) = \psi_1(\infty) = 0. \tag{3.230}$$

We shall look for the solution of the boundary value problem (3.227)–(3.230) in the form

$$\phi_3(z) = A\left(1 + \frac{z}{z_0}\right)^{\beta}, \quad \psi_1(z) = B\left(1 + \frac{z}{z_0}\right)^{\beta + j2\alpha''}. \tag{3.231}$$

By substituting expressions (3.231) into differential Eqs. (3.227)–(3.228), we find that these equations have solutions in the form (3.231) if β satisfies the following characteristic equation:

$$\left(\beta^2-\beta-j\chi_3 a z_0^2\right)\left[\left(\beta+j2\alpha''\right)^2-\left(\beta+j2\alpha''\right)-j\chi_1 a z_0^2\right]+\chi_1\chi_3 z_0^4=0. \tag{3.232}$$

By involving formulas (3.207) and (3.145), we obtain

$$\chi_1 a z_0^2=\frac{(n+1)(n+3)}{2(n-1)^2}\sqrt{2(n+1)}, \tag{3.233}$$

$$\chi_3 a z_0^2=\frac{3(n+1)(n+3)}{2(n-1)^2}\sqrt{2(n+1)}, \tag{3.234}$$

$$\chi_1\chi_3 z_0^4=\frac{3(n+3)^2(n+1)}{2(n-1)^2}. \tag{3.235}$$

Thus, the coefficients of characteristic Eqs. (3.232) depend only on n. Consequently, the roots of this equation depend only on n as well. To satisfy the boundary conditions (3.230), exponent β in formula (3.231) has to have a negative real part. It can be shown that characteristic Eq. (3.232) has two roots, β' and β'', with negative real parts. These roots have been computed as functions of n, and the results of computations are presented in Figs. 3.8 a and 3.8 b.

By using roots β' and β'' general solutions (3.231) to differential Eqs. (3.227) and (3.228) can be written in the form

$$\phi_3(z)=A'\left(1+\frac{z}{z_0}\right)^{\beta'}+A''\left(1+\frac{z}{z_0}\right)^{\beta''}, \tag{3.236}$$

$$\psi_1(z)=B'\left(1+\frac{z}{z_0}\right)^{\beta'+j2\alpha''}+B''\left(1+\frac{z}{z_0}\right)^{\beta''+j2\alpha''}. \tag{3.237}$$

By using the same line of reasoning as in Section 2.2, we find that functions (3.236) and (3.237) will satisfy boundary conditions (3.229) and differential Eqs.(3.227)–(3.228) if coefficients A', A'', B', and B'' are determined from the following simultaneous equations:

$$A'+A''=0, \tag{3.238}$$

$$B'+B''=H_m, \tag{3.239}$$

$$\left[\left(\beta'\right)^2-\beta'-j\chi_3 z_0^2 a\right]A'-j\chi_3 z_0^2 B'=0, \tag{3.240}$$

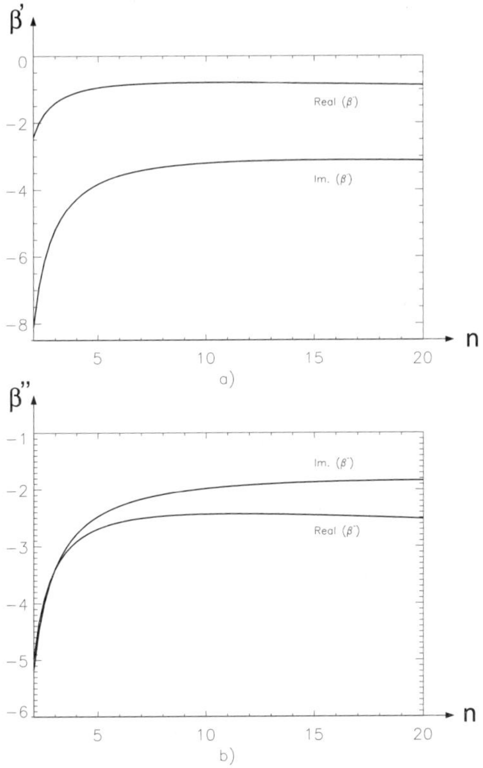

Fig. 3.8

$$\left[(\beta'')^2 - \beta'' - j\chi_3 z_0^2 a\right] A'' - j\chi_3 z_0^2 B'' = 0. \tag{3.241}$$

By solving the last equations, we find explicit formulas for the above coefficients:

$$A' = -A'' = \frac{j\chi_3 z_0^2}{\left[(\beta')^2 - (\beta')\right] - \left[(\beta'')^2 - \beta''\right]} H_m, \tag{3.242}$$

$$B' = \frac{(\beta')^2 - \beta' - j\chi_3 z_0^2 a}{[(\beta')^2 - \beta'] - [(\beta'')^2 - \beta'']} H_m, \tag{3.243}$$

$$B'' = -\frac{(\beta'')^2 - \beta'' - j\chi_3 z_0^2 a}{[(\beta')^2 - \beta'] - [(\beta'')^2 - \beta'']} H_m. \tag{3.244}$$

Having found the above coefficients and, consequently, functions $\phi_3(z)$ and $\psi_1(z)$, the first and third harmonics of perturbations can be determined. Indeed, from formulas (3.211), (3.212), and (3.225), we obtain

$$h_{x,3}(z) = \frac{1}{2}\phi_3(z), \quad h_{y_3}(z) = -\frac{j}{2}\phi_3(z), \tag{3.245}$$

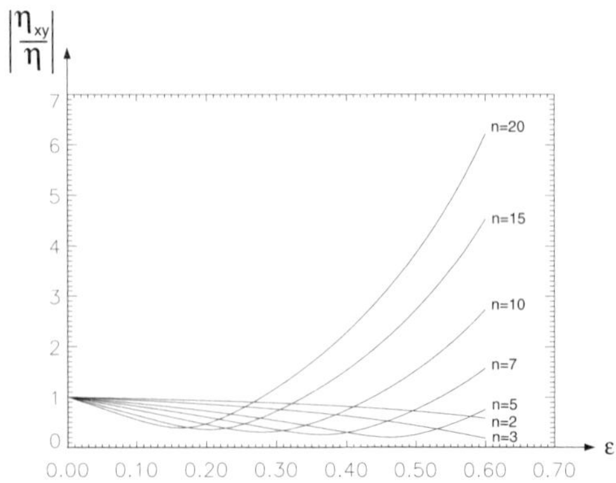

Fig. 3.9

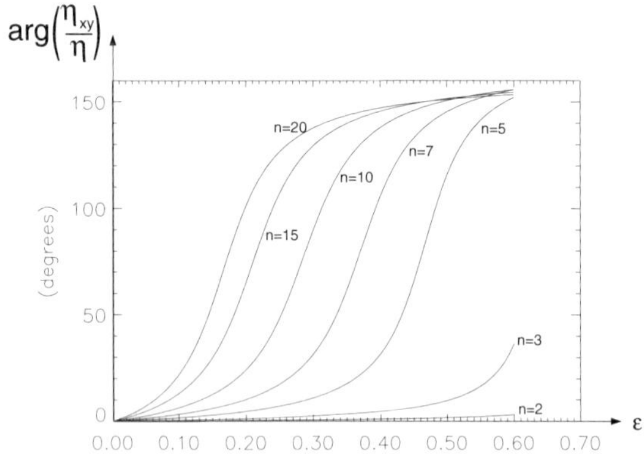

Fig. 3.10

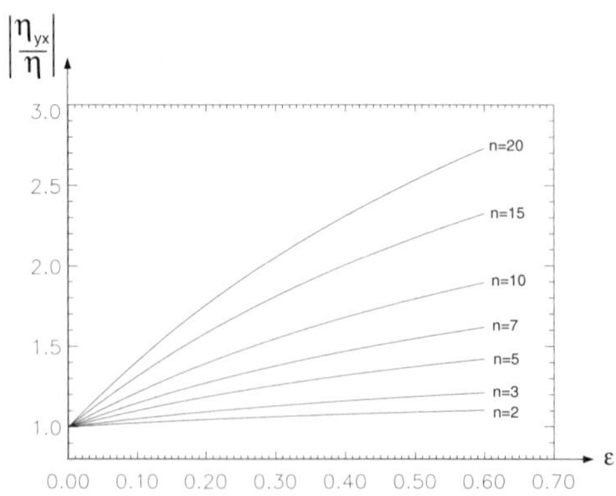

Fig. 3.11

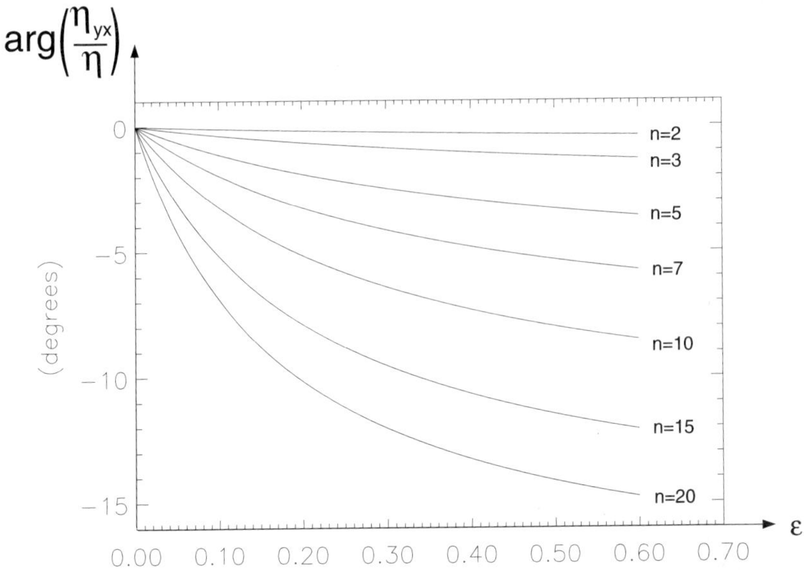

Fig. 3.12

$$h_{x,1}(z) = \frac{1}{2}\psi_1(z), \quad h_{y,1}(z) = \frac{j}{2}\psi_1(z). \tag{3.246}$$

By using $h_{x,1}(z)$ and $h_{y,1}(z)$, we can find the phasors of the first harmonic of the total field:

$$\hat{H}_{x,1}(z) = \hat{H}_x^0(z) + 2\epsilon h_{x,1}(z), \tag{3.247}$$

$$\hat{H}_{y,1}(z) = \hat{H}_y^0(z) + 2\epsilon h_{y,1}(z). \tag{3.248}$$

By taking into account formulas (3.236), (3.237), and (3.246), the last two expressions can be rewritten as follows:

$$\begin{aligned}\hat{H}_{x,1}(z) =& H_m \left(1 + \frac{z}{z_0}\right)^{-\alpha} + \epsilon \left[B' \left(1 + \frac{z}{z_0}\right)^{\beta' + j2\alpha''} \right. \\ &\left. + B'' \left(1 + \frac{z}{z_0}\right)^{\beta'' + j2\alpha''} \right],\end{aligned} \tag{3.249}$$

$$\begin{aligned}\hat{H}_{y,1}(z) =& -jH_m \left(1 + \frac{z}{z_0}\right)^{-\alpha} + j\epsilon \left[B' \left(1 + \frac{z}{z_0}\right)^{\beta' + j2\alpha''} \right. \\ &\left. + B'' \left(1 + \frac{z}{z_0}\right)^{\beta'' + j2\alpha''} \right].\end{aligned} \tag{3.250}$$

By using the last two formulas as well as formulas (3.160) and (3.161), we derive the following expressions for the surface impedance:

$$\frac{\eta_{xy}}{\eta} = \frac{\alpha H_m + \epsilon[B'(\beta' + j2\alpha'') + B''(\beta'' + j2\alpha'')]}{\alpha(1-\epsilon)H_m}, \tag{3.251}$$

$$\frac{\eta_{yx}}{\eta} = \frac{\alpha H_m - \epsilon\left[B'\left(\beta' + j2\alpha''\right) + B''\left(\beta'' + j2\alpha''\right)\right]}{\alpha(1+\epsilon)H_m}, \tag{3.252}$$

where η is the surface impedance in the case of circular polarization of the magnetic field.

Formulas (3.251) and (3.252) allows one to evaluate to what extent the surface impedance is affected by deviations from the circular polarizations. These deviations are accounted for by parameter ϵ in (3.251) and (3.252). It is instructive to note that the right-hand sides of formulas (3.251) and (3.252) do not depend on H_m. This is because, according to formulas (3.243)–(3.244), coefficients B' and B'' are directly proportional to H_m. Thus, the right-hand sides of (3.251) and (3.252) are functions only of n and ϵ. These functions have been computed for various values of n and ϵ and the results of computations are shown in Figs. 3.9, 3.10, 3.11, and 3.12.

3.4 NONLINEAR DIFFUSION IN ANISOTROPIC MEDIA

In this section, a two-parameter perturbation technique is applied to the analysis of nonlinear diffusion of elliptically polarized electromagnetic fields in anisotropic media. One parameter is used for the description of anisotropic magnetic properties of media as perturbations of isotropic properties, while another parameter is introduced to treat elliptical polarizations as perturbations of circular polarizations.

A convenient way to introduce the first perturbation parameter is through the potential function $U(\mathbf{H})$ of $\mathbf{B}$-field in $\mathbf{H}$-space (see Section 2.4). For isotropic media with constitutive relation (3.127), this potential function is given by the expression

$$U(\mathbf{H}) = \int_0^{\mathbf{H}} B(H')dH' = \frac{k}{n+1}\left|\mathbf{H}\right|^{n+1}. \tag{3.253}$$

The last expression can be rewritten in terms of H_x and H_y as follows:

$$U\left(H_x, H_y\right) = \frac{k}{n+1}\left(H_x^2 + H_y^2\right)^{\frac{n+1}{2}}. \tag{3.254}$$

The symmetry of the last expression with respect to H_x and H_y reflects the isotropicity of media. To generate the potential for anisotropic media, the above symmetry must be perturbed. A simple way to do this is to assign different "weight" coefficients for H_x and H_y. By using a perturbation parameter ϵ, this can be achieved as follows:

$$U_\epsilon(H_x, H_y) = \frac{k}{n+1}\left[(1+\epsilon)H_x^2 + (1-\epsilon)H_y^2\right]^{\frac{n+1}{2}}. \tag{3.255}$$

By taking the gradient of $U_\epsilon(H_x, H_y)$ with respect to H_x and H_y (see formula (2.317)), we end up with the following constitutive relations for magnetically anisotropic media:

$$B_x(H_x, H_y) = (1+\epsilon)kH_x\left(\sqrt{(1+\epsilon)H_x^2 + (1-\epsilon)H_y^2}\right)^{n-1}, \tag{3.256}$$

$$B_y(H_x, H_y) = (1-\epsilon)kH_y\left(\sqrt{(1+\epsilon)H_x^2 + (1-\epsilon)H_y^2}\right)^{n-1}. \tag{3.257}$$

Next, we introduce another parameter $\tilde{\epsilon}$ to describe elliptical polarizations as perturbations of circular polarizations. This is done as follows:

$$H_x(0,t) = H_m\cos(\omega t + \theta_0) + \tilde{\epsilon}H_m\cos(\omega t - \theta_0), \tag{3.258}$$

$$H_y(0,t) = H_m\sin(\omega t + \theta_0) - \tilde{\epsilon}H_m\sin(\omega t - \theta_0), \tag{3.259}$$

where

$$H_m = \frac{H_{mx'} + H_{my'}}{2}, \qquad \tilde{\epsilon} = \frac{H_{mx'} - H_{my'}}{H_{mx'} + H_{my'}}, \tag{3.260}$$

while $H_{mx'}$ and $H_{my'}$ are peak values of the magnetic field along the major (x') and minor (y') axes of the polarization ellipse, respectively, and θ_0 is the angle between the major axis x' and the anisotropy axis x.

In accordance with the general idea of the perturbation technique, we look for the magnetic field in the form of the following expansions:

$$H_x(z,t) = H_x^0(z,t) + \epsilon h_x(z,t) + \tilde{\epsilon}\tilde{h}_x(z,t) + \dots, \tag{3.261}$$

$$H_y(z,t) = H_y^0(z,t) + \epsilon h_y(z,t) + \epsilon\tilde{h}_y(z,t) + \dots. \tag{3.262}$$

Next, we substitute formulas (3.261) and (3.262) into boundary conditions (3.258) and (3.259) and nonlinear diffusion equations

$$\frac{\partial^2 H_x}{\partial z^2} = \sigma\frac{\partial B_x(H_x, H_y)}{\partial t}, \tag{3.263}$$

$$\frac{\partial^2 H_y}{\partial z^2} = \sigma\frac{\partial B_y(H_x, H_y)}{\partial t}. \tag{3.264}$$

Then we equate the terms of like powers of ϵ and $\tilde{\epsilon}$. As a result, we end up with the following boundary value problems for H_x^0 and H_y^0, h_x, and h_y, $\tilde{h}_x$, and $\tilde{h}_y$, respectively:

$$\frac{\partial^2 H_x^0}{\partial z^2} = \sigma \frac{\partial}{\partial t}\left[\mu\left(\sqrt{(H_x^0)^2 + \left(H_y^0\right)^2}\right) H_x^0\right], \tag{3.265}$$

$$\frac{\partial^2 H_y^0}{\partial z^2} = \sigma \frac{\partial}{\partial t}\left[\mu\left(\sqrt{(H_x^0)^2 + \left(H_y^0\right)^2}\right) H_y^0\right], \tag{3.266}$$

$$H_x^0(0,t) = H_m \cos(\omega t + \theta_0), \tag{3.267}$$

$$H_y^0(0,t) = H_m \sin(\omega t + \theta_0), \tag{3.268}$$

$$H_x^0(\infty,t) = H_y^0(\infty,t) = 0. \tag{3.269}$$

Next,

$$\frac{\partial^2 h_x}{\partial z^2} - \sigma \frac{\partial}{\partial t}\left[h_x \frac{\partial B_x^0}{\partial H_x}\left(H_x^0, H_y^0\right) + h_y \frac{\partial B_x^0}{\partial H_y}\left(H_x^0, H_y^0\right)\right] = \sigma \frac{\partial g_x}{\partial t}, \tag{3.270}$$

$$\frac{\partial^2 h_y}{\partial z^2} - \sigma \frac{\partial}{\partial t}\left[h_x \frac{\partial B_y^0}{\partial H_x}\left(H_x^0, H_y^0\right) + h_y \frac{\partial B_y^0}{\partial H_y}\left(H_x^0, H_y^0\right)\right] = \sigma \frac{\partial g_y}{\partial t}, \tag{3.271}$$

$$h_x(0,t) = h_y(0,t) = 0, \tag{3.272}$$

$$h_x(\infty,t) = h_y(\infty,t) = h_y(\infty,t) = 0, \tag{3.273}$$

where

$$B_x^0\left(H_x^0, H_y^0\right) = \mu(|\mathbf{H}^0|) H_x^0, \tag{3.274}$$

$$B_y^0\left(H_x^0, H_y^0\right) = \mu(|\mathbf{H}^0|) H_y^0, \tag{3.275}$$

$$g_x = B_x^0\left(H_x^0, H_y^0\right)\left[1 + \frac{n-1}{2} \cdot \frac{\left(H_x^0\right)^2 - \left(H_y^0\right)^2}{H^2}\right], \tag{3.276}$$

$$g_y = B_y^0\left(H_x^0, H_y^0\right)\left[1 - \frac{n-1}{2} \cdot \frac{\left(H_x^0\right)^2 - \left(H_y^0\right)^2}{H^2}\right], \tag{3.277}$$

and finally,

$$\frac{\partial^2 \tilde{h}_x}{\partial z^2} - \sigma \frac{\partial}{\partial t}\left[\tilde{h}_x \frac{\partial B_x^0}{\partial H_x}\left(H_x^0, H_y^0\right) + \tilde{h}_y \frac{\partial B_x^0}{\partial H_y}\left(H_x^0, H_y^0\right)\right] = 0, \tag{3.278}$$

$$\frac{\partial^2 \tilde{h}_y}{\partial z^2} - \sigma \frac{\partial}{\partial t}\left[\tilde{h}_x \frac{\partial B_y^0}{\partial H_x}\left(H_x^0, H_y^0\right) + h_y \frac{\partial B_y^0}{\partial H_y}\left(H_x^0, H_y^0\right)\right] = 0, \tag{3.279}$$

$$\tilde{h}_x(0,t) = H_m \cos(\omega t - \theta_0), \tag{3.280}$$

$$\tilde{h}_y(0,t) = -H_m \sin(\omega t - \theta_0), \tag{3.281}$$

$$\tilde{h}_x(\infty,t) = \tilde{h}_y(\infty,t) = 0. \tag{3.282}$$

The boundary value problem (3.265)–(3.269) must be solved first. Then, by having found H_x^0 and H_y^0, boundary value problems (3.270)–(3.277) and (3.278)–(3.282) can be tackled. It is apparent that the boundary value problem (3.265)–(3.269) is identical (up to the initial phase θ_0) to the boundary value problem (3.180)–(3.185). Consequently, its solution is given by formulas (3.194)–(3.195) with the only modification that phase $\theta(z)$ takes into account the initial phase θ_0:

$$\theta(z) = \theta_0 - \alpha'' \ln\left(1 + \frac{z}{z_0}\right). \tag{3.283}$$

By using formulas (3.194), (3.195), and (3.283) in the boundary value problem (3.270)-(3.277) and by literally repeating the same line of reasoning as in Sections 2.4 and 2.5, it can be shown that $h_x(z,t)$ and $h_y(z,t)$ have only the first and third time-harmonics. To find these harmonics, the following ordinary differential equations have to be solved for auxiliary functions ϕ_3 and ψ_1:

$$\left(1+\frac{z}{z_0}\right)^2 \frac{d^2\phi_3}{dz^2} - j\chi_3\left[a\phi_3 + e^{j2\theta_0}\left(1+\frac{z}{z_0}\right)^{-j2\alpha''}\cdot\psi_1\right] = j\zeta e^{j3\theta_0}\left(1+\frac{z}{z_0}\right)^{-\frac{2}{n-1}-j3\alpha''}, \tag{3.284}$$

$$\left(1+\frac{z}{z_0}\right)^2 \frac{d^2\psi_1}{dz^2} - j\chi_1\left[a\psi_1 + e^{-j2\theta_0}\left(1+\frac{z}{z_0}\right)^{j2\alpha''}\cdot\phi_3\right] = j\nu e^{j\theta_0}\left(1+\frac{z}{z_0}\right)^{-\frac{2}{n-1}-j\alpha''}, \tag{3.285}$$

subject to the boundary conditions

$$\phi_3(0) = \psi_1(0) = 0, \tag{3.286}$$

$$\phi_3(\infty) = \psi_1(\infty) = 0, \tag{3.287}$$

where

$$\zeta = 3\omega\sigma\mu_m H_m \frac{n-1}{4}, \quad \nu = \omega\sigma\mu_m H_m \frac{n+3}{4}. \tag{3.288}$$

Coupled differential Eqs. (3.284)–(3.285) are inhomogeneous. Their particular solution can be sought in the form

$$\phi_3^{(p)}(z) = e^{j3\theta_0} C_3 \left(1 + \frac{z}{z_0}\right)^{\lambda_3}, \tag{3.289}$$

$$\psi_1^{(p)}(z) = e^{j\theta_0} C_1 \left(1 + \frac{z}{z_0}\right)^{\lambda_1}. \tag{3.290}$$

By substituting functions (3.289) and (3.290) into differential equations (3.284) and (3.285), we find that these equations will be satisfied if

$$\lambda_1 = -\frac{2}{n-1} - j\alpha'' = -\alpha, \tag{3.291}$$

$$\lambda_3 = \lambda_1 - j2\alpha'', \tag{3.292}$$

and coefficients C_3 and C_1 are the solution of simultaneous linear equations identical to Eqs. (2.385) and (2.386). This means that these coefficients can be computed by using formulas (2.387) and (2.388), and these coefficients do not depend on θ_0.

Having found the particular solution (3.289) and (3.290), a general solution to coupled differential Eqs. (3.284)–(3.285) can be represented in the form

$$\phi_3(z) = e^{j3\theta_0} \left[A' \left(1 + \frac{z}{z_0}\right)^{\beta'} + A'' \left(1 + \frac{z}{z_0}\right)^{\beta''} + C_3 \left(1 + \frac{z}{z_0}\right)^{\lambda_3} \right], \tag{3.293}$$

$$\psi_1(z) = e^{j\theta_0} \left[B' \left(1 + \frac{z}{z_0}\right)^{\beta'} + B'' \left(1 + \frac{z}{z_0}\right)^{\beta''} + C_1 \left(1 + \frac{z}{z_0}\right)^{\lambda_1} \right], \tag{3.294}$$

where β' and β'' are the roots of the characteristic Eq. (3.232) with negative real parts.

By using literally the same line of reasoning as in Sections 2.4 and 2.5, we arrive at the conclusion that coefficients A', A'', B', and B'' can be found from simultaneous linear Eqs. (2.391)–(2.394). After these equations are solved and the above coefficients determined, the first and third time harmonics of perturbations $h_{x,1}$ and $h_{y,1}$ can be calculated according to the formulas

$$h_{x,1}(z) = \frac{1}{2}\psi_1(z), \quad h_{y,1}(z) = \frac{j}{2}\psi_1(z), \tag{3.295}$$

$$h_{x,3}(z) = \frac{1}{2}\phi_3(z), \quad h_{y,3}(z) = -\frac{j}{2}\phi_3(z). \tag{3.296}$$

Now we proceed to the solution of the boundary value problem (3.278)–(3.282). As before, it can be demonstrated that perturbations $\tilde{h}_x(z,t)$ and $\tilde{h}_y(z,t)$ have only the first and third time-harmonics, while all other harmonics are equal to zero. The calculation of the first and third time-harmonics can be reduced to the solution of the following coupled ordinary differential equations for auxiliary functions $\phi_3(z)$ and $\psi_1(z)$:

$$\left(1+\frac{z}{z_0}\right)^2 \frac{d^2\tilde{\phi}_3}{dz^2} - j\chi_3 \left[a\tilde{\phi}_3 + e^{j2\theta_0}\left(1+\frac{z}{z_0}\right)^{-j2\alpha''} \cdot \tilde{\psi}_1\right] = 0, \tag{3.297}$$

$$\left(1+\frac{z}{z_0}\right)^2 \frac{d\tilde{\psi}_1}{dz^2} - j\chi_1 \left[a\tilde{\psi}_1 + e^{-j2\theta_0}\left(1+\frac{z}{z_0}\right)^{j2\alpha''} \cdot \tilde{\phi}_3\right] = 0, \tag{3.298}$$

subject to the boundary conditions

$$\tilde{\phi}_3(0) = 0, \quad \tilde{\psi}_1(0) = H_m e^{-j\theta_0}, \tag{3.299}$$

$$\tilde{\phi}_3(\infty) = \tilde{\psi}_1(\infty) = 0. \tag{3.300}$$

A general solution of the above coupled differential equations can be represented in the form

$$\tilde{\phi}_3(z) = e^{j\theta_0}\left[\tilde{A}'\left(1+\frac{z}{z_0}\right)^{\beta'} + \tilde{A}''\left(1+\frac{z}{z_0}\right)^{\beta''}\right], \tag{3.301}$$

$$\tilde{\psi}_1(z) = e^{-j\theta_0}\left[\tilde{B}'\left(1+\frac{z}{z_0}\right)^{\beta'+j2\alpha''} + \tilde{B}''\left(1+\frac{z}{z_0}\right)^{\beta''+j2\alpha''}\right], \tag{3.302}$$

where, as before, β' and β'' are the roots of the characteristic Eq. (3.232) with negative real parts.

By invoking the same reasoning as in the previous section, it can be demonstrated that coefficients $\tilde{A}'$, $\tilde{A}''$, $\tilde{B}'$, and $\tilde{B}''$ can be computed by using formulas (3.242)–(3.244). Having found these coefficients, the first and third time-harmonics of perturbations $\tilde{h}_x(z,t)$ and $\tilde{h}_y(z,t)$ can be determined as follows:

$$\tilde{h}_{x,1}(z) = \frac{1}{2}\tilde{\psi}_1(z), \qquad \tilde{h}_{y,1}(z) = \frac{j}{2}\tilde{\psi}_1(z), \tag{3.303}$$

$$\tilde{h}_{x,3}(z) = \frac{1}{2}\tilde{\phi}_3(z), \qquad \tilde{h}_{y,3}(z) = -\frac{j}{2}\tilde{\phi}_3(z). \tag{3.304}$$

For the phasors of the first harmonic of the total field, we have

$$\hat{H}_{x,1}(z) = \hat{H}_x^0(z) + 2\epsilon\left[h_{x,1}(z) + \tilde{h}_{x,1}(z)\right], \tag{3.305}$$

$$\hat{H}_{y,1}(z) = \hat{H}_y^0(z) + 2\epsilon\left[h_{y,1}(z) + \tilde{h}_{y,1}(z)\right]. \tag{3.306}$$

This leads to the following expressions:

$$\begin{aligned}\hat{H}_{x,1}(z) = &(H_m + \epsilon C_1)\, e^{j\theta_0}\left(1+\frac{z}{z_0}\right)^{-\alpha}\\ &+ \epsilon e^{j\theta_0}\left[B'\left(1+\frac{z}{z_0}\right)^{\beta'+j2\alpha''} + B''\left(1+\frac{z}{z_0}\right)^{\beta''+j2\alpha''}\right]\\ &+ \tilde{\epsilon} e^{-j\theta_0}\left[\tilde{B}'\left(1+\frac{z}{z_0}\right)^{\beta'+j2\alpha''} + \tilde{B}''\left(1+\frac{z}{z_0}\right)^{\beta''+j2\alpha''}\right],\end{aligned} \tag{3.307}$$

$$\begin{aligned}\hat{H}_{y,1}(z) = &-j\,(H_m - \epsilon C_1)\, e^{j\theta_0}\left(1+\frac{z}{z_0}\right)^{-\alpha}\\ &+ j\epsilon e^{j\theta_0}\left[B'\left(1+\frac{z}{z_0}\right)^{\beta'+j2\alpha''} + B''\left(1+\frac{z}{z_0}\right)^{\beta''+j2\alpha''}\right]\\ &+ j\tilde{\epsilon} e^{-j\theta_0}\left[\tilde{B}'\left(1+\frac{z}{z_0}\right)^{\beta'+j2\alpha''} + \tilde{B}''\left(1+\frac{z}{z_0}\right)^{\beta''+j2\alpha''}\right].\end{aligned} \tag{3.308}$$

By using the last two formulas, the expressions for surface impedances similar to (2.448)–(2.449) can be derived. These expressions give explicit dependence of surface impedances on perturbation parameters ϵ and $\tilde{\epsilon}$ as well as on the "orientation" angle θ_0.

The discussion presented in this Chapter is an extended and modified version of our previous publications [1] and [2].

REFERENCES

[1] I.D. Mayergoyz, Izvestia VUZ, Electromechanika, No. 6, pp. 583–592, (1967).

[2] I.D. Mayergoyz and G. Friedman, IEEE Transactions on Magnetics, **22**, pp. 259–271, (1986).

CHAPTER 4

Nonlinear Diffusion in Superconductors

4.1 SUPERCONDUCTORS WITH SHARP RESISTIVE TRANSITIONS (THE BEAN MODEL FOR SUPERCONDUCTING HYSTERESIS AND ITS RELATION TO THE PREISACH MODEL)

It is well known that high field (hard) type-II superconductors are not actually ideal conductors of electric current. It is also known that these superconductors exhibit magnetic hysteresis. Finite resistivity and magnetic hysteresis in these superconductors appear because the motion of flux filaments is pinned by defects such as voids, normal inclusions, dislocations, grain boundaries, and compositional variations. This pinning results in the multiplicity of metastable states, which manifest themselves in hysteresis. When the flux filaments depin by thermal activation or because a current density exceeds some critical value, their motion induces an electric field. As a result, superconductors exhibit "current-voltage" laws $E(J)$, which are strongly nonlinear. Thus, the very phenomenon (pinning) that makes type-II superconductors useful in practical applications is also responsible for their magnetic hysteresis and nonzero resistivity.

From the point of view of phenomenological electrodynamics, type-II superconductors can be treated as electrically nonlinear conductors, and the process of electromagnetic field penetration in such superconductors is the process of nonlinear diffusion. Analysis of nonlinear diffusion in type-II superconductors is of practical and theoretical importance because it can be useful for the evaluation of magnetic hysteresis in these superconductors as well as for the study of creep phenomena.

Mathematically, this analysis has many features in common with the analysis of nonlinear diffusion in magnetically nonlinear conductors that has been carried out in the previous chapters. For this reason, our discussion of nonlinear diffusion in superconductors will inevitably contain some repetitions, however, it will be deliberately more concise and it will stress the points that are distinct to superconductors.

We begin with the case of a sharp (ideal) resistive transition shown in Fig. 4.1. This transition implies that persistent currents up to a critical current density J_c are always induced in superconductors. We consider nonlinear diffusion of linearly polarized electromagnetic fields in a lamination (slab) of thickness Δ. At first, it may seem natural to use the scalar nonlinear diffusion equation

$$\frac{\partial^2 E}{\partial z^2} = \mu_0 \frac{\partial J(E)}{\partial t}, \quad J(E) = J_c \operatorname{sign} E \tag{4.1}$$

for the analysis of this nonlinear diffusion. However, since the magnetic field at the slab boundary is usually specified, a simpler way to solve the problem at hand is to base our analysis on the equation

$$\operatorname{curl} \mathbf{H} = \mathbf{J}, \tag{4.2}$$

which in our one-dimensional case can be written as

$$\frac{dH}{dz} = -J_c. \tag{4.3}$$

Since the critical current density J_c is constant, the last equation implies linear profiles of the magnetic field within the slab.

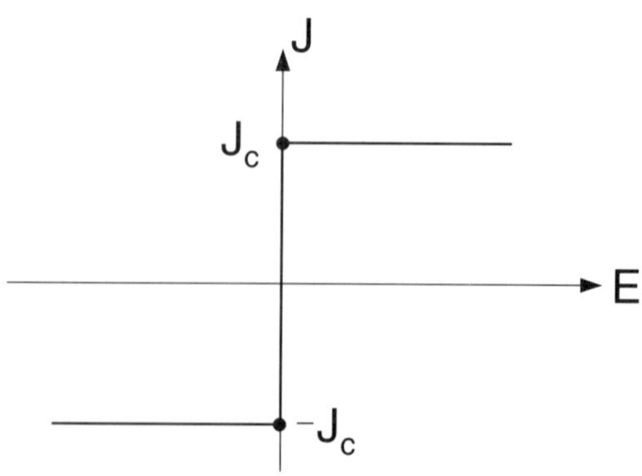

Fig. 4.1

The sharp (ideal) resistive transition (see Fig. 4.1) along with formula (4.3) form the basis for the critical state model for magnetic hysteresis of type-II superconductors. This model was first proposed by C.P. Bean [2], [3], (see also [16]) and then it was further generalized in [12] to take into account a dependence of critical current density on the magnetic field. The critical state type models have been tested experimentally and have proved to be fairly accurate for simple specimen geometries (plane slabs, circular cross-section cylinders). It was also realized that the critical state type models have some intrinsic limitations. First, these models do not take into account actual gradual resistive transitions of type-II superconductors. Second, even under the assumption of ideal resistive transitions, these models lead to explicit analytical results only for very simple specimen geometries.

Next, we shall briefly describe some basic facts concerning the critical state (Bean) model for superconducting hysteresis. Then, we shall demonstrate that the critical state type models are particular cases of the Preisach model of hysteresis discussed in detail in Section 1.9. By using this fact, we shall try to make the case for the Preisach model as an efficient tool for the description of superconducting hysteresis.

Consider a plane superconducting slab subject to an external time-varying magnetic field $H_0(t)$. We will be interested in the B vs. H_0 relation. Here, B is an average magnetic flux density that is defined as

$$B = \frac{\mu_0}{\Delta} \int_{-\frac{\Delta}{2}}^{\frac{\Delta}{2}} H(z) dz, \tag{4.4}$$

and $H(z)$ is the magnetic field within the slab.

In practice, B and H_0 are quantities that are experimentally measured and it is their relation that exhibits hysteresis.

It follows from formulas (4.4) that in order to compute B for any H_0, we have to find a magnetic field profile (magnetic field distribution) within the superconducting slab. This is exactly what we shall do next.

Suppose that no magnetic field was present prior to the instant of time t_0. It is assumed that for times $t > t_0$, the external magnetic field $H_0(t)$ is monotonically increased until it reaches some maximum value H_m. The monotonic increase in the external magnetic field induces persisting electric currents of density J_c. According to formula (4.3), this results in the formation of linear profiles of the magnetic field shown in Fig. 4.2. The corresponding distribution of persisting electric currents is shown in Fig. 4.3. It is easy to see that the instantaneous depth of penetration of the magnetic field is given by

$$z_0(t) = \frac{H_0(t)}{J_c}. \tag{4.5}$$

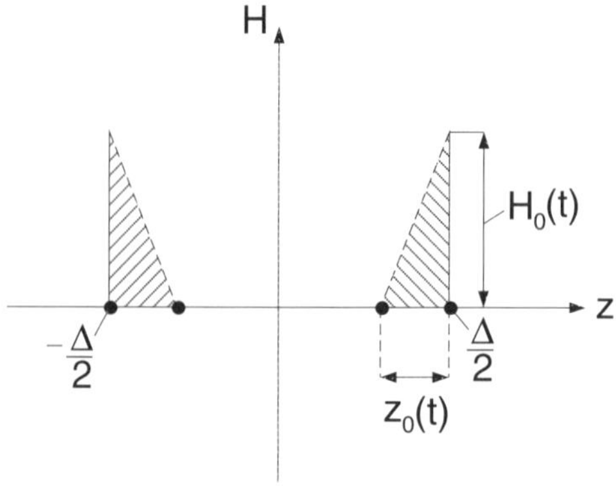

Fig. 4.2

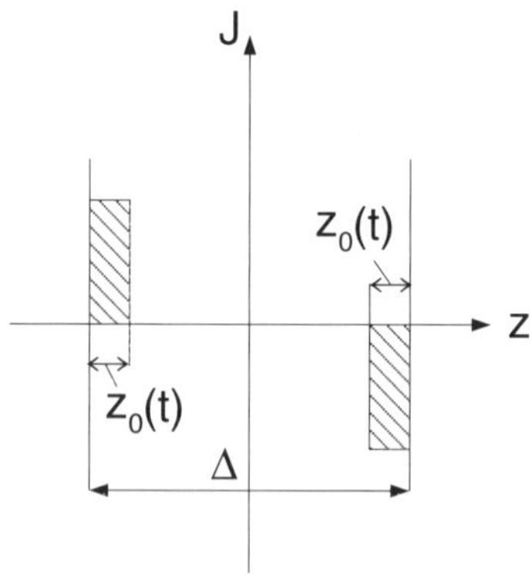

Fig. 4.3

It is also clear that

$$z_0(t) \leq \frac{\Delta}{2}, \tag{4.6}$$

if

$$H_0(t) \leq \frac{J_c \Delta}{2} = H^*. \tag{4.7}$$

By using Fig. 4.2 and formulas (4.4), (4.5), and (4.7), we find the average value of the magnetic flux density:

$$B(t) = \frac{\mu_0 H_0(t) z_0(t)}{\Delta} = \frac{\mu_0 (H_0(t))^2}{2H^*}. \tag{4.8}$$

Suppose now that after achieving the maximum value, H_m, the external magnetic field is monotonically decreased to zero. As soon as the maximum value H_m is achieved, the motion of the previous linear profile is terminated and a new moving linear profile of magnetic field is formed. Due to the previously induced persisting currents, the previous profile stays still and is partially wiped out by the motion of the new profile. The distribution of the magnetic field within the slab at the instant of time when the external magnetic field is reduced to zero is shown in Fig. 4.4. This figure shows that there is nonzero (positive) average magnetic flux density, which is given by:

$$\tilde{B} = \frac{\mu_0 H_m^2}{4H^*} > 0. \tag{4.9}$$

This clearly suggests that the B vs. H_e relation exhibits hysteresis. We next demonstrate the validity of this statement by computing the hysteresis loop for the case of back-and-forth variation of the external magnetic field between $-H_m$ and $+H_m$. For the sake of simplicity of our computations, we shall assume that

$$H_m \leq H^*. \tag{4.10}$$

We first consider the half-period when the external magnetic field is monotonically decreased. A typical magnetic field distribution for this half-period is shown in Fig. 4.5. For the penetration depths z_0 and δ, shown in this figure, we have

$$\delta = \frac{H_m}{J_c}, \qquad z_0(t) = \frac{H_m - H_0(t)}{2J_c}. \tag{4.11}$$

By using Fig. 4.5 and formula (4.11), we find the increment ΔB of the average magnetic flux density:

$$\Delta B = \frac{2\mu_0}{\Delta} \cdot \frac{(H_m - H_0)z_0(t)}{2} = \mu_0 \frac{(H_m - H_0)^2}{4H^*}. \tag{4.12}$$

This leads to the following expression for the average magnetic flux density on the descending branch of the hysteresis loop:

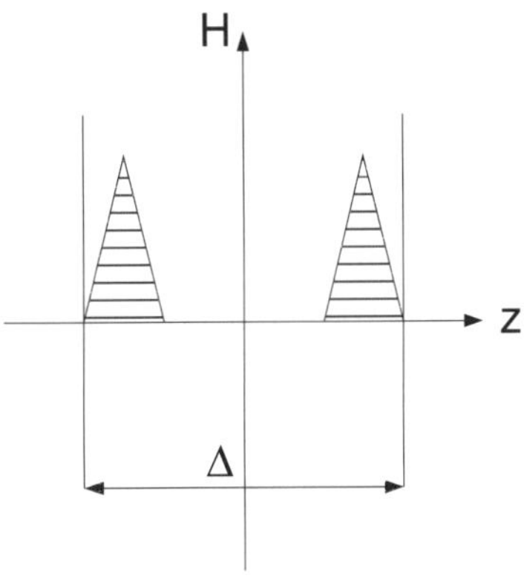

Fig. 4.4

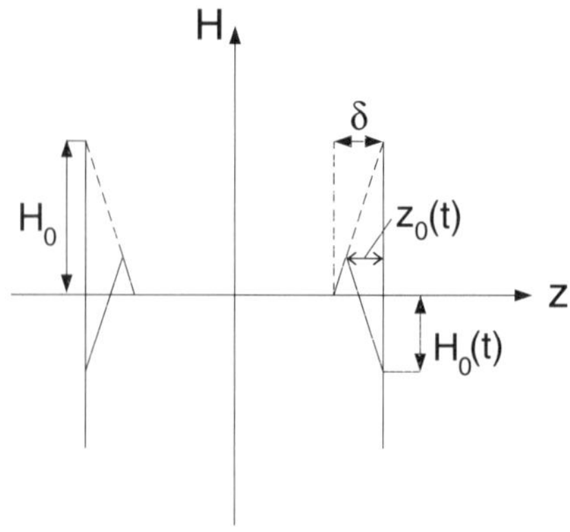

Fig. 4.5

$$B = B_m - \Delta B - \frac{\mu_0 H_m^2}{2H^*} - \frac{\mu_0 (H_m - H_0)^2}{4H^*}. \tag{4.13}$$

Consider now the half-period during which the external magnetic field is monotonically increased from $-H_m$ to $+H_m$. A typical magnetic field distribution for this half-period is shown in Fig. 4.6. By using this figure, as before we find

$$\Delta B = \frac{\mu_0 (H_m + H_0)^2}{4H^*}, \tag{4.14}$$

and

$$B = -B_m + \Delta B = -\frac{\mu_0 H_m^2}{2H^*} + \frac{\mu_0 (H_m + H_0)^2}{4H^*}. \tag{4.15}$$

The expressions (4.13) and (4.15) can be combined into one formula:

$$B = \pm \mu_0 \left[\frac{H_m^2}{2H^*} - \frac{(H_m \mp H_0)^2}{4H^*} \right], \tag{4.16}$$

where the upper signs correspond to the descending branch of the loop, while the lower signs correspond to the ascending branch.

On the basis of the previous discussion, the essence of the Bean model can now be summarized as follows. Each reversal of the magnetic field $H_0(t)$ at the boundary of the superconducting slab results in the formation of a linear profile of the magnetic field. This profile extends inward into the superconductor until another reversal value of the magnetic field at the boundary is reached. At this point, the motion of the previous profile is terminated and a new moving linear profile is formed. Due to the previously induced persisting currents, the previous linear profiles stay still and they represent past history, which leaves its mark upon future values of average

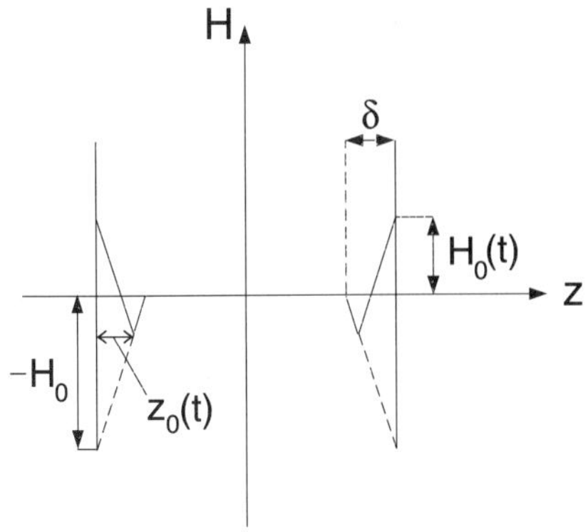

Fig. 4.6

magnetic flux density. These persisting linear profiles of the magnetic field may be partially or completely wiped out by new moving profiles.

Next, we shall establish the connection between the critical state (Bean) model for superconducting hysteresis and the Preisach model described in Section 1.9. To do this, we shall establish that the wiping-out property and congruency property hold for the Bean model. Indeed, a moving linear profile of the magnetic field will wipe out those persisting linear profiles if they correspond to the previous extremum values of $H_0(t)$, which are exceeded by a new extremum value. In this way, the effect of those previous extremum values of $H_0(t)$ on the future average values of magnetic flux density B will be completely eliminated. This means that the wiping-out property holds. It can also be shown that the congruency property of minor loops corresponding to the same reversal values of $H_0(t)$ holds as well. Indeed, consider two variations of external magnetic field $H_0^{(1)}(t)$ and $H_0^{(2)}(t)$. Suppose that these external fields may have different past histories, but starting from some instant of time t_0 they vary back-and-forth between the same reversal values. It is apparent from the previous description of the Bean model that these back-and-forth variations will affect in the *same* way *equal* surface layers of superconductors. Consequently, these variations will result in equal increments of B, which is tantamount to the congruency of the corresponding minor loops.

In the case of generalized critical state models [12], the linear profiles of the magnetic field within superconductors are replaced by curved profiles. However, the creation and motion of these profiles are basically governed by the same rules as in the case of the Bean model. As a result, the previous reasoning holds, and, consequently, the wiping-out property and the congruency property are valid for the generalized critical state models as well. It was established in Chapter 1 that the wiping-out property and congruency property constitute necessary and sufficient conditions for the representation of actual hysteresis nonlinearity by the Preisach model. Thus, we conclude that the Bean model and generalized critical state models are particular cases of the Preisach model:

$$B(t) = \iint_{\alpha \geq \beta} \mu(\alpha, \beta)\hat{\gamma}_{\alpha\beta} H_0(t) d\alpha d\beta. \tag{4.17}$$

It is instructive to find such a function $\mu(\alpha, \beta)$ for which the Preisach model coincides with the Bean model. To do this, consider a "major" loop formed when the external magnetic field varies back-and-forth between $+H_m$ and $-H_m$. Consider first-order transition curves $B_{\alpha\beta}$ attached to the ascending branch of the previously mentioned loop. We recall that the curves $B_{\alpha\beta}$ are formed when, after reaching the value $-H_m$, the external magnetic field is monotonically increased to the value α and subsequently monotonically decreased to the value β. Depending on particular values of α and β, we may have three typical field distributions shown in Fig. 4.7, 4.8, and 4.9. We will use these figures to evaluate the function

$$F(\alpha, \beta) = \frac{1}{2}(B_\alpha - B_{\alpha\beta}). \tag{4.18}$$

Figure 4.7 is valid under the condition:

$$H_m + \alpha \leq 2H^*. \tag{4.19}$$

From this figure we find

$$F(\alpha, \beta) = \frac{\mu_0(\alpha - \beta)^2}{8H^*}. \tag{4.20}$$

Figure 4.8 holds when

$$H_m + \alpha \geq 2H^*, \qquad \alpha - \beta \leq 2H^*. \tag{4.21}$$

By using this figure, we derive

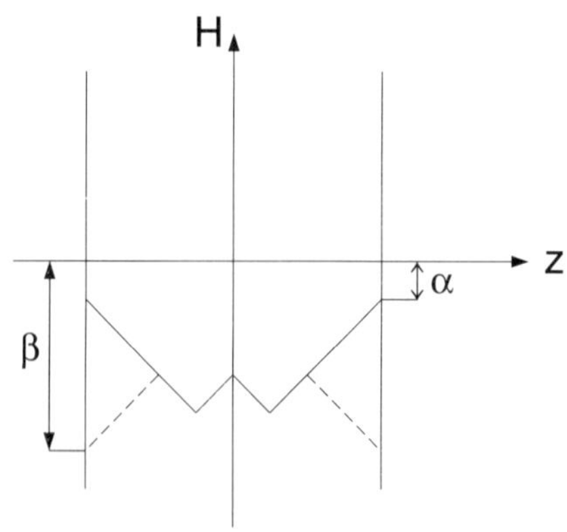

Fig. 4.7

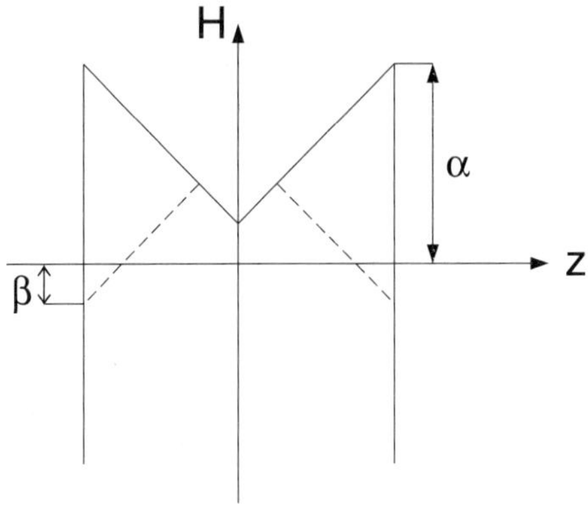

Fig. 4.8

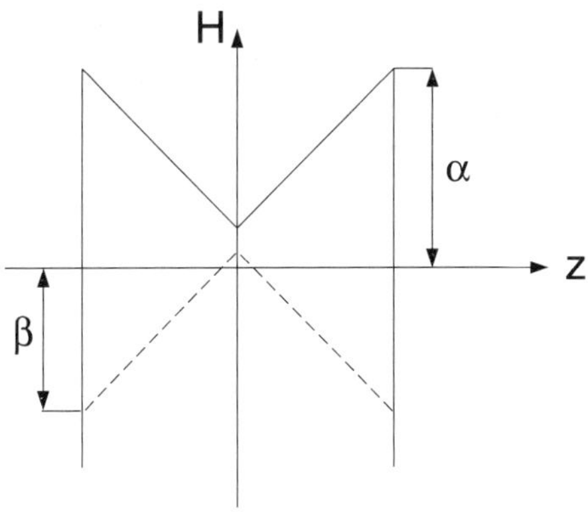

Fig. 4.9

$$F(\alpha, \beta) = \frac{\mu_0(\alpha - \beta)^2}{8H^*}. \tag{4.22}$$

Finally, the distribution of the magnetic field shown in Fig. 4.9 occurs when

$$H_m + \alpha \geq 2H^* \quad \text{and} \quad \alpha - \beta \geq 2H^*. \tag{4.23}$$

From Fig. 4.9, we obtain

$$F(\alpha, \beta) = \frac{\mu_0}{2}(\alpha + \beta - H^*). \tag{4.24}$$

The expressions (4.20), (4.22), and (4.24) can be combined into one formula:

$$F(\alpha, \beta) = \begin{cases} \frac{\mu_0(\alpha-\beta)^2}{8H^*}, & \text{if } 0 < \alpha - \beta \leq 2H^*, \quad |\alpha| \leq H_m, |\beta| \leq H_m, \\ \frac{\mu_0}{2}(\alpha + \beta - H^*), & \text{if } \alpha - \beta \geq 2H^*, \quad |\alpha| \leq H_m, |\beta| \leq H_m. \end{cases} \tag{4.25}$$

By using formula (4.25) as well as the formula (see Chapter 1):

$$\mu(\alpha, \beta) = -\frac{\partial^2 F(\alpha, \beta)}{\partial\alpha\partial\beta}, \tag{4.26}$$

we find

$$\mu(\alpha, \beta) = \begin{cases} \frac{\mu_0}{4H^*}, & \text{if } 0 < \alpha - \beta \leq 2H^*, \quad |\alpha| \leq H_m, \quad |\beta| \leq H_m, \\ 0, & \text{otherwise.} \end{cases} \tag{4.27}$$

The trapezoidal support of $\mu(\alpha, \beta)$ given by (4.27) is illustrated in Fig. 4.10.

Thus, it has been shown that the critical state model for superconducting hysteresis is a very particular case of the Preisach model. This result has been established for one-dimensional flux distributions and specimens of simple shapes (plane slabs). For these cases, explicit analytical expressions for magnetic field distributions within the superconductors are readily available, and they have been instrumental in the discussion just presented.

Next, we demonstrate that the critical state model is a particular case of the Preisach model for specimens of arbitrary shapes and complex flux distributions. For these specimens, analytical machinery for the calculation of magnetic fields within the superconductors does not exist. Nevertheless, it will be shown next that the superconducting hysteresis (as described by the critical state model) still exhibits the wiping-out property and the congruency property of minor hysteresis loops.

To start the discussion, consider a superconducting cylinder of arbitrary cross-section subject to the uniform external field $\mathbf{B}(t)$ whose direction does not change with time and lies in the plane of superconductor cross-section (Fig. 4.11). We will choose this direction as the direction of axis

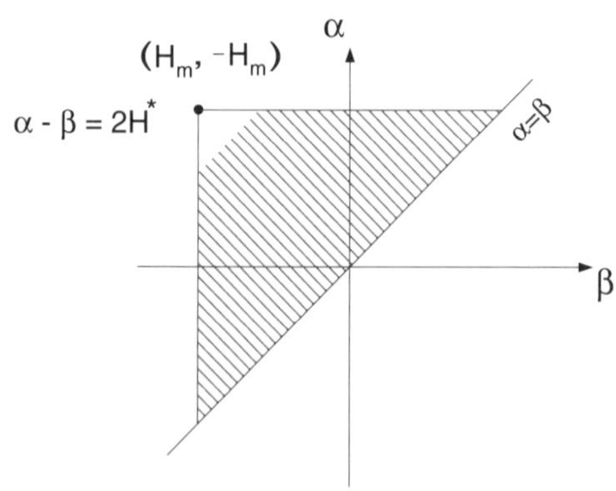

Fig. 4.10

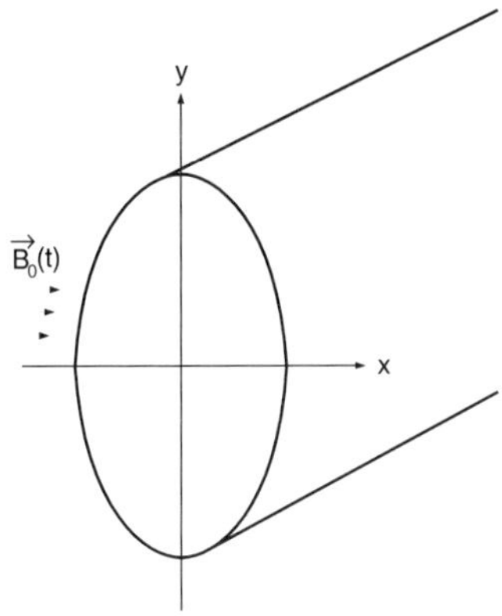

Fig. 4.11

x. As the time-varying flux enters the superconductor, it induces screening (shielding) currents of density $\pm J_c$. The distribution of these superconducting screening currents is such that they create the magnetic field, which at any instant of time completely compensates for the change in the external field $\mathbf{B}_0(t)$. Mathematically, this can be expressed as follows:

$$\delta B_0(t) + B_i(t) = 0. \tag{4.28}$$

Here $\delta B_0(t)$ is the change in $B_0(t)$, while $B_i(t)$ is the field created by superconducting screening currents, and equality (4.28) holds in the region interior to these currents.

It is clear that $\delta B_0(t) \geq 0$ when $B_0(t)$ is monotonically increased, and $\delta B_0(t) \leq 0$ when $B_0(t)$ is monotonically decreased. By using this fact and (4.28), it can be concluded that there is a reversal in the direction (polarity) of superconducting screening currents as $B_0(t)$ goes through its maximum or minimum values.

With these facts in mind, consider how the distribution of superconducting currents is generically modified in time by temporal variations of the external magnetic field. Suppose that, starting from zero value, the external field is monotonically increased until it reaches its maximum value M_1 at some time $t = t_1^+$. This monotonic variation of $B_0(t)$ induces a surface layer of superconducting screening currents. The interior boundary of this current layer extends inwards as $B_0(t)$ is increased [see Fig. 4.12 a], and at any instant of time this boundary is uniquely determined by the instantaneous values of $B_0(t)$. Next, we suppose that this monotonic increase is followed by a monotonic decrease until $B_0(t)$ reaches its minimum value m_1 at some time $t = t_1^-$. For the time being it is assumed that $|m_1| < M_1$. As soon as the maximum value M_1 is achieved, the inward progress of the previous current layer is terminated and a new surface current layer of reversed polarity (direction) is induced [see Fig. 4.12 b]. This new current

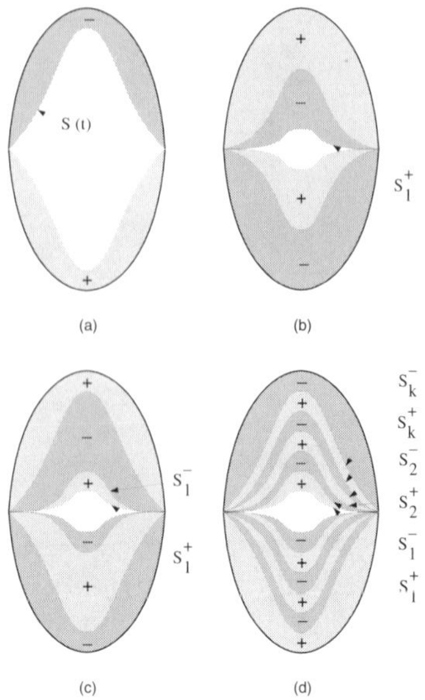

Fig. 4.12

layer creates field $B_i(t)$, which compensates for monotonic decrease in $B_0(t)$ in the region interior to this current layer. For this reason, it is clear that the interior boundary of the new current layer extends inward as $B_0(t)$ is monotonically decreased. It is also clear that this boundary is uniquely determined by the instantaneous value of $\delta B_0(t)$, and, consequently, by the instantaneous value of $B_0(t)$ for any specific (given) value of M_1. Now suppose that the monotonic decrease is followed by a monotonic increase until $B_0(t)$ reaches its new maximum value M_2 at some time $t = t_2^+$. For the time being, it is assumed that $M_2 < |m_1|$. As soon as the minimum value m_1 is achieved, the inward progress of the second layer of superconducting screening currents of reversed polarity is introduced to counteract the monotonic increase of the external field [see Fig. 4.12 c]. This current layer progresses inward until the maximum value M_2 is achieved; at this point the inward progress of the current layer is terminated. As before, the instantaneous position of the interior boundary of this layer is uniquely determined by the instantaneous value of $\delta B_0(t)$, and, consequently, by the instantaneous value of $B_0(t)$ for a specific (given) value of m_1.

Thus, it can be concluded that at any instant of time there exist several (many) layers of persisting superconducting currents [see Fig. 4.12 d]. These persisting currents have opposite polarities (directions) in adjacent layers. The interior boundaries S_k^+ and S_k^- of all layers (except the last one) remain still and they are uniquely determined by the past extremum values M_k and m_k of $B_0(t)$, respectively. The last induced current layer extends inward as the external field changes in time monotonically.

The magnetic moment $\mathbf{M}$ of the superconductor is related to the distribution of the superconducting screening currents as follows:

$$\mathbf{M}(t) = \int_S [\mathbf{r} \times \mathbf{j}(t)] ds, \tag{4.29}$$

where the integration is performed over the superconductor cross-section.

In general, this magnetic moment has x and y components. According to (4.29), these components are given by the expressions

$$M_x(t) = \int_S y j(t) ds, \tag{4.30}$$

$$M_y(t) = -\int_S x j(t) ds. \tag{4.31}$$

It is clear that if the superconductor cross-section is symmetric with respect to the x-axis, then only the x component of the magnetic moment is present. In the absence of this symmetry, two components of the magnetic moments exist.

It is apparent from the previous discussion that the instantaneous values of $M_x(t)$ and $M_y(t)$ depend not only on the current instantaneous value of the external field $B_0(t)$ but on the past extremum values of $B_0(t)$ as well. This is because the overall distribution of persisting superconducting currents depends on the past extrema of $B_0(t)$. Thus, it can be concluded that relationships $M_x(t)$ vs. $B_0(t)$ and $M_y(t)$ vs. $B(t)$ exhibit *discrete* memories that are characteristic and intrinsic for the rate-independent hysteresis. It is worthwhile to note that it is the hysteretic relationship $M_x(t)$ vs. $B_0(t)$ that is typically measured in experiments by using, for instance, a vibrating sample magnetometer (VSM) with a one pair of pickup coils. By using a VSM equipped with two pairs of orthogonal pickup coils, the hysteretic relation between $M_y(t)$ and $B_0(t)$ can be measured as well.

It is important to stress here that the origin of rate independence of superconducting hysteresis can be traced back to the assumption of ideal (sharp) resistive transitions. This connection is especially apparent for superconducting specimens of simple shapes (plane slabs). For such specimens, the explicit and single-valued relations between the increments of the external field and the location of inward boundaries of superconducting layers can be found by resorting only to Ampére's law. It is also clear that there is a strong mathematical similarity and a close formal parallel between the superconducting hysteresis and the eddy current hysteresis discussed in the Section 1.9. The main distinction, however, is that the eddy current hysteresis is rate dependent. This distinction appears because in the case of superconductors we deal with electrically nonlinear conducting media, whereas in the case of eddy currents we deal with magnetically nonlinear conducting media whereas we look at hysteresis between the same variables B and H.

It is clear from the presented discussion that a newly induced and inward-extending layer of superconducting currents will wipe out (replace) some layers of persisting superconducting currents if they correspond to the previous extremum values of $B_0(t)$, which are exceeded by a new extremum value. In this way, the effect of those previous extremum values of $B_0(t)$ on the overall future current distributions will be completely eliminated. According to formulas (4.30) and (4.31), the effect of those past extremum values of the external magnetic field on the magnetic moment will be eliminated as well. This is the wiping-out property of the superconducting hysteresis as described by the critical state model.

Next, we proceed with the discussion of the congruency property. Consider two distinct variations of the external field, $B_0^{(1)}(t)$ and $B_0^{(2)}(t)$. Suppose that these two external fields have different past histories and, consequently, different sequences of local past extreme, $\{M_k^{(1)}, m_k^{(1)}\}$ and $\{M_k^{(2)}, m_k^{(2)}\}$. However, starting from some instant of time they vary back-and-forth between the same reversal values. It is apparent from the description of the critical state model and expressions (4.30) and (4.31) that these two identical back-and-forth variations of the external field will result in the formation of two minor loops for the hysteretic relation $M_x(t)$ vs. $B_0(t)$ [or $M_y(t)$ vs. $B_0(t)$]. It is also apparent from the same description of the critical state model that these two back-and-forth variations of the external field will affect in the *identical* way the *same* surface layers of a superconductor. Unaffected layers of the persistent superconducting currents will be different because of different past histories of $B_0^{(1)}(t)$ and $B_0^{(2)}(t)$. However, according to (4.30) and (4.31), these unaffected layers of persistent currents result in constant-in-time ("background") components of the magnetic moment. Consequently, it can be concluded that the same *incremental* variations of $B_0^{(1)}(t)$ and $B_0^{(2)}(t)$ will result in equal *increments* of M_x (and M_y). This is tantamount to the congruency of the corresponding minor loops. Thus, the congruency property is established for the superconducting hysteresis as described by the critical state model.

It has been previously established that the wiping-out property and the congruency property constitute the necessary and sufficient conditions for the representation of actual hysteresis nonlinearities by the Preisach model. Thus, the description of the superconducting hysteresis by the critical state model is equivalent to the description of the same hysteresis by the Preisach model.

The question can be immediately asked, "What is to be gained from this result?" The answer to this question can be stated as follows: There is no readily available analytical machinery for the calculation of the interior

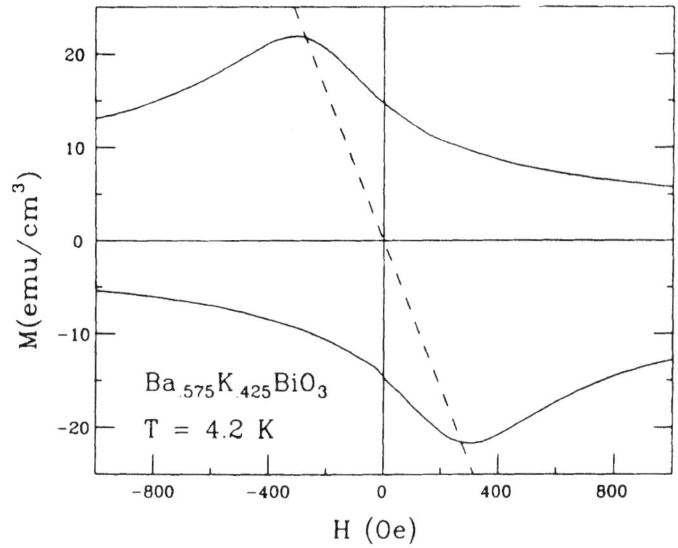

Fig. 4.13

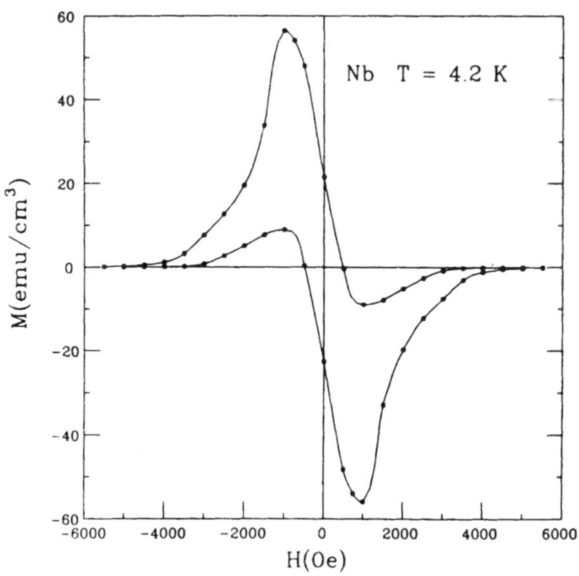

Fig. 4.14

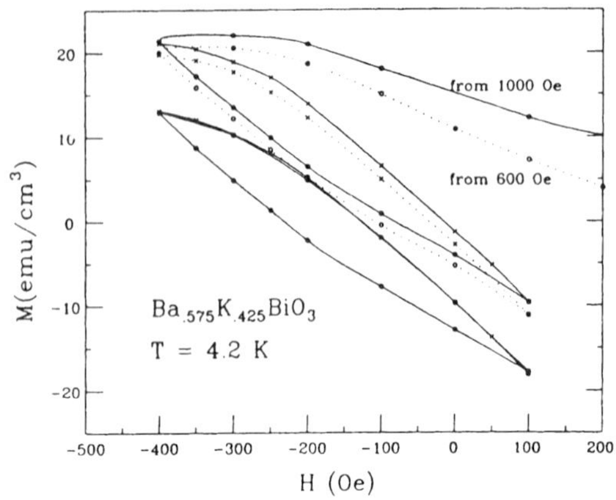

20
10
0
-10
-20
M(emu/cm^3)
from 1000 Oe
from 600 Oe
$Ba_{.575}K_{.425}BiO_3$
T = 4.2 K
-500
-400
-300
-200
-100
0
100
200
H (Oe)

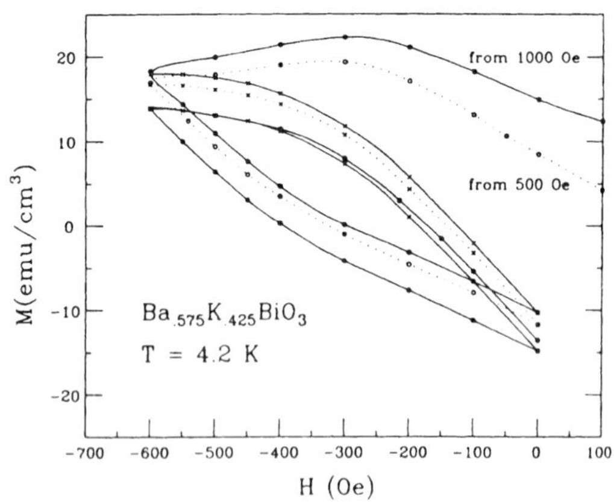

20
10
0
-10
-20
M(emu/cm^3)
from 1000 Oe
from 500 Oe
$Ba_{.575}K_{.425}BiO_3$
T = 4.2 K
-700
-600
-500
-400
-300
-200
-100
0
100
H (Oe)

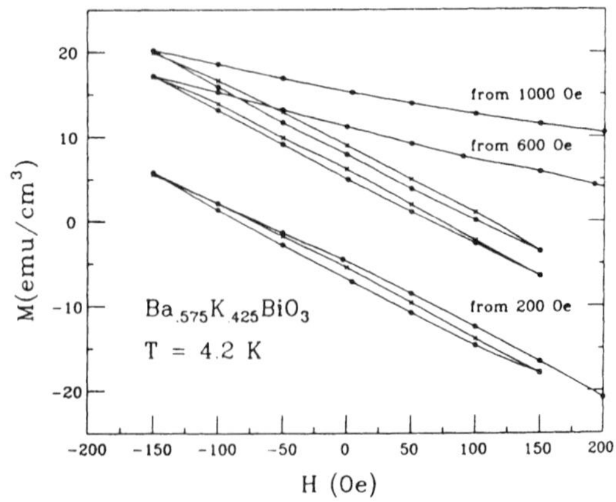

20
10
0
-10
-20
M(emu/cm^3)
from 1000 Oe
from 600 Oe
from 200 Oe
$Ba_{.575}K_{.425}BiO_3$
T = 4.2 K
-200
-150
-100
-50
0
50
100
150
200
H (Oe)

Fig. 4.15

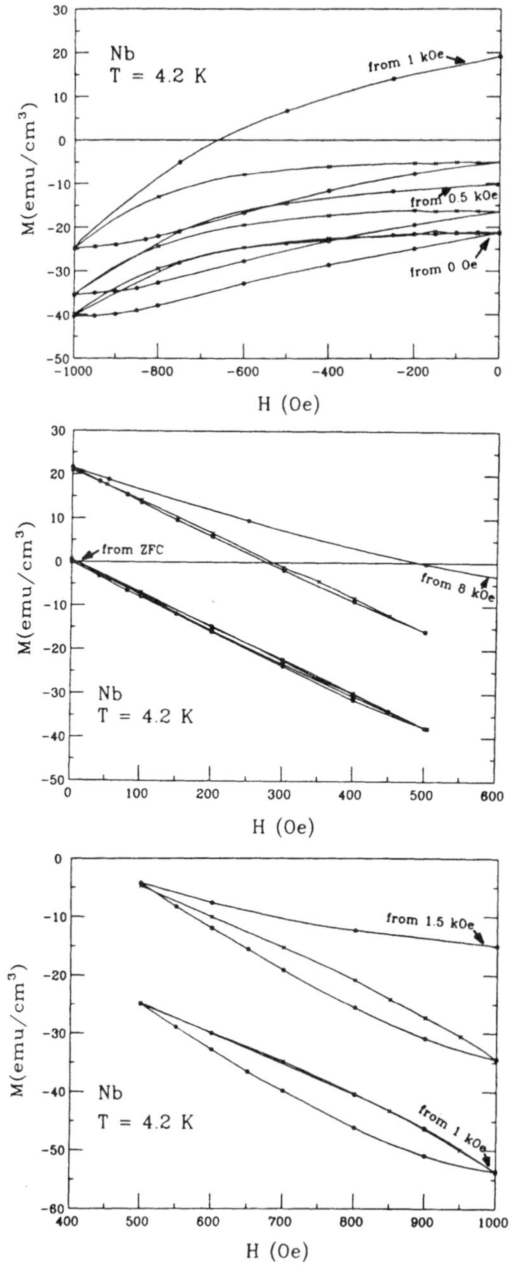
Nb
T = 4.2 K
from 1 kOe
from 0.5 kOe
from 0 Oe
M(emu/cm³)
H (Oe)
from ZFC
from 8 kOe
Nb
T = 4.2 K
M(emu/cm³)
H (Oe)
from 1.5 kOe
from 1 kOe
Nb
T = 4.2 K
M(emu/cm³)
H (Oe)

Fig. 4.16

boundaries of superconducting current layers for specimens of arbitrary shapes. For this reason, the critical state model does not lead to mathematically explicit results. The application of the Preisach model allows one to circumvent these difficulties by using some experimental data. Namely, for any superconducting specimen, the "first-order transition" curves can be measured and used for the identification of the Preisach model for the given specimen. By using these curves, complete prediction of hysteretic behavior of the specimen can be given at least at the same level of accuracy and physical legitimacy as in the case of the critical state model. In particular, cyclic and "ramp" losses can be explicitly expressed in terms of the first-order transition curves (see [17]).

Experimental testing of the congruency and wiping-out properties (and with them the applicability of the Preisach model) has been recently carried out by G. Friedman, L. Liu, and J.S. Kouvel [9]. In the reported experiments, two superconducting samples were used. One was a high-temperature superconductor $Ba_{0.575}K_{0.425}\ B_iO_3$. The other was niobium (Nb). Their major loops are shown in Figs. 4.13 and 4.14, respectively. The wiping-out property was checked by observing closure of minor loops at the end of the first cycle of the magnetic field. To examine the congruency property, minor hysteresis loops were formed by cycling magnetic fields with different prior histories. The results of these experiments for the $BaKB_iO_3$ sample and the Nb sample are shown in Figs. 4.15 and 4.16, respectively. These experimental results suggest that the Preisach model is fairly accurate for these particular superconductors.

As an aside, we point out that the presented discussion can also be useful whenever numerical implementation of the Bean model is attempted. Indeed, the numerical implementation of the Bean model can be appreciably simplified by computing only the "first-order transition" curves and then by using these curves for the prediction of hysteretic behavior for arbitrary piecewise monotonic variations of the external field. The latter is possible because, whenever the congruency and wiping-out properties are valid, all hysteretic data can be compressed (collapsed) into the "first-order transition" curves.

4.2 PREISACH MODEL WITH STOCHASTIC INPUT AS A MODEL FOR CREEP (AFTEREFFECT)*

It is well known that the physical origin of hysteresis is due to the multiplicity of metastable states exhibited by hysteretic materials or

*This section is not conceptually related to the subsequent sections of this chapter.

systems. At equilibrium, large deviations of random (thermal) perturbations may cause a hysteretic system to move from one metastable state to another This may result in gradual (slow) changes of an output variable (magnetization or magnetic flux density). This temporal loss of memory of hysteretic systems is generally referred to as "creep" in the literature on superconductors and as "aftereffect" or "viscosity" in the literature on magnetics. In our discussion, we shall mostly use the term "viscosity."

Traditionally, the phenomenological modeling of hysteresis and viscosity has been pursued along two quite distinct lines. In modeling of hysteresis, the Preisach approach has been prominent, whereas creep, aftereffect, and viscosity have been studied by using thermal activation-type models. It is desirable to develop a uniform approach to the modeling of hysteresis and viscosity. In this section, we shall explore such an approach to the modeling of viscosity. The central idea of this approach is to model random thermal agitations by a stochastic input to the Preisach model.

Consider a deterministic input $u(t)$, which at time $t = 0$ assumes some value u and remains constant thereafter. In a purely deterministic situation, the output $f(t)$ would remain constant for $t \geq 0$ as well. However, in order to model viscosity phenomenon, we assume that some noise described by the stochastic process X_t is superimposed on constant input u. Consequently, the Preisach model is driven by the following random process:

$$x_t = u + X_t, \qquad \overline{X}_t = 0, \tag{4.32}$$

where $\overline{X}_t$ stands for the expected value of X_t.

The output will also be a random process given by

$$f_t = \iint_{\alpha \geq \beta} \mu(\alpha, \beta) \hat{\gamma}_{\alpha\beta} x_t \, d\alpha \, d\beta. \tag{4.33}$$

It is instructive to note that adding noise X_t to the deterministic input $u(t)$ is mathematically equivalent to subtracting the same noise from switching thresholds α and β. This is because

$$\hat{\gamma}_{\alpha - X_t, \beta - X_t} u(t) = \hat{\gamma}_{\alpha\beta}(u(t) + X_t).$$

Imposing noise on the switching thresholds (barriers) may be more transparent from the physical point of view.

To simplify the problem, we shall first model the noise by a discrete-time independent identically distributed (i.i.d.) random process X_n. The continuous-time noise will be discussed afterwards. In the discrete-time case, it can be assumed that process samples remain constant at time intervals $\Delta t = t_{n+1} - t_n$ and undergo monotonic step changes at times t_{n+1}. Accordingly, the Preisach model is driven by the process

$$f_n = \iint_{\alpha \geq \beta} \mu(\alpha, \beta) \hat{\gamma}_{\alpha\beta} x_n d\alpha d\beta. \tag{4.35}$$

We will be interested in the time evolution of the expected value of the output process. Since integration is a linear operation, from (4.35) we derive

$$\overline{f}_n = E\{f_n\} = \iint_{\alpha \geq \beta} \mu(\alpha, \beta) E\{\hat{\gamma}_{\alpha\beta} x_n\} d\alpha d\beta. \tag{4.36}$$

Thus, the whole problem is reduced to the evaluation of the expected value, $E\{\hat{\gamma}_{\alpha\beta} x_n\}$. Since $\hat{\gamma}_{\alpha\beta} x_n$ may assume only two values +1 and −1, we find

$$E\{\hat{\gamma}_{\alpha\beta} x_n\} = P\{\hat{\gamma}_{\alpha\beta} x_n = +1\} - P\{\hat{\gamma}_{\alpha\beta} x_n = -1\}. \tag{4.37}$$

It is also clear that

$$P\{\hat{\gamma}_{\alpha\beta} x_n = +1\} + P\{\hat{\gamma}_{\alpha\beta} x_n = -1\} = 1. \tag{4.38}$$

By introducing the notation

$$P\{\hat{\gamma}_{\alpha\beta} x_n = +1\} = q_{\alpha\beta}(n), \tag{4.39}$$

from (4.37) and (4.38) we obtain

$$E\{\hat{\gamma}_{\alpha\beta} x_n\} = 2q_{\alpha\beta}(n) - 1. \tag{4.40}$$

We next derive the finite difference equation for $q_{\alpha\beta}(n)$. According to the total probability theorem, we have

$$\begin{aligned} &P\{\hat{\gamma}_{\alpha\beta} x_{n+1} = +1\} \\ &\quad = P\{\hat{\gamma}_{\alpha\beta} x_{n+1} = +1 \,|\, \hat{\gamma}_{\alpha\beta} x_n = +1\} P\{\hat{\gamma}_{\alpha\beta} x_n = +1\} \\ &\quad + P\{\hat{\gamma}_{\alpha\beta} x_{n+1} = +1 \,\Big|\, \hat{\gamma}_{\alpha\beta} x_n = -1\} P\{\hat{\gamma}_{\alpha\beta} x_n = -1\}. \end{aligned} \tag{4.41}$$

It is convenient now to introduce the switching probabilities:

$$P_{\alpha\beta}^{++}(n) = P\{\hat{\gamma}_{\alpha\beta} x_{n+1} = +1 \,|\, \hat{\gamma}_{\alpha\beta} x_n = +1\}, \tag{4.42}$$

$$P_{\alpha\beta}^{-+}(n) = P\{\hat{\gamma}_{\alpha\beta} x_{n+1} = +1 \,|\, \hat{\gamma}_{\alpha\beta} x_n = -1\}. \tag{4.43}$$

Similar meanings hold for the switching probabilities $P^{--}_{\alpha\beta}(n)$ and $P^{+-}_{\alpha\beta}(n)$. It is clear that

$$P^{++}_{\alpha\beta}(n) + P^{+-}_{\alpha\beta}(n) = 1, \tag{4.44}$$

$$P^{-+}_{\alpha\beta}(n) + P^{--}_{\alpha\beta}(n) = 1, \tag{4.45}$$

$$P\{\hat{\gamma}_{\alpha\beta}x_n = -1\} = 1 - q_{\alpha\beta}(n). \tag{4.46}$$

By using (4.39) and (4.40)–(4.46), we can transform (4.41) to

$$q_{\alpha\beta}(n+1) = q_{\alpha\beta}(n)[1 - (P^{-+}_{\alpha\beta}(n) + P^{+-}_{\alpha\beta}(n))] + P^{-+}_{\alpha\beta}(n). \tag{4.47}$$

We next proceed to the evaluation of switching probabilities $P^{-+}_{\alpha\beta}(n)$ and $P^{+-}_{\alpha\beta}(n)$. It is clear that these probabilities can also be defined as

$$P^{-+}_{\alpha\beta}(n) = P\{x_{n+1} > \alpha \,|\, \hat{\gamma}_{\alpha\beta}x_n = -1\}, \tag{4.48}$$

$$P^{+-}_{\alpha\beta}(n) = P\{x_{n+1} < \beta \,|\, \hat{\gamma}_{\alpha\beta}x_n = +1\}. \tag{4.49}$$

In general, these probabilities are difficult to evaluate because multidimensional conditional probability density functions are required. However, the problem is significantly simplified if the noise is modeled by the i.i.d. process (independent identically distributed process). In this case we find

$$P^{-+}_{\alpha\beta}(n) = P^{-+}_{\alpha} = P\{x_{n+1} > \alpha\} = \int_{\alpha}^{\infty} \rho(x)dx, \tag{4.50}$$

$$P^{+-}_{\alpha\beta}(n) = P^{+-}_{\beta} = P\{x_{n+1} < \beta\} = \int_{-\infty}^{\beta} \rho(x)dx, \tag{4.51}$$

where $\rho(x)$ is a probability density function.

By using (4.50) and (4.51), the finite difference Eq. (4.47) can be represented as follows:

$$q_{\alpha\beta}(n+1) = r_{\alpha\beta}q_{\alpha\beta}(n) + P^{-+}_{\alpha}, \tag{4.52}$$

where

$$r_{\alpha\beta} = 1 - (P^{-+}_{\alpha} + P^{+-}_{\beta}) = \int_{\beta}^{\alpha} \rho(x)dx. \tag{4.53}$$

If the probability density function $\rho(x)$ is strictly positive, then

$$0 < r_{\alpha\beta} < 1. \tag{4.54}$$

Equation (4.52) is a constant coefficient first-order finite difference equation whose general solution has the form:

$$q_{\alpha\beta}(n) = Ar_{\alpha\beta}^n + B \tag{4.55}$$

By substituting (4.55) into (4.52), we find

$$B = \frac{P_\alpha^{-+}}{1 - r_{\alpha\beta}} = \frac{P_\alpha^{-+}}{P_\alpha^{-+} + P_\beta^{+-}}. \tag{4.56}$$

From the initial condition we obtain

$$q_{\alpha\beta}(0) = A + B, \tag{4.57}$$

where

$$q_{\alpha\beta}(0) = \begin{cases} 1, & \text{if } (\alpha, \beta) \in S^+(0), \\ 0, & \text{if } (\alpha, \beta) \in S^-(0), \end{cases} \tag{4.58}$$

and $S^+(0)$ and $S^-(0)$ are positive and negative sets on the (α, β)-diagram, respectively, at the instant of time $t = 0$. From (4.55)–(4.57) we derive

$$q_{\alpha\beta}(n) = [q_{\alpha\beta}(0) - q_{\alpha\beta}(\infty)]r_{\alpha\beta}^n + \frac{P_\alpha^{-+}}{P_\alpha^{-+} + P_\beta^{+-}}, \tag{4.59}$$

where

$$q_{\alpha\beta}(\infty) = \lim_{n\to\infty} q_{\alpha\beta}(n) = \frac{P_\alpha^{-+}}{P_\alpha^{-+} + P_\beta^{+-}}. \tag{4.60}$$

By substituting (4.59) into (4.40), we obtain

$$E\{\hat{\gamma}_{\alpha\beta}x_n\} = 2[q_{\alpha\beta}(0) - q_{\alpha\beta}(\infty)]r_{\alpha\beta}^n + \frac{P_\alpha^{-+} - P_\beta^{+-}}{P_\alpha^{-+} + P_\beta^{+-}}. \tag{4.61}$$

We next introduce the functions

$$\zeta(\alpha, \beta) = \frac{P_\alpha^{-+} - P_\beta^{+-}}{P_\alpha^{-+} + P_\beta^{+-}}, \tag{4.62}$$

$$\vartheta(\alpha, \beta) = \begin{cases} 1, & \text{if } (\alpha, \beta) \in S^+(0), \\ -1, & \text{if } (\alpha, \beta) \in S^-(0). \end{cases} \tag{4.63}$$

By using (4.58), (4.62), and (4.63), we can transform (4.61) to

$$E\{\hat{\gamma}_{\alpha\beta}x_n\} = [\vartheta(\alpha, \beta) - \zeta(\alpha, \beta)]r_{\alpha\beta}^n + \zeta(\alpha, \beta). \tag{4.64}$$

By substituting (4.64) into (4.36), we finally find

$$\overline{f}_n = \overline{f}_\infty + \iint\limits_{\alpha \geq \beta} \mu(\alpha,\beta)[\vartheta(\alpha,\beta) - \zeta(\alpha,\beta)] r^n_{\alpha\beta} d\alpha d\beta, \tag{4.65}$$

where

$$\overline{f}_\infty = \lim_{n\to\infty} \overline{f}_n = \iint\limits_{\alpha \geq \beta} \mu(\alpha,\beta)\zeta(\alpha,\beta) d\alpha d\beta. \tag{4.66}$$

It is apparent from (4.65) and (4.66) that the limiting expected value of output $\overline{f}_\infty$ does not depend on the history of input variations prior to time $t = 0$. In this respect, the value, $\overline{f}_\infty$, bears some resemblance to the anhysteretic output value. This resemblance is enhanced by the fact that $\overline{f}_\infty = 0$ if the expected value u of x_n is equal to zero. To prove this fact, we assume that $\rho(x) = \rho(X)$ is an even function. In this case, according to (4.50) and (4.51), we find

$$P^{-+}_{\alpha} = P^{+-}_{-\alpha}. \tag{4.67}$$

By using (4.67), it is easy to check that

$$\zeta(\alpha,\beta) = -\zeta(-\beta,-\alpha). \tag{4.68}$$

On the other hand, we recall that

$$\mu(\alpha,\beta) = \mu(-\beta,-\alpha). \tag{4.69}$$

From (4.68) and (4.69), we conclude that $\mu(\alpha,\beta)\zeta(\alpha,\beta)$ is an odd function with respect to the line $\alpha = -\beta$. From the last fact and (4.66), we find that

$$\overline{f}_\infty = 0. \tag{4.70}$$

If u is not equal to zero, then $\overline{f}_\infty$ is not equal to zero as well, and its value depends on noise characteristics. In the case when $\rho(x)$ is Gaussian, $\overline{f}_\infty$ depends only on the variance σ^2 of the noise. This dependence has been numerically analyzed and it has been found that $\overline{f}_\infty$ is not very sensitive to σ^2. This fact is another reason to identify $\overline{f}_\infty$ with anhysteretic output values.

Next, we shall compare the result (4.65) with thermal activation type models for viscosity. For this purpose, we replace discrete time n by continuous time t and rewrite (4.65) as follows:

$$\overline{f}_t = \overline{f}_\infty + \iint\limits_{\alpha \geq \beta} \chi(\alpha,\beta) e^{-\xi(\alpha,\beta)t} d\alpha d\beta, \tag{4.71}$$

where

$$\chi(\alpha,\beta) = \mu(\alpha,\beta)[\vartheta(\alpha,\beta) - \zeta(\alpha,\beta)], \tag{4.72}$$

$$\xi(\alpha,\beta) = -\ln r_{\alpha\beta}. \tag{4.73}$$

In thermal activation type models, it is assumed that metastable equilibrium states of hysteretic systems are separated by energy barriers E_B. It is also assumed that there is a continuum of these energy barriers and it is postulated (with some physical justification) that the viscosity phenomenon is described by the model

$$f(t) = f(\infty) + A\int_0^\infty g(E_B)e^{-\lambda(E_B)t}dE_B, \tag{4.74}$$

where $g(E_B)$ is some density of states,

$$\lambda(E_B) = \lambda_0 e^{\frac{E_B}{kT}}, \tag{4.75}$$

k is Boltzmann's constant, T is the absolute temperature, while A and λ_0 are some constants.

It is clear by inspection that there is formal similarity between our result (4.71) and the thermal activation model (4.74). Actually, our model (4.71) can be reduced to (4.74) in the particular case when only symmetrical loops (operators) $\hat{\gamma}_{\alpha,-\alpha}$ are used in the Preisach model. In this case

$$\mu(\alpha,\beta) = \theta(\alpha,\beta)\delta(\alpha+\beta), \tag{4.76}$$

where $\delta(\alpha+\beta)$ is the Dirac delta function.

By substituting (4.76) into (4.72) and (4.71), after simple transformations we can represent (4.71) as

$$\overline{f}_t = \overline{f}_\infty + \int_0^\infty \tilde{\chi}(\alpha)e^{-\tilde{\xi}(\alpha)t}d\alpha. \tag{4.77}$$

Now, by using the change of variables:

$$\tilde{\xi}(\alpha) = \lambda(E_B), \qquad \alpha = \tilde{\xi}^{-1}(\lambda(E_B)), \tag{4.78}$$

the expression (4.77) can be reduced to (4.74). This shows that the Preisach model of viscosity (4.71) is reduced to the thermal activation model in a very particular case. This case occurs when only symmetrical rectangular loops are used in the Preisach model. Since it is generally believed that nonsymmetrical loops in the Preisach model account for "particle interactions," the last reduction is consistent with the generally held opinion that the thermal activation model (4.74) is a "noninteracting particle" model.

In comparison with the thermal activating type models of the form (4.74), the model (4.65) (or (4.71)) has certain attractive features. First, the model (4.65) explicitly accounts for the specific hysteretic nature of the system as well as for specific input histories. Second, stochastics characteristics of thermal noise explicitly appear in the model (4.65), whereas the thermal activation type models are formulated in purely deterministic terms. Third, thermal activation type models (4.74) are intrinsically scalar models, whereas the model (4.65) can be generalized to the vector case (see, for instance, [10]).

The model (4.65) has been experimentally tested by C.E. Korman and P. Rugkwamsook [13] for $\gamma - Fe_3O_2$ magnetic recording material. Thermal viscosity for this material (and magnetic materials in general) has many features in common with thermal creep for superconductors. By using a vibrating sample magnetometer (VSM), first-order reversal curves were measured. These experimental data were used for the identification of the Preisach model (determination of $\mu(\alpha, \beta)$). It was also found from these data that the coercivity of the sample was approximately 290 Oe. Then viscosity measurements were performed under the conditions that the sample was first brought to positive saturation and afterward the applied field was reversed, held constant at some fixed value, and gradual decay of magnetization was observed. The viscosity measurements were performed at constant fields in the range between -800 Oe and 100 Oe. These measurements are demonstrated in Figs. 4.17 and 4.18.

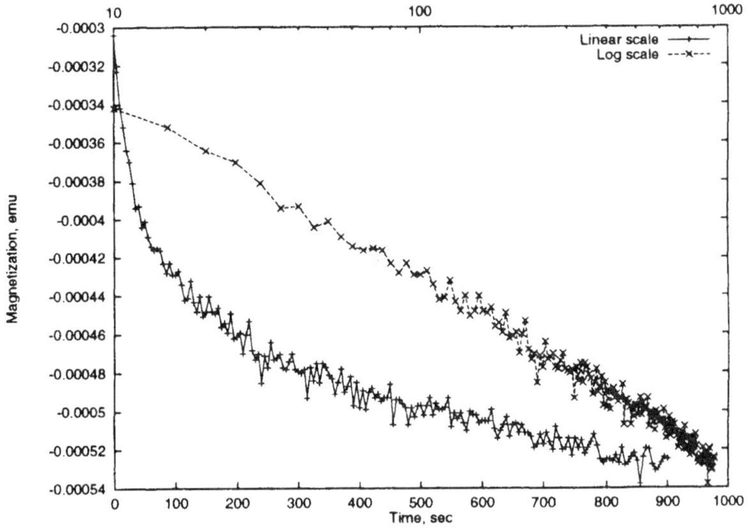

Fig. 4.17. (©1997 IEEE)

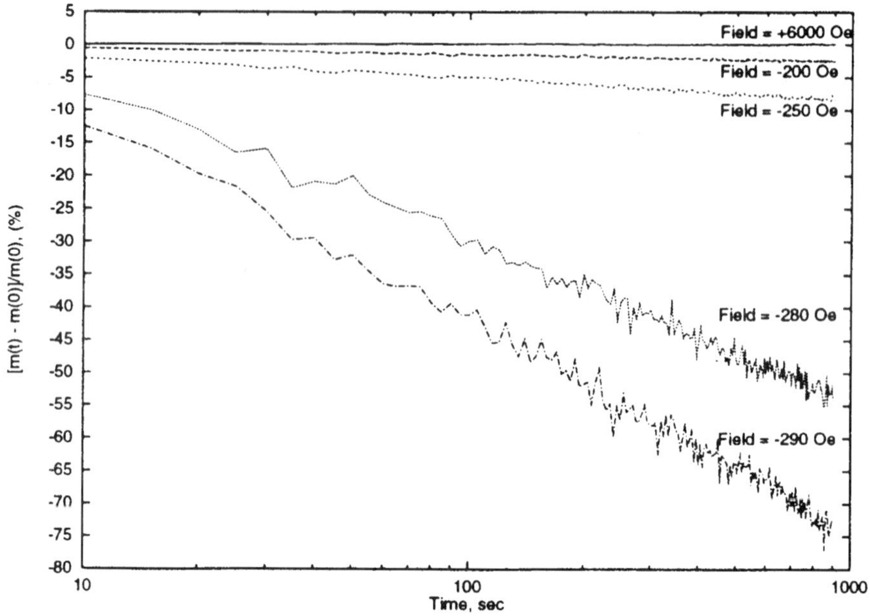

Fig. 4.18. (©1997 IEEE)

Figure 4.17 shows temporal changes in magnetization in the linear and logarithmic time scales when the magnetic field was held constant at the value of 290 *Oe*. On the logarithmic time scale temportal variations of average magnetization can be approximated by a straight line for sufficiently long times. This supports the following well-known **intermediate** asymptotics:

$$\overline{f}_t \approx -S(u) \ln t + c. \tag{4.79}$$

Figure 4.18 shows temporal changes in magnetization on the logarithmic time-scale measured at various fixed values of the magnetic field. It is apparent from this figure that slope S reaches its maximum near the coercivity 290 *Oe*.

The described viscosity experiments were numerically simulated by using model (4.65) and experimentally measured first-order transition curves. The noise was assumed to be Gaussian and calculations we performed for different values of variances σ^2. The calculations revealed the same "$\ln t$"-intermediate asymptotics as those given by formula (4.79). The slopes of these asymptotics were computed by means of the formula

$$S = -\frac{\Delta M(t)}{\Delta \ln t} = -\frac{M_{n'} - M_n}{\ln \frac{n'}{n}} \tag{4.80}$$

then they were normalized and it was observed that the normalized curves $S(u)/S_{\max}$ computed for different variances practically collapsed into one curve. This computed universal curve is almost identical to the normalized curve $S(u)/S_{\max}$ found from the experimental data previously described (see Fig. 4.19). This suggests that the variance σ^2 of the noise can be found by matching $S_{\max}$, and then the model (4.65) predicts the same curve $S(u)$ as observed in experiments. It would be interesting to examine the existence of the collapse of the normalized curves $S(u)/S_{\max}$ experimentally observed for different temperatures.

Next, we turn to the discussion of continuous-time noise. From the mathematical point of view, this makes the problem quite complicated. It is shown next that these difficulties can be largely overcome by using the mathematical machinery of the "exit problem."

The noise X_t in equations (4.32) and (4.33) will be modeled by a (continuous time and continuous samples) diffusion process, which is a solution to the Ito stochastic differential equation ([11]):

$$dX_t = b(X_t)dt + \sigma(X_t)dW_t. \tag{4.81}$$

In this equation, W_t is the Wiener process, and its formal time derivative is the white noise. Formula (4.81) can be construed as a generic equation for dynamical systems driven by the white noise, and trajectories of such dynamical systems can be viewed as samples of stochastic diffusion

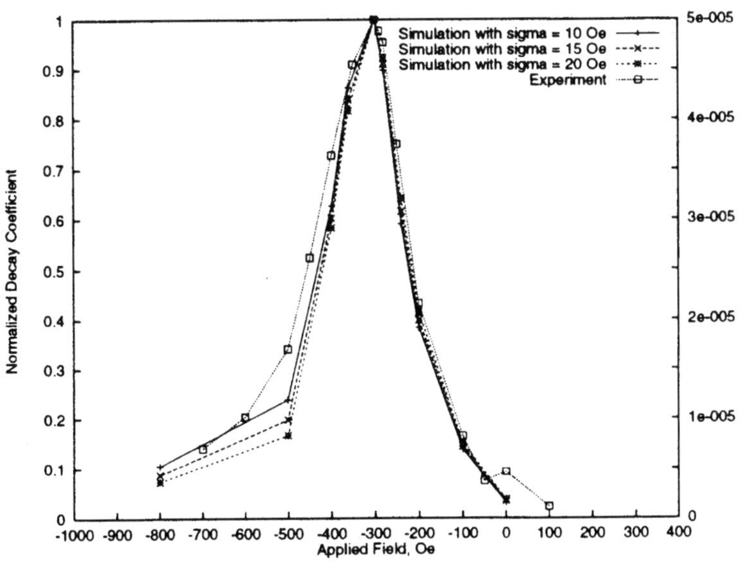

Fig. 4.19. (©1997 IEEE)

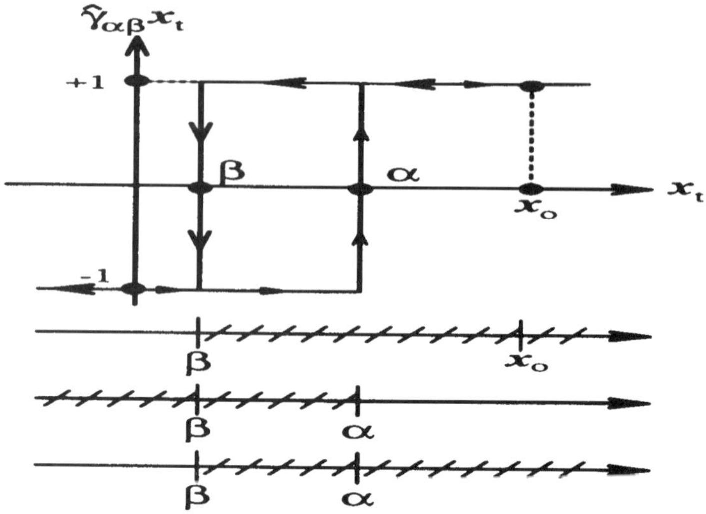

Fig. 4.20

processes. From the purely mathematical point of view, the Ito stochastic differential equation generates complicated diffusion processes by using the Wiener process, which is one of the simplest and most studied stochastic processes.

Now we shall return to Eq. (4.33). Since integration is a linear operation, from (4.33) we derive:

$$\overline{f}_t = \iint_{\alpha \geq \beta} \mu(\alpha, \beta) E\{\hat{\gamma}_{\alpha,\beta} X_t\} d\alpha d\beta. \tag{4.82}$$

Thus, the problem is reduced to the evaluation of the expected value, $E\{\hat{\gamma}_{\alpha,\beta} X_t\}$.

Let

$$q_{\alpha,\beta}(t) = \text{Prob}\{\hat{\gamma}_{\alpha,\beta} x_t = +1\}. \tag{4.83}$$

Since $\hat{\gamma}_{\alpha,\beta} x_t$ may assume only two values +1 and −1, we find:

$$E\{\hat{\gamma}_{\alpha,\beta} x_t\} = 2q_{\alpha,\beta}(t) - 1. \tag{4.84}$$

In this way, the problem is reduced to the calculation of $q_{\alpha,\beta}(t)$. The last quantity can be expressed in terms of switching probabilities $P_k^+(t)$ and $P_k^-(t)$, which are defined as follows:

$$P_k^+(t) = \text{Prob}\left\{ \begin{array}{c} k \text{ switchings of } \hat{\gamma}_{\alpha,\beta} \text{ during} \\ \text{time interval } (0,t) \mid \hat{\gamma}_{\alpha,\beta}\, x_0 = +1 \end{array} \right\} \tag{4.85}$$

$$P_k^-(t) = \text{Prob}\left\{ \begin{array}{c} k \text{ switchings of } \hat{\gamma}_{\alpha,\beta} \text{ during} \\ \text{time interval } (0,t) \mid \hat{\gamma}_{\alpha,\beta}\, x_0 = -1. \end{array} \right\} \tag{4.86}$$

By using these switching probabilities, we derive:

$$q_{\alpha,\beta}(t) = \begin{cases} \sum_{k=0}^{\infty} P_{2k}^+(t), & \text{if } \hat{\gamma}_{\alpha,\beta} x_0 = +1, \\ \sum_{k=0}^{\infty} P_{2k+1}^-(t), & \text{if } \hat{\gamma}_{\alpha,\beta} x_0 = -1. \end{cases} \tag{4.87}$$

The last expression is valid because occurrences of different numbers of switchings are nonintersecting (disjoint) events.

Next, we shall discuss the mechanism of switching. It is clear from Fig. 4.20 that the first switching occurs at the moment when the stochastic process x_t starting from the point x_0 exits the semi-infinite interval (β, ∞). Then, the second switching occurs at the moment when the process x_t starting from the point $x = \beta$ exits the semi-infinite interval $(-\infty, \alpha)$. The third switching takes place at the moment when the process x_t starting from the point $x = \alpha$ exits the semi-infinite interval (β, ∞). It is apparent that the mechanism of all subsequent even switchings is identical to the mechanism of the second switching, while all subsequent odd switchings occur in the same manner as the third switching. Thus, switchings of rectangular loops $\hat{\gamma}_{\alpha,\beta}$ are closely related to the exit problem for stochastic processes. This problem is one of the most studied problems in the theory of diffusion processes and the mathematical machinery developed for the solution of this problem will be utilized in the calculation of probabilities $P_k^{\pm}(t)$.

The exit problems just described can be characterized by exit times $\tau_x^{\pm}$, which are random variables. In the above notation for the exit times, subscript "x" means that process x_t starts from point x, while superscripts "$\pm$" mean that upward and downward switchings, respectively, occur at these exit times. Next, we introduce the functions

$$v^{\pm}(t,x) = \text{Prob}\{\tau_x^{\pm} \geq t\}, \tag{4.88}$$

$$V^{\pm}(t,x) = e(t) - v^{\pm}(t,x), \tag{4.89}$$

where $e(t)$ is a unit step-function.

It is clear that

$$V^{\pm}(t,x) = \text{Prob}\{\tau_x^{\pm} \leq t\}, \tag{4.90}$$

which means that $V^{\pm}(t,x)$ has the meaning of a cumulative distribution function for the random variable $\tau_x^{\pm}$. This, in turn, implies that

$$\rho^{\pm}(t,x) = \frac{\partial V^{\pm}(t,x)}{\partial t} \tag{4.91}$$

is the probability density function for the random variable $\tau_x^{\pm}$.

It is apparent from (4.89)–(4.91) that $\rho^{\pm}(t,x)$ can be easily computed if $v^{\pm}(t,x)$ are somehow found. It turns out (and this is a well-known result from the theory of stochastic processes) that $v^{\pm}(t,x)$ is the solution to the following initial boundary value problem for the backward Kolmogorov equation:

$$\frac{\partial v^{\pm}}{\partial t} = \frac{\sigma^2(x)}{2}\frac{\partial^2 v^{\pm}}{\partial x^2} + b(x)\frac{\partial v^{\pm}}{\partial x}, \tag{4.92}$$

$$v(0,x) = 1, \qquad v(t,c^{\pm}) = 0, \tag{4.93}$$

where $c^{\pm}$ are the exit points for the process X_t, which are equal to $\alpha - u_0$ and $\beta - u_0$, respectively.

Next, we shall show that switching probabilities $P_k^{\pm}(t)$ can be expressed in terms of $v^{\pm}(t)$ and $\rho^{\pm}(t)$. Note that, according to (4.89)–(4.91), $\rho^{\pm}(t)$ are related to $v^{\pm}(t)$ as follows:

$$\rho^{\pm}(t) = \frac{\partial}{\partial t}[e(t) - v^{\pm}(t)]. \tag{4.94}$$

It is clear from the very definition of $v^{\pm}(t,x)$ that

$$P_o^{\pm}(t) = v^{\pm}(t,0). \tag{4.95}$$

It is apparent from Fig. 4.21 that the occurrence of exactly one downward switching is the union of the following disjoint elementary events: downward switching occurs in the time interval $(\lambda, \lambda + d\lambda)$ and then no upward switching occurs up to the time t. Due to the strong Markov property of X_t, the probability of this elementary event is given by

$$\rho^{-}(\lambda,0)v^{+}(t-\lambda, \beta - u_o)d\lambda. \tag{4.96}$$

Now the probability $P_1^{+}(t)$ of exactly one downward switching can be found by integrating (4.96) from 0 to t:

$$P_1^{+}(t) = \int_0^t \rho^{-}(\lambda,0)v^{+}(t-\lambda, \beta - u_o)d\lambda. \tag{4.97}$$

In other words, $P_1^{+}(t)$ is the convolution of $\rho^{-}(t,0)$ and $v^{+}(t, \beta - u_o)$:

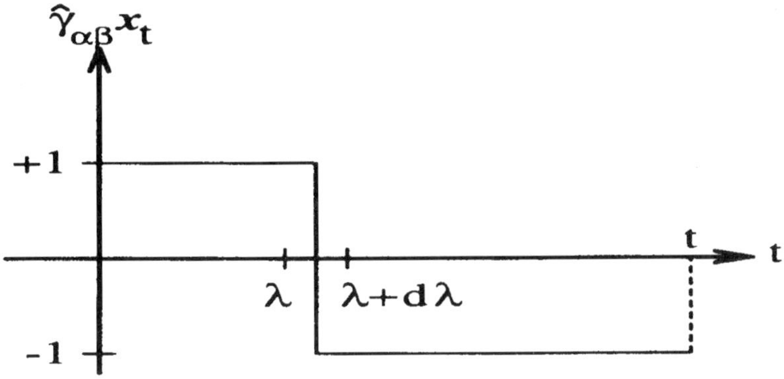

Fig. 4.21

$$P_1^+(t) = \rho^-(t,0) * v^+(t, \beta - u_o). \tag{4.98}$$

By using similar reasoning, we can derive

$$P_1^-(t) = \rho^+(t,0) * v^-(t, \alpha - u_o). \tag{4.99}$$

Next, consider the probability $P_2^+(t)$ of the occurrence of exactly two switchings starting from the initial state $\hat{\gamma}_{\alpha\beta}x_o = 1$. According to Fig. 4.22, this occurrence can be considered as the union of the following disjoint elementary events: downward switching occurs in the time interval $(\lambda, \lambda + d\lambda)$ and then exactly one upward switching occurs up to the time t. The probability of these elementary events is given by

$$\rho^-(\lambda, 0) P_1^-(t - \lambda) d\lambda. \tag{4.100}$$

Now, by integrating (4.100), we find

$$P_2^+(t) = \int_0^t \rho^-(\lambda, 0) P_1^-(t, \lambda) d\lambda. \tag{4.101}$$

From (4.99) and (4.101) we obtain

$$P_2^+(t) = \rho^-(t,0) * \rho^+(t, \beta - u_o) * v^-(t, \alpha - u_o). \tag{4.102}$$

By using the same line of reasoning, we derive

$$P_2^-(t) = \rho^+(t,0) * \rho^-(t, \alpha - u_o) * v^+(t, \beta - u_o). \tag{4.103}$$

For the sake of conciseness, we introduce the notations:

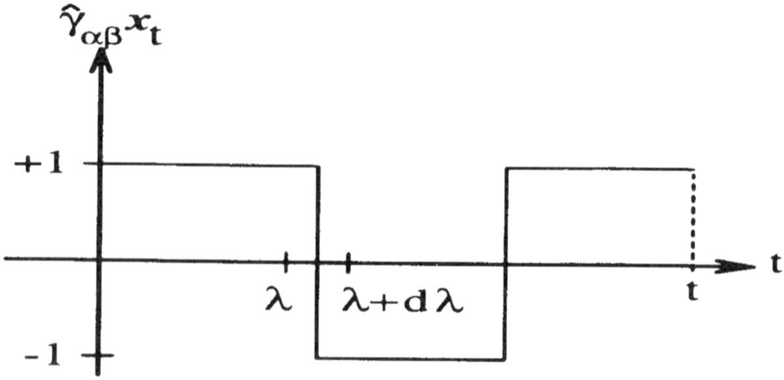

Fig. 4.22

$$\rho^{\pm}(t,0) = \rho_o^{\pm}(t), \quad \rho^{+}(t,\beta - u_o) = \rho^{+}(t), \quad \rho^{-}(t,\alpha - u_o) = \rho^{-}(t), \tag{4.104}$$

$$v^{\pm}(t,0) = v_o^{\pm}(t), \quad v^{+}(t,\beta - u_o) = v^{+}(t), \quad v^{-}(t,\alpha - u_o) = v^{-}(t). \tag{4.105}$$

Now, by using the same line of reasoning as before and the induction argument, we can easily derive the following expressions for the switching probabilities:

$$P_{2k}^{+}(t,u_o) = \rho_o^{-}(t) * \rho^{+}(t) * \overbrace{\rho^{-}(t) * \rho^{+}(t) * \cdots * \rho^{-}(t) * \rho^{+}(t)}^{2k-2 \text{ terms}} \ldots * v^{-}(t), \tag{4.106}$$

$$P_{2k+1}^{-}(t,u_o) = \rho_o^{+}(t) * \overbrace{\rho^{-}(t) * \rho^{+}(t) * \cdots * \rho^{-}(t) * \rho^{+}(t)}^{2k \text{ terms}} * v^{-}(t). \tag{4.107}$$

By substituting (4.106) and (4.107) into (4.87), we obtain the expression for $q_{\alpha\beta}(t)$ in terms of infinite series of iterated convolutions. These series can be reduced to geometric ones by employing Laplace transforms:

$$\tilde{\rho}(s) = \int_0^{\infty} \rho(t) e^{-st} dt, \quad (\text{Re } s > 0), \tag{4.108}$$

$$\tilde{v}(s) = \int_0^{\infty} v(t) e^{-st} dt. \tag{4.109}$$

It is clear that

$$|\, \tilde{\rho}(s) \,| < 1. \tag{4.110}$$

By using these Laplace transforms, from (4.95), (4.106), and (4.107) we obtain

$$\tilde{P}_o^{\pm}(s) = \tilde{v}_o^{\pm}(s), \tag{4.111}$$

$$\tilde{P}^+_{2k}(s) = \tilde{\rho}^-_o(s)\tilde{\rho}^+(s)\tilde{v}^-(s)[\tilde{\rho}^-(s)\tilde{\rho}^+(s)]^{k-1}, \tag{4.112}$$

$$\tilde{P}^-_{2k+1}(s) = \tilde{\rho}^+_o(s)\tilde{v}^-(s)[\tilde{\rho}^-(s)\tilde{\rho}^+(s)]^k. \tag{4.113}$$

From (4.113) and (4.87), we derive

$$\tilde{q}_{\alpha\beta}(s) = \frac{\tilde{\rho}^+_o(s)\tilde{v}^-(s)}{1-\tilde{\rho}^-(s)\tilde{\rho}^+(s)}, \quad \text{if } \hat{\gamma}_{\alpha\beta}x_o = -1. \tag{4.114}$$

A similar expression can be derived for the case $\hat{\gamma}_{\alpha\beta}x_o = +1$.

According to (4.94),

$$\tilde{\rho}^{\pm}(s) = 1 - s\tilde{v}^{\pm}(s). \tag{4.115}$$

Thus, the problem of computing $\tilde{q}_{\alpha\beta}$ is reduced to the problem of determining $\tilde{v}^{\pm}(s)$. This can be accomplished by using the initial boundary value problem (4.92)–(4.93). The complexity of this task will depend on the nature of the stochastic process X_t, which models the noise in hysteretic systems. It is natural to require that the stochastic process that models the noise must be a stationary Gaussian Markov process. According to the Doob theorem [11], the only process that satisfies these requirements is the Ornstein-Uhlenbeck process. This process is the solution to the following Ito stochastic differential equation:

$$dX_t = -bX_t dt + \sigma dW_t, \tag{4.116}$$

where $1/b$ has the meaning of the correlation time. (This means that X_t and $X_{t'}$ are only significantly correlated if $\mid t - t' \mid \leq 1/b$).

The backward Kolmogorov equation for the Ornstein-Uhlenbeck process has the form:

$$\frac{\partial v^{\pm}}{\partial t} = \frac{\sigma^2}{2}\frac{\partial^2 v^{\pm}}{\partial x^2} - bx\frac{\partial v^{\pm}}{\partial x}. \tag{4.117}$$

This equation should be considered jointly with initial and boundary conditions (4.93). By applying the Laplace transform to (4.117) and (4.93), we arrive at the following boundary value problem for $\tilde{v}^{\pm}(s)$:

$$\frac{\sigma^2}{2}\frac{d^2\tilde{v}^{\pm}(s,x)}{dx^2} - bx\frac{d\tilde{v}^{\pm}(s,x)}{dx} - s\tilde{v}^{\pm}(s,x) = -1, \tag{4.118}$$

$$\tilde{v}^{\pm}(s,c^{\pm}) = 0, \quad \tilde{v}^{\pm}(s,\infty) = 1/s. \tag{4.119}$$

The solution to the boundary value problem (4.118)–(4.119) can be written in the form:

$$\tilde{v}^{\pm}(s,x) = \frac{1}{s}\left(1 - e^{[x^2-(c^{\pm})^2]/4\lambda^2}\frac{\mathcal{D}_{-s/b}(x/\lambda)}{\mathcal{D}_{-s/b}(c^{\pm}/\lambda)}\right), \tag{4.120}$$

where $\mathcal{D}_{-s/b}(x/\lambda)$ are parabolic cylinder functions, while

$$\lambda = \sigma/\sqrt{2b}.$$

Expressions (4.115), (4.114), and (4.120) jointly with (4.82) and (4.84) outline the main steps of computing $\overline{f}_t$.

It is apparent that the case of continuous-time noise is computationally more expensive. Some sample examples of computations for this case can be found in [14], [15].

4.3 NONLINEAR DIFFUSION IN SUPERCONDUCTORS WITH GRADUAL RESISTIVE TRANSITIONS (LINEAR POLARIZATION)

In the previous sections of this chapter, nonlinear diffusion in superconductors with sharp (ideal) resistive transitions was discussed. However, actual resistive transitions are gradual and it is customary to describe them by the following power law:

$$E = \left(\frac{J}{k}\right)^n \cdot \operatorname{sign} J, \qquad (n > 1), \tag{4.121}$$

where E is electric field, J is current density, and k is some parameter that coordinates the dimensions of both sides in expression (4.121).

The exponent "n" is a measure of the sharpness of the resistive transition and it may vary in the range 7–1000. Initially, the power law was regarded only as a reasonable empirical description of the resistive transition. Recently, there has been a considerable research effort to justify this law theoretically. As a result, models based on Josephson-junction coupling [27], "sausaging" [8], and spatial distribution of critical current [28] have been proposed. However, the most plausible explanation for the power law came from the thermal activation theory [6], [7], [29]. According to this theory, the electric field E induced by thermally activated drift of flux filaments (vortices) can be written in the form of the Arrhenius law:

$$E = E_c \exp\left[-U(J)/kT\right], \tag{4.122}$$

where $U(J)$ is a current-dependent flux creep potential barrier, which supposedly vanishes at some critical current J_c; E_c is an electric field at $J = J_c$. If a logarithmic dependence of activation barrier U on current J

$$U(J) = U_c \ln\left(\frac{J_c}{J}\right) \tag{4.123}$$

is assumed, then from formula (4.122) we readily obtain the power law (4.121) for the resistive transition.

Whatever the theoretical rationale may be behind the power law, this law has been observed in numerous experiments. For this reason, in our subsequent discussions, this law will be used as a constitutive relation for hard superconductors. By using this constitutive relation and Maxwell's equations, it is easy to show that nonlinear diffusion of linearly polarized electromagnetic fields for monotonically increasing boundary conditions is described by the following nonlinear partial differential equation:

$$\frac{\partial^2 J^n}{\partial z^2} = \mu_0 k^n \frac{\partial J}{\partial t}. \tag{4.124}$$

We shall first consider the solution of this equation for the following boundary and initial conditions:

$$J(0,t) = ct^p, \qquad (t \geq 0, p > 0), \tag{4.125}$$

$$J(z,0) = 0 \qquad (z > 0). \tag{4.126}$$

It may seem at first that these boundary conditions are of a very specific nature. However, it can be remarked that these boundary conditions do describe a wide class of monotonically increasing functions as p varies from 0 to ∞ (see Fig. 4.23). It will be shown below that for all these boundary conditions the profile of electric current density as a function of z remains practically the same. This observation will suggest using the same profile of electric current density for arbitrary monotonically increasing boundary conditions. This will lead to very simple analytical solutions.

The initial boundary value problem (4.124)–(4.126) is mathematically identical to the model problem discussed in Sections 1.3 and 1.4. As a result, the solution of the initial boundary value problem (4.124)–(4.126) closely parallels the solution of the model problem mentioned above. For this reason, we shall only outline the main steps of this solution.

The initial boundary value problem (4.124)–(4.126) can be reduced to the boundary value problem for an ordinary differential equation. This reduction is based on the dimensional analysis of Eqs. (4.124) and (4.125). This analysis leads to the conclusion that the following variable is dimensionless:

$$\zeta = \frac{z}{t^m \sqrt{k^{-n} \mu_0^{-1} c^{n-1}}}, \tag{4.127}$$

where

$$m = \frac{p(n-1)+1}{2}. \tag{4.128}$$

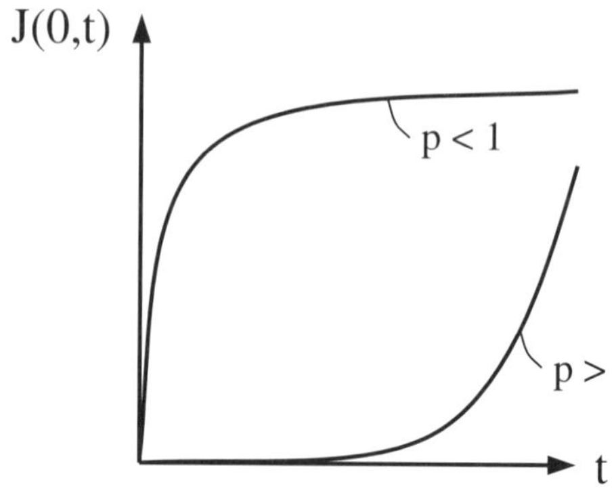

Fig. 4.23

By using this dimensionless variable, we look for the self-similar solution of initial boundary value problem (4.125)–(4.124) in the form:

$$J(z,t) = ct^p f(\zeta), \tag{4.129}$$

where $f(\zeta)$ is a dimensionless function of ζ. By substituting formula (4.129) into Eq. (4.124), after simple transformations we end up with the following ordinary differential equation:

$$\frac{d^2 f^n}{d\zeta^2} + m\zeta \frac{df}{d\zeta} - pf = 0. \tag{4.130}$$

It is apparent that $J(z,t)$ given by expression (4.129) will satisfy boundary and initial conditions (4.125) and (4.126) if f satisfies the boundary conditions:

$$f(0) = 1, \tag{4.131}$$

$$f(\infty) = 0. \tag{4.132}$$

Thus, the initial boundary value problem (4.124)–(4.126) is reduced to the boundary value problem (4.130)–(4.132) for the ordinary differential Eq. (4.130). It can be proven that this nonlinear differential equation has the following group property: if $f(\zeta)$ is a solution to Eq. (4.130), then

$$F(\zeta) = \lambda^{-2/(n-1)} f(\lambda\zeta) \tag{4.133}$$

is also a solution to this equation for any constant λ. This property can be utilized as follows. Suppose we have solution $f(\zeta)$ to Eq. (4.130), which satisfies the boundary condition (4.132), however

$$f(0) = a \neq 1. \tag{4.134}$$

Then, by using $\lambda = a^{n-1/2}$, we find that

$$f(\zeta) = \frac{1}{a} f(a^{(n-1)/2}\zeta) \tag{4.135}$$

is the solution to Eq. (4.130), which satisfied (4.132) as well as the boundary condition (4.131). Thus, we can first find a solution to Eq. (4.130) subject to boundary condition (4.132), then, by using transformation (4.132), we can map this solution into the solution that also satisfies the boundary condition (4.131).

It can be shown that a solution to Eq. (4.130) satisfying the boundary condition (4.132) has the form:

$$f(\zeta) = \begin{cases} b(1-\zeta)^{1/(n-1)}[1 + b_1(1-\zeta) + b_2(1-\zeta)^2 + \ldots], & \text{if } 0 \le \zeta \le 1. \\ 0, & \text{if } \zeta > 1. \end{cases} \tag{4.136}$$

By substituting formula (4.136) into Eq. (4.130), after simple but lengthy transformations, we find

$$b = \left[\frac{m(n-1)}{n}\right]^{1/(n-1)}, \tag{4.137}$$

$$b_1 = \frac{p(n-1) - m}{2mn(n-1)}, \tag{4.138}$$

$$b_2 = -b_1 \frac{1 + \frac{1}{2} b_1 [(2n-1)(3n-2) - 4n]}{3(2n-1)}. \tag{4.139}$$

It is clear that

$$f(0) = b(1 + b_1 + b_2 + \ldots) \neq 1. \tag{4.140}$$

This difficulty is overcome by using transformation (4.133) with

$$\lambda = [b(1 + b_1 + b_2 + \ldots)]^{(n-1)/2}. \tag{4.141}$$

This leads to the following solution of the boundary value problem (4.130)–(4.132):

$$f(\zeta) = \begin{cases} (1-\lambda\zeta)^{1/(n-1)} \frac{1 + b_1(1-\lambda\zeta) + b_2(1-\lambda\zeta)^2 + \ldots}{1 + b_1 + b_2 + \ldots}, & \text{if } 0 \le \lambda\zeta < 1, \\ 0, & \text{if } \lambda\zeta > 1. \end{cases} \tag{4.142}$$

The last expression can be simplified by exploiting the fact that the exponent n in the power law is usually greater than 7. This simplification can be accomplished by using the following inequalities for b_1 and b_2, which can be easily derived from Eqs. (4.128), (4.138), and (4.139):

$$|b_1| \leq \frac{1}{2n(n-1)}, \tag{4.143}$$

$$|b_2| \leq \frac{1}{6(n-1)(2n-1)n} + \frac{1}{8(n-1)n^2}. \tag{4.144}$$

From the above inequalities, for $n \geq 7$ we find

$$|b_1| \leq 0.012, \quad |b_2| \leq 0.00075. \tag{4.145}$$

This suggests the following simplification of solution (4.142):

$$f(\zeta) = \begin{cases} \left(1 - \sqrt{m(n-1)/n}\zeta\right)^{1/(n-1)}, & \text{if } 0 \leq \zeta < \sqrt{n/m(n-1)}, \\ 0, & \text{if } \zeta > \sqrt{n/m(n-1)}. \end{cases} \tag{4.146}$$

By substituting formula (4.146) into expression (4.129) and taking into account Eq. (4.127), we end up with the following analytical expression for the current density:

$$J(z,t) = \begin{cases} ct^p \left(1 - \frac{z}{dt^m}\right)^{1/(n-1)}, & \text{if } z \leq dt^m, \\ 0, & \text{if } z \geq dt^m, \end{cases} \tag{4.147}$$

where

$$d = \sqrt{(nc^{n-1})/[\mu_0 k^n m(n-1)]}. \tag{4.148}$$

The brief examination of self-similar solutions (4.147) leads to the following conclusion: the profile of electric current density $J(z,t)$ remains approximately the same in spite of wide-ranging variations of boundary conditions (4.125) (see Fig. 4.23). For typical values of n (usually $n \geq 7$), this profile is very close to a rectangular one. This suggests that the actual profile of electric current density will be close to a rectangular one for other boundary conditions as well. Thus, we arrive at the following generalization of the critical state model.

Current density $J(z,t)$ has a rectangular profile with the height equal to the instantaneous value $J_0(t)$ of electric current density on the boundary of the superconductor (see Fig. 4.24). Magnetic field $H(z,t)$ has a linear profile with a slope determined by the instantaneous value of $J_0(t)$.

To better appreciate this generalization, we recall that in the critical state model the current has a rectangular profile of constant (in time) height, while the magnetic field has a linear profile with constant (in time) slope.

For the zero front of the current profile we have

$$z_0(t) = \frac{H_0(t)}{J_0(t)}. \tag{4.149}$$

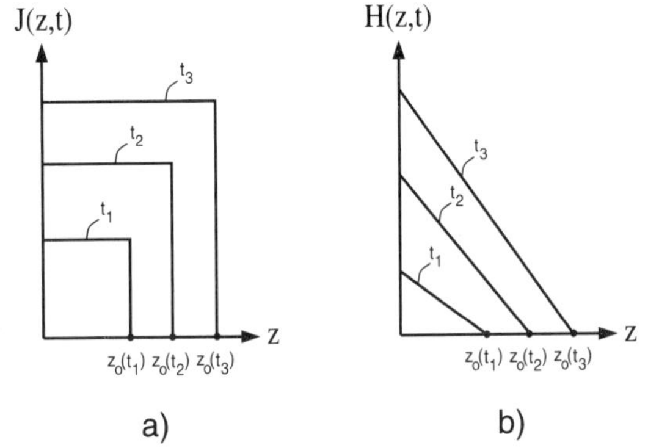

Fig. 4.24

However, $H_0(t)$ and $J_0(t)$ are not simultaneously known. For this reason, we intend to find $J_0(t)$ in terms of $H_0(t)$. To this end, we multiply Eq. (4.124) by z and integrate from 0 to $z_0(t)$ with respect to z and from 0 to t with respect to t. By literally repeating the same transformations as in Section 1.5, we arrive at the following equation:

$$\mu_0 k^n \int_0^{z_0(t)} zJ(z,t)dz = \int_0^t J_0^n(\tau)d\tau. \tag{4.150}$$

By using in the last equation the rectangular profile approximation for $J(z,t)$, we obtain

$$\frac{\mu_0 k^n}{2} J_0(t) z_0^2(t) = \int_0^t J_0^n(\tau)d\tau. \tag{4.151}$$

By substituting formula (4.149) into Eq. (4.151), we find

$$\frac{\mu_0 k^n}{2} \frac{d}{dt}\left[\frac{H_0^2(t)}{J_0(t)}\right] = J_0^n(t). \tag{4.152}$$

By introducing a new variable:

$$\gamma(t) = \frac{H_0^2(t)}{J_0(t)}, \tag{4.153}$$

we can represent formula (4.152) as the following differential equation with respect to $\gamma(t)$:

$$\frac{d\gamma^{n+1}}{dt} = \frac{2(n+1)}{\mu_0 k^n} H_0^{2n}(t). \tag{4.154}$$

By integrating Eq. (4.154) and by using Eq. (4.153), we arrive at the following expression for $J_0(t)$:

$$J_0(t) = \frac{H_0^2(t)}{\{[(2(n+1)/\mu_0 k^n)] \int_0^t H_0^{2n}(\tau)d\tau\}^{1/(n+1)}}. \tag{4.155}$$

By substituting formula (4.155) into Eq. (4.149), we find the following expression for zero front $z_0(t)$ in terms of the magnetic field, $H_0(t)$, at the boundary of the superconductor:

$$z_0(t) = \frac{1}{H_0(t)} \left[\frac{2(n+1)}{\mu_0 k^n} \int_0^t H_0^{2n}(\tau)d\tau \right]^{1/(n+1)}. \tag{4.156}$$

Up to this point, nonlinear diffusion of electromagnetic fields in semi-infinite superconducting half-space has been discussed. However, the above results can be directly extended to the case of a slab of finite thickness Δ. This can be done due to the finite speed of propagation of zero front $z_0(t)$. As a result, if $z_0(t) < \frac{\Delta}{2}$, nonlinear diffusion at both sides of a superconducting slab occurs in the same way as in the case of superconducting half-space. This is illustrated by Figs. 4.25 a and 4.25 b.

After the instant of time t^*, when two fronts meet at the middle of the slab, formula (4.155) is not valid anymore. To find the appropriate formula for $J_0(t)$ in the case $t > t^*$, we shall again use the first moment relation for the nonlinear diffusion equation. However, this moment relation should be somewhat modified (in comparison with (4.150)). To find this modification, we start with the nonlinear diffusion equation:

$$\frac{\partial^2 E}{\partial z^2} = \mu_0 \frac{\partial J}{\partial t}. \tag{4.157}$$

We multiply both sides of this equation by z and integrate with respect to z from the boundary $z = 0$ to $z = \frac{\Delta}{2}$:

$$\int_0^{\frac{\Delta}{2}} z \frac{\partial^2 E}{\partial z^2} dz = \mu_0 \int_0^{\frac{\Delta}{2}} z \frac{\partial J}{\partial t} dz. \tag{4.158}$$

Next, we shall transform the first integral in formula (4.158) by using integration by parts:

$$\int_0^{\frac{\Delta}{2}} z \frac{\partial^2 E}{\partial z^2} dz = z \frac{\partial E}{\partial z}\bigg|_0^{\frac{\Delta}{2}} - \int_0^{\frac{\Delta}{2}} \frac{\partial E}{\partial z} dz = \frac{\Delta}{2} \frac{\partial E}{\partial z}\left(\frac{\Delta}{2}, t\right) + E_0(t). \tag{4.159}$$

From equation curl $\mathbf{E} = -\mu_0 \frac{\partial \mathbf{H}}{\partial t}$, we find

$$\frac{\partial E}{\partial z}\left(\frac{\Delta}{2}, t\right) = -\mu_0 \frac{\partial H}{\partial t}\left(\frac{\Delta}{2}, t\right). \tag{4.160}$$

By using the last equation in formula (4.159), as well as the power law (4.121), we obtain

$$\int_0^{\frac{\Delta}{2}} z \frac{\partial^2 E}{\partial z^2} dz = -\frac{\mu_0 \Delta}{2} \frac{\partial H}{\partial t}\left(\frac{\Delta}{2}, t\right) + \frac{J_0^n(t)}{k^n}. \tag{4.161}$$

By substituting formula (4.161) into the moment relation (4.158), we derive

$$J_0^n(t) - \frac{\mu_0 \Delta k^n}{2} \frac{\partial H}{\partial t}\left(\frac{\Delta}{2}, t\right) = \mu_0 k^n \frac{\partial}{\partial t} \int_0^{\frac{\Delta}{2}} z J(z,t) dz. \tag{4.162}$$

Since it is assumed (in our generalization of the critical state model) that

$$J(z,t) = J_0(t), \qquad (0 \le z < \frac{\Delta}{2}), \tag{4.163}$$

we can transform formula (4.162) as follows:

$$J_0^n(t) - \frac{\mu_0 \Delta k^n}{2} \frac{\partial H}{\partial t}\left(\frac{\Delta}{2}, t\right) = \frac{\mu_0 \Delta^2 k^n}{8} \frac{dJ_0(t)}{dt}. \tag{4.164}$$

Relation (4.163) implies that

$$H_0(t) - H\left(\frac{\Delta}{2}, t\right) = \frac{\Delta}{2} J_0(t), \tag{4.165}$$

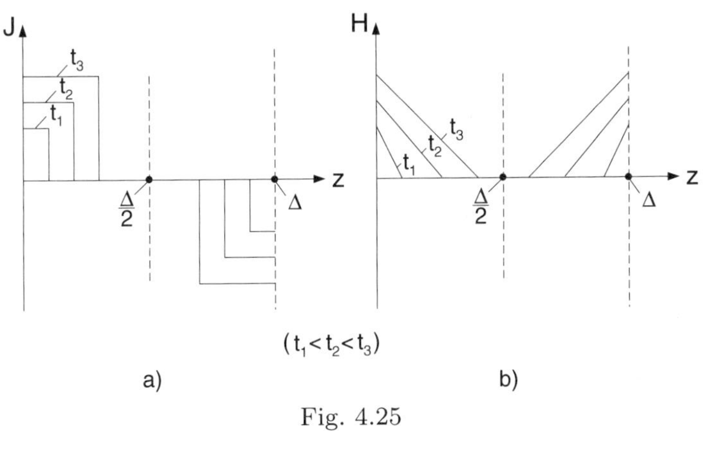

Fig. 4.25

which yields

$$\frac{\partial H}{\partial t}\left(\frac{\Delta}{2}, t\right) = -\frac{\Delta}{2}\frac{dJ_0(t)}{dt} + \frac{dH_0(t)}{dt}. \tag{4.166}$$

By substituting the last formula into expression (4.164), we arrive at the following ordinary differential equation for $J_0(t)$:

$$\boxed{\tau_0 \frac{dJ_0(t)}{dt} + J_0^n(t) = \frac{4\tau_0}{\Delta} \cdot \frac{dH_0(t)}{dt},} \tag{4.167}$$

where

$$\boxed{\tau_0 = \frac{\mu_0 \Delta^2 k^n}{8}.} \tag{4.168}$$

Thus, in order to find $J_0(t)$ for $t \geq t^*$, the solution to differential Eq. (4.167) subject to the initial condition

$$J_0(t^*) = J_0^* \tag{4.169}$$

must be found. Here J_0^* is the value of the current density immediately prior to the instant t^*, and this value can be computed by using formula (4.155).

As an example, consider an important case when

$$H_0(t) = H_0 = \text{const}, \qquad (t > 0). \tag{4.170}$$

For this case, Eq. (4.167) is reduced to

$$\tau_0 \frac{dJ_0(t)}{dt} = -J_0^n(t). \tag{4.171}$$

This equation can be integrated by employing the following separation of variables:

$$\frac{dJ_0}{J_0^n} = -\frac{dt}{\tau_0}, \tag{4.172}$$

which leads to

$$J_0(t) = \left(\frac{\tau_0'}{t + t'}\right)^{\frac{1}{n-1}}, \tag{4.173}$$

where

$$\tau_0' = \frac{\tau_0}{n-1} = \frac{\mu_0 \Delta^2 k^n}{8(n-1)}, \tag{4.174}$$

and t' is determined from initial condition (4.169).

It is interesting to note that formula (4.173) coincides with the long-time (intermediate) asymptotics found in [29] (see also [6], [7]). These asymptotics are used to describe the phenomenon of flux creep in superconductors. To better appreciate this, we shall rewrite formula (4.173) in the form:

$$J_0(t) = \exp\left\{-\frac{1}{n-1}\ln\frac{t+t'}{\tau_0'}\right\}. \tag{4.175}$$

By assuming that

$$n \gg 1, \quad t \gg t', \quad \frac{1}{n-1}\ln\frac{t+t'}{\tau_0'} \ll 1, \tag{4.176}$$

and by using only two terms of the Taylor expansion in formula (4.175), we find

$$J_0(t) \approx 1 - \frac{1}{n-1}\ln\frac{t}{\tau_0'}. \tag{4.177}$$

This is the well-known logarithmic intermediate asymptotics, which is a characteristic of creep phenomena. Thus, it can be concluded that long-time solutions to the nonlinear diffusion Eq. (4.124) are instrumental for the description of creep. The idea of using nonlinear diffusion equations for the description of flux creep can be traced back to the landmark papers of P.W. Anderson and Y.B. Kim [1] and M.R. Beasley, R. Labusch, and W.W. Webb [5].

Typical distributions of the electric current density and the magnetic field computed by using the described generalization of the critical state model for the case $H_0(t) = H_0 =$ const are shown in Figs. 4.26 a and 4.26 b.

Next, we intend to show that electromagnetic field diffusion in superconductors with gradual resistive transitions may exhibit a peculiar (anomalous) mode that does not exist in superconductors with ideal resistive transition. This is a standing mode. In the case of this mode, the electromagnetic field on a superconductor boundary increases with time, while the region occupied by the electromagnetic field does not expand. We shall first find the condition for the existence of this mode by using the "rectangular profile" approximation for the current density. Then we shall derive the exact expressions for the standing mode through the analytical solution of the nonlinear diffusion equation, that is, without resorting to the rectangular profile approximation. Finally, we shall compare these two results.

To start the discussion, we turn to equation (4.150) and try to find such a monotonically increasing boundary condition $J_0(t)$ for which the zero front, $z_0(t)$, stands still. To this end, we assume that $z_0(t) = z_0 = \text{const}$, and, by differentiating both sides of (4.150), we arrive at

$$J_0^n(t) = \frac{\mu_0 k^n}{2} z_0^2 \frac{dJ_0(t)}{dt}. \tag{4.178}$$

The last expression can be transformed as follows:

$$\frac{2}{\mu_0 k^n z_0^2} dt = \frac{dJ_0(t)}{J_0^n(t)}. \tag{4.179}$$

By integrating both parts of (4.179), we obtain

$$\frac{2}{\mu_0 k^n z_0^2} t = \frac{1}{n-1}\left[J_0^{1-n}(0) - J_0^{1-n}(t)\right]. \tag{4.180}$$

From (4.180), we derive

$$J_0(t) = \frac{1}{\left(J_0^{1-n}(0) - \frac{2(n-1)}{\mu_0 k^n z_0^2}(t)\right)^{\frac{1}{n-1}}}. \tag{4.181}$$

The last expression can be represented in the form:

$$J_0(t) = \frac{c}{(t_0 - t)^{\frac{1}{n-1}}}, \tag{4.182}$$

where

$$c = \left(\frac{\mu_0 k^n z_0^2}{2(n-1)}\right)^{\frac{1}{n-1}}, \qquad t_o = \frac{\mu_0 k^n z_0^2 J_0^{1-n}(0)}{2(n-1)}. \tag{4.183}$$

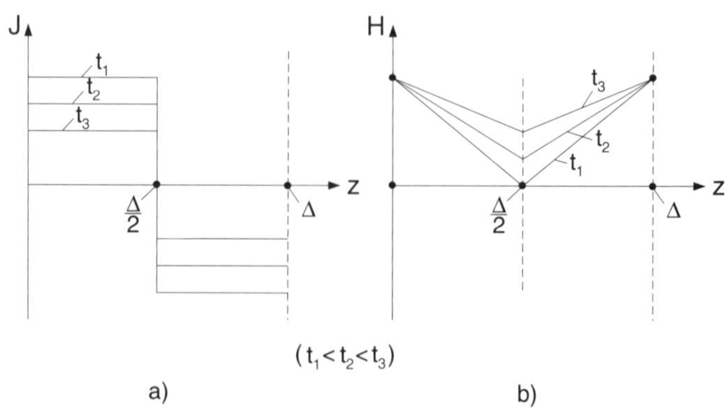

Fig. 4.26

Thus, we have established that, if the current density on the boundary of superconducting half-space varies with time according to the expressions (4.182)–(4.183), then the zero front, $z_0(t)$, of the current density stands still during the time interval $0 \leq t < t_0$. In other words, during this time interval the electromagnetic field diffusion exhibits a standing mode. This mode is illustrated by Fig. 4.27.

It is desirable to express the boundary condition for the standing mode in terms of magnetic field $H_0(t)$ at the superconductor boundary. This can be easily accomplished by using (4.182) and Ampére's law, which lead to:

$$H_0(t) = \frac{cz_0}{(t_0 - t)^{\frac{1}{n-1}}}. \tag{4.184}$$

Our previous derivation has been based on the rectangular profile approximation for the electric current density. Next, we shall derive the expressions for the standing mode solution without resorting to the above approximation, but instead through analytical solution of the nonlinear diffusion Eq. (4.124). It is remarkable that the standing mode solution can be obtained by using the method of separation of variables. Actually, this is the only "short-time" solution that can be obtained by this method. According to the method of separation of variables, we look for the solution of Eq. (4.185) in the form:

$$J(z,t) = \varphi(z)\psi(t). \tag{4.185}$$

By substituting (4.185) into (4.124), after simple transformations we derive

$$\frac{1}{\varphi(z)}\frac{d^2\varphi_(z)^n}{dz^2} = \frac{\mu_o k^n}{\psi^n(t)}\frac{d\psi(t)}{dt}. \tag{4.186}$$

This means that

$$\frac{\mu_0 k^n}{\psi^n(t)}\frac{d\psi(t)}{dt} = \lambda, \tag{4.187}$$

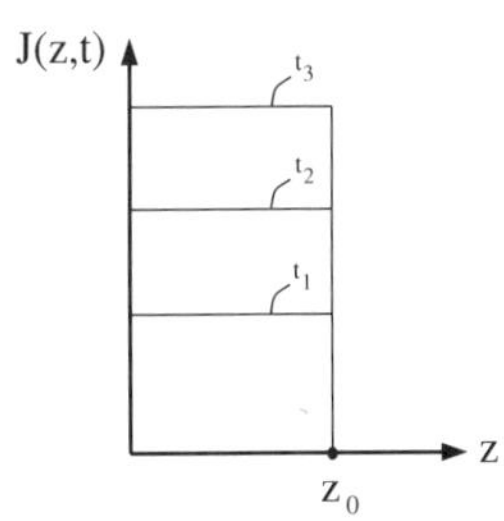

Fig. 4.27

$$\frac{1}{\varphi(z)}\frac{d^2\varphi^n(z)}{dz^2} = \lambda, \tag{4.188}$$

where λ is some constant.

By integrating Eq. (4.187), we easily obtain

$$\psi(t) = \left[\frac{\mu_0 k^n}{(n-1)\lambda(t_0 - t)}\right]^{\frac{1}{n-1}}, \tag{4.189}$$

where t_0 is a constant of integration.

Equation (4.188) is somewhat more complicated than Eq. (4.187) and its integration is somewhat more involved. To integrate Eq. (4.188), we introduce the following auxiliary functions:

$$\varphi^n(z) = \theta(z), \qquad R(z) = \frac{d\theta(z)}{dz}. \tag{4.190}$$

From (4.190) and (4.188), we derive

$$\frac{dR}{dz} = \frac{d^2\theta}{dz^2} = \frac{d^2\varphi^n}{dz^2} = \lambda\varphi(z) = \lambda\theta^{\frac{1}{n}}(z). \tag{4.191}$$

On the other hand,

$$\frac{dR}{dz} = \frac{dR}{d\theta}\cdot\frac{d\theta}{dz} = R\frac{dR}{d\theta} = \frac{1}{2}\frac{d(R^2)}{d\theta}. \tag{4.192}$$

By equating the right-hand sides of (4.191) and (4.192), we obtain

$$\frac{d(R^2)}{d\theta} = 2\lambda\theta^{\frac{1}{n}}. \tag{4.193}$$

By integrating equation (1.93), we find

$$R(z) = \sqrt{\frac{2n}{n+1}\lambda}\,[\theta(z)]^{\frac{n+1}{2n}}. \tag{4.194}$$

In formula (4.194), a constant of integration was set to zero. This can be justified on the physical grounds. Indeed, the magnetic field should vanish at the zero front, that is, at the same point where $J(z,t)$ vanishes. By using (4.185) and (4.190), it can be shown that the magnetic field and $J(z,t)$ are proportional to $R(z)$ and $\theta^{\frac{1}{n}}(z)$, respectively. This means that these two functions should vanish simultaneously. This is only possible if the integration constant in (4.194) is set to zero.

Next, by using (4.190) in (4.194), we find

$$\frac{d\theta(z)}{dz} = \sqrt{\frac{2n}{n+1}\lambda}\,[\theta(z)]^{\frac{n+1}{2n}}. \tag{4.195}$$

By integrating (4.195), we derive

$$[\theta(z)]^{\frac{n-1}{n}} = \frac{(n-1)^2}{2(n+1)n}\lambda(z_0 - z)^2, \tag{4.196}$$

where z_0 is a constant of integration.

From (4.190) and (4.196), we obtain

$$\varphi(z) = \left[\frac{(n-1)^2\lambda}{2(n+1)n}(z_0 - z)^2\right]^{\frac{1}{n-1}}. \tag{4.197}$$

Now, by substituting (4.189) and (4.197) into (4.185), we find the following analytical (and *exact*) solution of nonlinear diffusion:

$$J(z,t) = \left[\frac{(n-1)\mu_0 k^n (z_0 - z)^2}{2(n+1)n(t_0 - t)}\right]^{\frac{1}{n-1}}. \tag{4.198}$$

It is remarkable that, as a result of substitution, the "separation" constant λ cancels out, and the solution (4.198) does not depend on λ at all.

The obtained solution (4.198) can be physically interpreted as follows. Suppose that at time $t = 0$ the electric current density satisfies the following initial condition:

$$J(z,0) = \begin{cases} \left[\frac{(n-1)\mu_0 k^n (z_0 - z)^2}{2(n+1)nt_0}\right]^{\frac{1}{n-1}}, & \text{if } 0 \leq z \leq z_0, \\ 0, & \text{if } z \geq z_0. \end{cases} \tag{4.199}$$

Suppose also that the current density satisfies the following boundary condition during the time interval $0 \leq t < t_0$:

$$J_0(t) = J(0,t) = \left[\frac{(n-1)\mu_0 k^n z_0^2}{2(n+1)n(t_0 - t)}\right]^{\frac{1}{n-1}}. \tag{4.200}$$

Then, according to (4.198), the exact solution to the initial boundary value problem (4.199)–(4.200) for the nonlinear diffusion Eq. (4.124) can be written as follows:

$$J(z,t) = \begin{cases} \left[\frac{(n-1)\mu_0 k^n (z_0 - z)^2}{2(n+1)n(t_0 - t)}\right]^{\frac{1}{n-1}}, & \text{if } 0 \leq z \leq z_0 \\ 0, & \text{if } z > z_0. \end{cases} \tag{4.201}$$

This solution is illustrated by Fig. 4.28 and it is apparent that it has the physical meaning of the standing mode. It is also clear from formula (4.201) (and Fig. 4.28) that the above solution has the following self-similarity property: the profiles of electric current density for different instants of time can be obtained from one another by dilation (or contraction) along the J-axis. In other words, these profiles remain similar to one another. This suggests that solution (4.201) can be derived by using the dimensionality analysis. However, we shall not delve further into this matter.

From the practical point of view, it is desirable to express the boundary condition (4.200) for the standing mode in terms of magnetic field $H_0(t)$ on the superconductor boundary. According to Ampére's law, we have

$$H_0(t) = \int_0^{z_0} J(z,t)dz. \tag{4.202}$$

By substituting (4.201) into (4.202) and performing integration, we obtain

$$H_0(t) = \frac{n-1}{n+1} z_0 \left[\frac{(n-1)\mu_0 k^n z_0^2}{2(n+1)n(t_0 - t)} \right]^{\frac{1}{n-1}}. \tag{4.203}$$

It is also instructive to compare the above exact standing mode solution with the standing mode expressions derived on the basis of rectangular profile approximation. First, it is clear from formula (4.201) (and Fig. 4.28) that, for sufficiently large n, the actual current density profiles for the standing mode are almost rectangular. Second, it is apparent that the boundary condition (4.200) can be written in the form (4.182) with c and t_0 defined as follows:

$$c = \left[\frac{(n-1)\mu_0 k^n z_0^2}{2(n+1)n} \right]^{\frac{1}{n-1}}, \tag{4.204}$$

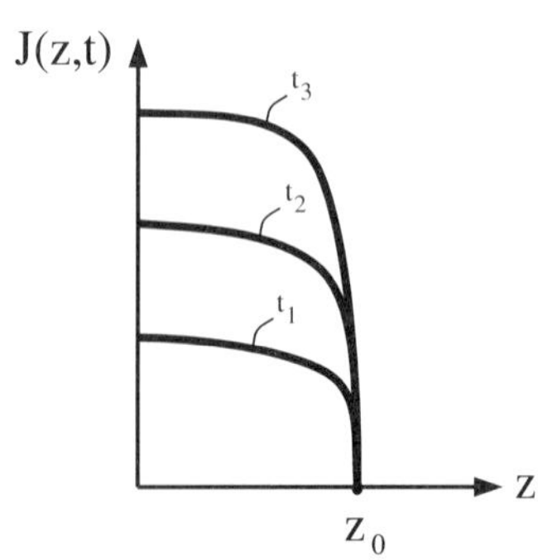

Fig. 4.28

$$t_0 = \frac{(n-1)\mu_0 k^n z_0^2 J_0^{1-n}(0)}{2(n+1)n}. \tag{4.205}$$

By comparing (4.204)–(4.205) with (4.183), it can be observed that for sufficiently large n these expressions are practically identical. Thus, the rectangular profile approximation is fairly accurate as far as the prediction of the standing mode diffusion is concerned. This brings further credence to the rectangular profile approximation.

The origin of the standing mode can be elucidated on physical grounds as follows. Under the boundary condition (4.200), the electromagnetic energy entering the superconducting material at any instant of time is just enough to affect the almost uniform increase in electric current density in the region $(0 \leq z \leq z_0)$ already occupied by the field but insufficient to affect the further diffusion of the field in the material.

In the discussion presented above, the method of separation of variables has been used in order to find the "short-time" solution, which describes the standing mode of nonlinear diffusion. It turns out that the same method of separation of variables can be used in order to study "long-time" solutions, which describe the phenomenon of flux creep. As has been demonstrated by E.H. Brandt [7], these "long-time" solutions can be found in quite general situations. In our presentation below, we closely follow the paper [7] of E.H. Brandt.

Consider a long superconducting cylinder of an arbitrary cross-section subject to uniform magnetic field $\mathbf{B}_0$ directed along the y-axis (see Fig. 4.29). This magnetic field induces currents $\mathbf{J}$ in the superconductor and these currents are directed along the z-axis. The vector magnetic potential $\mathbf{A}$ is also directed along the z-axis and it is given by the following formula:

$$A(\mathbf{r}, t) = -\frac{\mu_0}{2\pi} \int_S J(\mathbf{r}', t) \ln |\mathbf{r} - \mathbf{r}'| \, d^2\mathbf{r}' + A_0(\mathbf{r}), \tag{4.206}$$

where A_0 is the vector magnetic potential of the external field:

$$A_0(\mathbf{r}) = -x B_0. \tag{4.207}$$

From equations curl $\mathbf{E} = -\frac{\partial \mathbf{B}}{\partial t}$ and curl $\mathbf{A} = \mathbf{B}$ we conclude that

$$E(\mathbf{r}, t) = -\frac{\partial}{\partial t} A(\mathbf{r}, t). \tag{4.208}$$

Combining formulas (4.206), (4.207), and (4.208), we derive

$$E(\mathbf{r}, t) = \frac{\mu_0}{2\pi} \int_S \frac{\partial J(\mathbf{r}', t)}{\partial t} \ln |\mathbf{r} - \mathbf{r}'| \, d^2\mathbf{r}' + x \frac{\partial B_0}{\partial t}. \tag{4.209}$$

Now we consider "long-time" conditions when the external magnetic field is maintained constant:

$$B_0 = \text{const}, \qquad \frac{\partial B_0}{\partial t} = 0. \tag{4.210}$$

By substituting (4.210) into Eq. (4.209) and taking into account power law (4.121), we arrive at the following nonlinear integro-differential equation:

$$E(\mathbf{r}, t) = \frac{\mu_0 k}{2\pi} \int_S \frac{\partial E^{\frac{1}{n}}(\mathbf{r}', t)}{\partial t} \cdot \ln|\mathbf{r} - \mathbf{r}'| d^2\mathbf{r}'. \tag{4.211}$$

We look for a nonzero solution of this equation in the form:

$$E(\mathbf{r}, t) - \varphi(\mathbf{r}) \left(\frac{t}{\tau}\right)^{\alpha}. \tag{4.212}$$

By substituting formula (4.212) into Eq. (4.211), we find that the above equation is satisfied if

$$\alpha = \frac{\alpha}{n} - 1, \tag{4.213}$$

which yields

$$\alpha = \frac{n}{1-n}. \tag{4.214}$$

This means that solution (4.212) takes the form:

$$E(\mathbf{r}, t) = \varphi(\mathbf{r}) \left(\frac{t}{\tau}\right)^{\frac{n}{1-n}}, \tag{4.215}$$

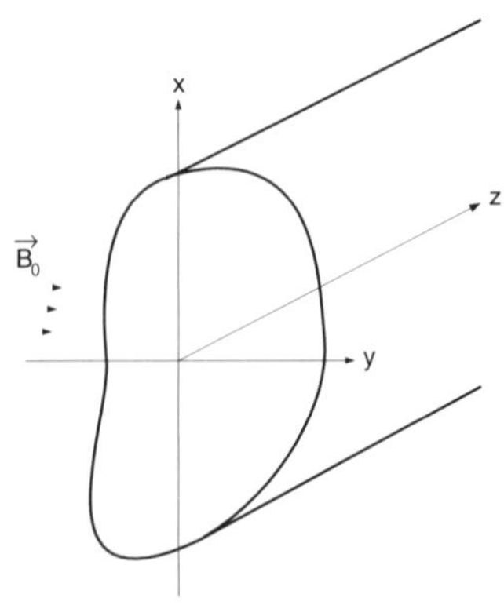

Fig. 4.29

and Eq. (4.211) is reduced to the following nonlinear integral equation:

$$\varphi(\mathbf{r}) = \frac{-\mu_0 k}{2\pi\tau(n-1)} \int_S \left[\varphi(\mathbf{r}')\right]^{\frac{1}{n}} \ln|\mathbf{r}-\mathbf{r}'|\, d^2\mathbf{r}'. \tag{4.216}$$

As $n \gg 1$ and τ can be chosen sufficiently large, Eq. (4.216) can be solved numerically by using contraction mapping iterations [7].

Having computed $\varphi(\mathbf{r})$, we can find the electric current density:

$$J(\mathbf{r},t) = k\left[\varphi(\mathbf{r})\right]^{\frac{1}{n}} \left(\frac{t}{\tau}\right)^{\frac{1}{1-n}} \text{sign}\,\varphi(\mathbf{r}). \tag{4.217}$$

To investigate the intermediate asymptotical behavior of the current density, we shall employ the formula

$$\left(\frac{t}{\tau}\right)^{\frac{1}{1-n}} = \exp\left\{-\frac{1}{n-1}\ln\frac{t}{\tau}\right\}. \tag{4.218}$$

Since $n \gg 1$, there are always such times that

$$\frac{1}{n-1}\ln\frac{t}{\tau} \ll 1. \tag{4.219}$$

For such times, we can retain only two terms of the Taylor expansion of the exponent in formula (4.218). This leads to the following intermediate logarithmic asymptotics:

$$J(\mathbf{r},t) \approx k\left[\varphi(\mathbf{r})\right]^{\frac{1}{n}} \left[1-\frac{1}{n-1}\ln\frac{t}{\tau}\right] \text{sign}\,\varphi(\mathbf{r}), \tag{4.220}$$

which is typical of creep phenomena.

4.4 NONLINEAR DIFFUSION IN ISOTROPIC SUPERCONDUCTORS WITH GRADUAL RESISTIVE TRANSITIONS (CIRCULAR POLARIZATION)

In the previous sections of this chapter, nonlinear diffusion of linearly polarized electromagnetic fields was discussed. Our analysis was based on the solution of scalar nonlinear diffusion equations. Now we turn to the discussion of more complicated and challenging problems, where electromagnetic fields are not linearly polarized. This will require the solution of vector nonlinear diffusion equations. We begin with the simplest case when electromagnetic fields are circularly polarized and superconducting media are assumed to be electrically isotropic. We shall still use the power law as the constitutive relation for the isotropic superconducting media. In the vector case, the power law (4.121) can be written as follows:

$$J_x(E_x, E_y) = \sigma\left(\sqrt{E_x^2 + E_y^2}\right) E_x, \tag{4.221}$$

$$J_y(E_x, E_y) = \sigma\left(\sqrt{E_x^2 + E_y^2}\right) E_y, \tag{4.222}$$

where

$$\sigma\left(\sqrt{E_x^2 + E_y^2}\right) = \sigma\left(|\mathbf{E}|\right) = k\left|\mathbf{E}\right|^{\frac{1}{n}-1}. \tag{4.223}$$

By using the above formulas, the problem of nonlinear diffusion of circularly polarized electromagnetic fields in the superconducting half-space $z \geq 0$ can be framed as the following boundary value problem: find the periodic solution of the coupled nonlinear diffusion equations

$$\frac{\partial^2 E_x}{\partial z^2} = \mu_0 \frac{\partial}{\partial t}\left[\sigma\left(\sqrt{E_x^2 + E_y^2}\right) E_x\right], \tag{4.224}$$

$$\frac{\partial^2 E_y}{\partial z^2} = \mu_0 \frac{\partial}{\partial t}\left[\sigma\left(\sqrt{E_x^2 + E_y^2}\right) E_y\right], \tag{4.225}$$

subject to the following boundary conditions:

$$E_x(0,t) = E_m \cos(\omega t + \theta_0), \tag{4.226}$$

$$E_y(0,t) = E_m \sin(\omega t + \theta_0), \tag{4.227}$$

$$E_x(\infty,t) = E_y(\infty,t) = 0. \tag{4.228}$$

It is apparent that the mathematical structure of nonlinear differential Eqs. (4.224)–(4.225) and boundary conditions (4.226)–(4.228) is invariant with respect to rotations of the x- and y-axes around axis z. This suggests that the solution of the boundary value problem (4.224)–(4.228) should also be invariant with respect to the above rotations. The latter implies that the electric field is circularly polarized everywhere within the superconducting media:

$$E_x(z,t) = E(z)\cos[\omega t + \theta(z)], \tag{4.229}$$

$$E_y(z,t) = E(z)\sin[\omega t + \theta(z)]. \tag{4.230}$$

Now we formally demonstrate that the circularly polarized solution (4.229)–(4.230) is consistent with the mathematical structure of the boundary value problems (4.224)–(4.228). First, it is clear from the above formulas that

$$|\mathbf{E}(z,t)| = \sqrt{E_x^2(z,t) + E_y^2(z,t)} = E(z). \tag{4.231}$$

This means that the magnitude of the electric field, as well as the conductivity $\sigma(|\mathbf{E}|)$, do not change with time at every point of the superconducting media.

Next, we shall write formulas (4.229) and (4.230) in the phasor form:

$$\hat{E}_x(z) = E(z)e^{j\theta(z)}, \tag{4.232}$$

$$\hat{E}_y(z) = -jE(z)e^{j\theta(z)}. \tag{4.233}$$

It is apparent that

$$|\mathbf{E}(z,t)| = \left|\hat{E}_x(z)\right| = \left|\hat{E}_y(z)\right|, \tag{4.234}$$

and

$$\sigma(|\mathbf{E}(z,t)|) = \sigma(\left|\hat{E}_x(z)\right|) = \sigma(\left|\hat{E}_y(z)\right|). \tag{4.235}$$

Now, by using phasors (4.232) and (4.233) as well as the formula (4.235), we can **exactly** transform the boundary value problem (4.224)–(4.228) into the following boundary value problems for $\hat{E}_x(z)$ and $\hat{E}_y(z)$, respectively:

$$\frac{d^2\hat{E}_x}{dz^2} = j\omega\mu_0\sigma(|\hat{E}_x|)\hat{E}_x, \tag{4.236}$$

$$\hat{E}_x(0) = \hat{E}_m, \quad \hat{E}_x(\infty) = 0, \tag{4.237}$$

and

$$\frac{d^2\hat{E}_y}{dz^2} = j\omega\mu_0\sigma(|\hat{E}_y|)\hat{E}_y, \tag{4.238}$$

$$\hat{E}_y(0) = -j\hat{E}_m, \quad \hat{E}_y(\infty) = 0, \tag{4.239}$$

where

$$\hat{E}_m = E_m e^{j\theta_0}. \tag{4.240}$$

The above **exact** transformation can be construed as a mathematical proof that the circular polarization of the incident electromagnetic field is preserved everywhere within the superconducting media. This also proves that there are no higher-order time-harmonics of the electric field anywhere within the media despite its nonlinear properties.

From the purely mathematical point of view, the achieved simplification of the boundary value problem (4.224)–(4.228) is quite remarkable. First, partial differential Eqs. (4.224)–(4.225) are exactly reduced to the ordinary differential Eqs. (4.236) and (4.238), respectively. Second, these ordinary differential equations are completely decoupled. Finally, these decoupled equations have identical mathematical structures. As a result, the same solution technique can be applied to both of them.

To solve Eqs. (4.236) and (4.238), we shall first slightly transform them. According to formulas (4.223) and (4.235), we have

$$\sigma(|\hat{E}_x|) = k|\hat{E}_x|^{\frac{1}{n}-1}, \tag{4.241}$$

$$\sigma(|\hat{E}_y|) = k|\hat{E}_y|^{\frac{1}{n}-1}, \tag{4.242}$$

$$\sigma_m = \sigma(|\hat{E}_m|) = k|\hat{E}_m|^{\frac{1}{n}-1}. \tag{4.243}$$

From the last three expressions, we derive

$$\sigma(|\hat{E}_x|) = \sigma_m \left|\frac{\hat{E}_x}{\hat{E}_m}\right|^{\frac{1}{n}-1}, \tag{4.244}$$

$$\sigma(|\hat{E}_y|) = \sigma_m \left|\frac{\hat{E}_y}{\hat{E}_m}\right|^{\frac{1}{n}-1}. \tag{4.245}$$

By substituting formulas (4.244) and (4.245) into Eqs. (4.236) and (4.238), respectively, we find

$$\frac{d^2\hat{E}_x}{dz^2} = j\omega\mu_0\sigma_m \left|\frac{\hat{E}_x}{\hat{E}_m}\right|^{\frac{1}{n}-1} \hat{E}_x, \tag{4.246}$$

$$\frac{d^2\hat{E}_y}{dz^2} = j\omega\mu_0\sigma_m \left|\frac{\hat{E}_y}{\hat{E}_m}\right|^{\frac{1}{n}-1} \hat{E}_y. \tag{4.247}$$

The solution to Eq. (4.246), subject to the boundary conditions (4.237), can be sought in the form:

$$\boxed{\hat{E}_x(z) = \hat{E}_m \left(1 - \frac{z}{z_0}\right)^{\alpha},} \tag{4.248}$$

where

$$\boxed{\alpha = \alpha' + j\alpha''.} \tag{4.249}$$

Here, z_0, α' and α'' are parameters, which will be appropriately chosen to guarantee that $\hat{E}_x(z)$ given by formula (4.248) satisfies Eq. (4.246).

It is important to keep in mind that expression (4.248) is an abbreviated form of the solution. In other words, it is tacitly understood that this form is valid for $0 \le z \le z_0$, while for $z \ge z_0$ the solution is equal to zero.

From formulas (4.248) and (4.249), we derive

$$\left|\frac{\hat{E}_x(z)}{\hat{E}_m}\right| = \left(1 - \frac{z}{z_0}\right)^{\alpha'}, \tag{4.250}$$

which leads to

$$\left|\frac{\hat{E}_x(z)}{\hat{E}_m}\right|^{\frac{1}{n}-1} = \left(1 - \frac{z}{z_0}\right)^{-\frac{\alpha'(n-1)}{n}}. \tag{4.251}$$

By substituting formulas (4.248) and (4.251) into Eq. (4.246), we find

$$\alpha(\alpha-1)\hat{E}_m\left(1 - \frac{z}{z_0}\right)^{\alpha-2} = j\omega\mu_0\sigma_m z_0^2 \hat{E}_m\left(1 - \frac{z}{z_0}\right)^{\alpha-\frac{\alpha'(n-1)}{n}}. \tag{4.252}$$

It is clear that equality (4.252) will hold, if the following two conditions are satisfied:

$$2 = \frac{\alpha'(n-1)}{n}, \tag{4.253}$$

and

$$\alpha(\alpha-1) = j\omega\mu_0\sigma_m z_0^2. \tag{4.254}$$

From the first condition, we immediately find

$$\alpha' = \frac{2n}{n-1}. \tag{4.255}$$

The second condition (4.254) can be construed as a characteristic equation, which can be used for determination of α'' and z_0. This is an equation in terms of complex variable α. It can be reduced to the following two real equations:

$$\alpha'(\alpha'-1) - (\alpha'')^2 = 0, \tag{4.256}$$

$$\alpha''(2\alpha'-1) = \omega\mu_0\sigma_m z_0^2. \tag{4.257}$$

By using formulas (4.255) and (4.256), we derive

$$\alpha'' = \frac{\sqrt{2n(n+1)}}{n-1}. \tag{4.258}$$

By substituting expressions (4.255) and (4.258) into (4.257), we arrive at the following expression for z_0:

$$z_0 = \frac{[2n(n+1)(3n+1)^2]^{\frac{1}{4}}}{(n-1)\sqrt{\omega\mu_0\sigma_m}}. \tag{4.259}$$

Formulas (4.248), (4.249), (4.255), (4.258), and (4.259) completely define the phasor $\hat{E}_x(z)$ as the solution of the boundary value problem (4.236)–(4.237). The boundary value problem (4.238)–(4.239) has the same mathematical form (structure) as the boundary value problem (4.236)–(4.237). For this reason, the solution to the boundary value problem (4.238)-(4.239) can be written as follows:

$$\hat{E}_y(z) = -j\hat{E}_m \left(1 - \frac{z}{z_0}\right)^{\alpha}, \tag{4.260}$$

where, as before, α and z_0 are given by formulas (4.249), (4.255), (4.258), and (4.259).

Expressions (4.248) and (4.260) can be converted from the phasor form into the time-domain forms. This yields

$$E_x(z,t) = E_m \left(1 - \frac{z}{z_0}\right)^{\frac{2n}{n-1}} \cos\left[\omega t + \theta_0 + \frac{\sqrt{2n(n+1)}}{n-1} \ln\left(1 - \frac{z}{z_0}\right)\right], \tag{4.261}$$

$$E_y(z,t) = E_m \left(1 - \frac{z}{z_0}\right)^{\frac{2n}{n-1}} \sin\left[\omega t + \theta_0 + \frac{\sqrt{2n(n+1)}}{n-1} \ln\left(1 - \frac{z}{z_0}\right)\right], \tag{4.262}$$

The above formulas give the **exact** analytical solution to the boundary value problem (4.224)–(4.228) for coupled nonlinear diffusion equations. This is the high symmetry solution, which is invariant (up to a choice of initial phase θ_0) with respect to rotations of axes x and y.

By using the last two formulas along with expressions (4.221)–(4.223), we obtain the following relations for the electric current densities:

$$J_x(z,t) = J_m \left(1 - \frac{z}{z_0}\right)^{\frac{2}{n-1}} \cos\left[\omega t + \theta_0 + \frac{\sqrt{2n(n+1)}}{n-1} \ln\left(1 - \frac{z}{z_0}\right)\right], \tag{4.263}$$

$$J_y(z,t) = J_m \left(1 - \frac{z}{z_0}\right)^{\frac{2}{n-1}} \sin\left[\omega t + \theta_0 + \frac{\sqrt{2n(n+1)}}{n-1} \ln\left(1 - \frac{z}{z_0}\right)\right], \tag{4.264}$$

where

$$J_m = \sigma_m E_m. \tag{4.265}$$

From the above relations, we find

$$|\mathbf{J}(z)| = J_{mx}(z) = J_{my}(z) = J_m \left(1 - \frac{z}{z_0}\right)^{\frac{2}{n-1}}. \tag{4.266}$$

We see that for $n \gg 1$ profile of $|\mathbf{J}(z)|$ is almost rectangular. A typical plot of this profile is shown in Fig. 4.30. We can also observe the logarithmic variation of phase with respect to z. As a result, for any fixed time t, electric current densities J_x and J_y (as well as electric fields E_x and E_y) have infinite numbers of zeros (infinite numbers of oscillations) in the interval $0 \leq z \leq z_0$.

Up to this point, it has been assumed that the electric field components are specified on the boundary of the superconducting half-space $z \geq 0$. However, in applications it is more convenient to specify the boundary values of the magnetic field components. For this reason, we shall express the above solutions (4.261)–(4.262) in terms of the magnetic field at the boundary. To do this, we shall invoke the equations

$$\frac{d\hat{E}_x}{dz} = -j\omega\mu_0 \hat{H}_y, \tag{4.267}$$

$$\frac{d\hat{E}_y}{dz} = j\omega\mu_0 \hat{H}_x, \tag{4.268}$$

as well as formulas (4.248) and (4.260). As a result, we obtain

$$\hat{H}_x(z) = \hat{H}_m \left(1 - \frac{z}{z_0}\right)^{\alpha-1}, \tag{4.269}$$

$$\hat{H}_y(z) = -j\hat{H}_m \left(1 - \frac{z}{z_0}\right)^{\alpha-1}, \tag{4.270}$$

where

$$\hat{H}_m = \frac{\alpha}{\omega\mu_0 z_0} \hat{E}_m. \tag{4.271}$$

If $\hat{H}_m$ is given, then the last relation can be construed as a **nonlinear** equation for $\hat{E}_m$. This equation is nonlinear, because z_0 depends on σ_m, which is a nonlinear function of E_m. To make this nonlinearity manifest, we use formula (4.259) in (4.271) and derive

$$H_m = |\alpha| \sqrt{\frac{\sigma_m}{\omega\mu_0} \frac{n-1}{[2n(n+1)(3n+1)^2]^{\frac{1}{4}}}} E_m. \tag{4.272}$$

Now, by recalling formula (4.243), we transform the last equation as follows:

$$H_m = |\alpha| \sqrt{\frac{k}{\omega\mu_0} \frac{n-1}{[2n(n+1)(3n+1)^2]^{\frac{1}{4}}}} E_m^{\frac{1+n}{2n}}. \tag{4.273}$$

From expressions (4.249), (4.255), and (4.258), we find

$$|\alpha| = \frac{\sqrt{2n(3n+1)}}{n-1}. \tag{4.274}$$

By substituting formula (4.274) into (4.273), we end up with the nonlinear equation for E_m:

$$H_m = \sqrt{\frac{k}{\omega\mu_0}} \left(\frac{2n}{n+1}\right)^{\frac{1}{4}} E_m^{\frac{1+n}{2n}} \tag{4.275}$$

$$E_m = \left(\frac{\omega\mu_0}{k}\right)^{\frac{n}{n+1}} \left(\frac{n+1}{2n}\right)^{\frac{n}{2(n+1)}} H_m^{\frac{2n}{n+1}}, \tag{4.276}$$

and

$$\sigma_m = k^{\frac{2n}{n-1}} \left(\omega\mu_0\right)^{\frac{1-n}{n+1}} \left(\frac{n+1}{2n}\right)^{\frac{1-n}{2(n+1)}} H_m^{\frac{2(1-n)}{n+1}}. \tag{4.277}$$

Thus, for any given H_m we can find σ_m from formula (4.277), and then z_0 from formula (4.259). Having found z_0, we can use formulas (4.269), (4.270), and (4.271) for calculations of $\hat{H}_x(z)$, $\hat{H}_y(z)$, and $\hat{E}_m$.

Next, we consider the surface impedance of the superconducting half-space. This impedance is defined as follows:

$$\eta = \frac{\hat{E}_x(0)}{\hat{H}_y(0)} = -\frac{\hat{E}_y(0)}{\hat{H}_x(0)}. \tag{4.278}$$

From formulas (4.248), (4.270), (4.271), and (4.278), we find

$$\eta = \frac{j\omega\mu_0 z_0}{\alpha}. \tag{4.279}$$

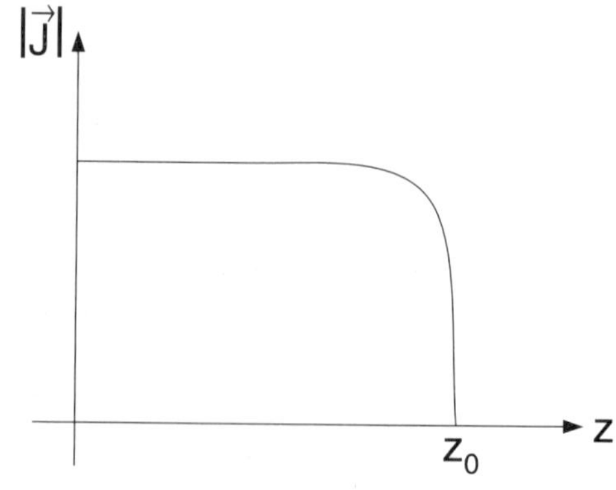

Fig. 4.30

By using the polar form of the impedance

$$\eta = |\eta| e^{j\varphi}, \tag{4.280}$$

from Eq. (4.279), we derive

$$|\eta| = \frac{\omega \mu_0 z_0}{|\alpha|}, \tag{4.281}$$

$$\tan \varphi = \frac{\alpha'}{\alpha''}. \tag{4.282}$$

By invoking formulas (4.259) and (4.274), from (4.281) we find

$$\boxed{|\eta| = \left(\frac{n+1}{2n} \right)^{\frac{1}{4}} \sqrt{\frac{\omega \mu_0}{\sigma_m}}.} \tag{4.283}$$

Similarly, from formulas (4.255), (4.258), and (4.282), we arrive at

$$\boxed{\tan \varphi = \sqrt{\frac{2n}{n+1}}.} \tag{4.284}$$

The last two formulas are remarkably simple. In the particular case of $n = 1$ (linear media), these formulas lead to the well-known expressions:

$$|\eta| = \sqrt{\frac{\omega \mu_0}{\sigma}}, \qquad \varphi = \frac{\pi}{4}. \tag{4.285}$$

It is important to note that the magnitude of the surface impedance is field dependent. This is clearly seen from formula (4.277). In contrast, the phase φ of the surface impedance is not field dependent. It is determined only by the sharpness of the resistive transition. Figs. 4.31 and 4.32 show respectively the dependence of $|\eta| / \sqrt{\frac{\omega \mu_0}{\sigma_m}}$ and $\tan \varphi$ on the exponent n, which is the natural measure of the sharpness of the resistive transition.

In another limiting case $n = \infty$ (sharp transition), from formulas (4.283) and (4.284) we derive

$$\boxed{|\eta| = \sqrt{\frac{\omega \mu_0}{\sqrt{2} \sigma_m}}, \qquad \tan \varphi = \sqrt{2}.} \tag{4.286}$$

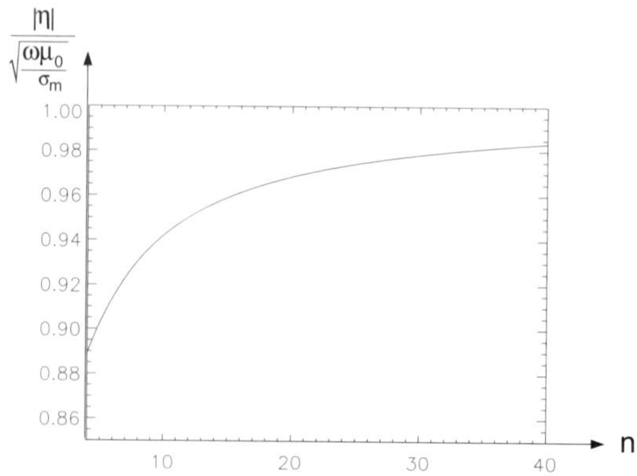

Fig. 4.31

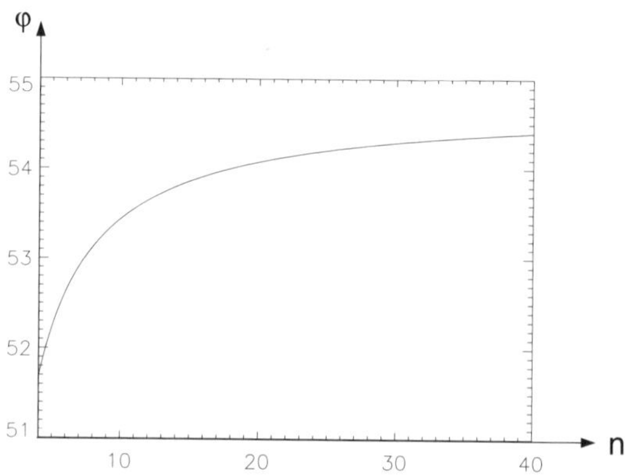

Fig. 4.32

Now we turn back to formulas (4.269) and (4.270) and convert them into time-domain form. As a result, we obtain

$$H_x(z,t) = H_m \left(1 - \frac{z}{z_0}\right)^{\frac{n+1}{n-1}} \cos\left[\omega t + \frac{\sqrt{2n(n+1)}}{n-1} \ln\left(1 - \frac{z}{z_0}\right)\right], \tag{4.287}$$

$$H_y(z,t) = H_m \left(1 - \frac{z}{z_0}\right)^{\frac{n+1}{n-1}} \sin\left[\omega t + \frac{\sqrt{2n(n+1)}}{n-1} \ln\left(1 - \frac{z}{z_0}\right)\right], \tag{4.288}$$

where, for the sake of simplicity, it is assumed that the initial phase of the magnetic field at the boundary is equal to zero.

Next, we shall show that, in the limiting case of sharp resitive transitions ($n = \infty$), the last two expressions are reduced to those that were asserted by C. Bean in the paper [4]. To this end, from formula (4.271), we find

$$H_m = \frac{|\alpha|}{\omega\mu_0\sigma_m z_o} J_m. \tag{4.289}$$

From the last equation as well as formulas (4.259) and (4.274), we obtain

$$\sqrt{\omega\mu_0\sigma_m} = \left(\frac{2n}{n+1}\right)^{\frac{1}{4}} \frac{J_m}{H_m}. \tag{4.290}$$

By substituting the last expression into formula (4.259), we arrive at

$$z_0 = \frac{\sqrt{(n+1)(3n+1)}}{n-1} \frac{H_m}{J_m}. \tag{4.291}$$

In the case of sharp resistive transitions,

$$J_m = J_c, \tag{4.292}$$

and the ratio

$$\delta = \frac{H_m}{J_m} = \frac{H_m}{J_c} \tag{4.293}$$

can be construed as the field dependent (Bean) penetration depth, which we dealt with for linear polarizations of magnetic fields (see formula (4.11)). By using expression (4.293) in formula (4.291), in the limiting case of $n \to \infty$, we obtain

$$z_0 = \sqrt{3}\delta. \tag{4.294}$$

Thus, the penetration depth in the case of the circular polarization is $\sqrt{3}$ times larger than in the case of linear polarization.

Finally, by substituting formula (4.294) into expressions (4.287) and (4.288) and letting n go to infinity, we derive

$$H_x(z,t) = H_m \left(1 - \frac{z}{\sqrt{3}\delta}\right) \cos\left[\omega t + \sqrt{2}\ln\left(1 - \frac{z}{z_0}\right)\right], \tag{4.295}$$

$$H_y(z,t) = H_m \left(1 - \frac{z}{\sqrt{3}\delta}\right) \sin\left[\omega t + \sqrt{2}\ln\left(1 - \frac{z}{z_0}\right)\right]. \tag{4.296}$$

The last expressions are identical to those published in the paper [4].

The results discussed up to this point were obtained for the superconducting half-space. However, they can be easily extended to the case of a superconducting slab of finite thickness Δ if the penetration depth z_0 satisfies the following inequality:

$$z_0 \leq \frac{\Delta}{2}. \tag{4.297}$$

Under condition (4.297), nonlinear diffusion of electromagnetic fields at each side of the superconducting slab occurs in the same way as in the case of the superconducting half-space. As a result, we obtain the following formulas:

$$\hat{H}_x(z) = \begin{cases} \hat{H}_m \left(1 - \frac{\frac{\Delta}{2}+z}{z_0}\right)^{\alpha-1}, & \text{if } -\frac{\Delta}{2} \leq z \leq -\frac{\Delta}{2} + z_0, \\ 0, & \text{if } -\frac{\Delta}{2} + z_0 \leq z \leq \frac{\Delta}{2} - z_0, \\ \hat{H}_m \left(1 - \frac{\frac{\Delta}{2}-z}{z_0}\right)^{\alpha-1}, & \text{if } \frac{\Delta}{2} - z_0 \leq z \leq \frac{\Delta}{2}, \end{cases} \tag{4.298}$$

$$\hat{H}_y(z) = \begin{cases} -j\hat{H}_m \left(1 - \frac{\frac{\Delta}{2}+z}{z_0}\right)^{\alpha-1}, & \text{if } -\frac{\Delta}{2} \leq z \leq -\frac{\Delta}{2} + z_0, \\ 0, & \text{if } -\frac{\Delta}{2} + z_0 \leq z \leq \frac{\Delta}{2} - z_0, \\ -j\hat{H}_m \left(1 - \frac{\frac{\Delta}{2}-z}{z_0}\right)^{\alpha-1}, & \text{if } \frac{\Delta}{2} - z_0 \leq z \leq \frac{\Delta}{2}. \end{cases} \tag{4.299}$$

Similar generalizations can be given for the electric field and current density. It would be interesting to find the solution to the nonlinear diffusion problem in the case when $z_0 > \frac{\Delta}{2}$.

4.5 NONLINEAR DIFFUSION IN THE CASE OF ELLIPTICAL POLARIZATIONS AND ANISOTROPIC MEDIA

In this section, we shall use the perturbation technique in order to extend the results from the previous section to more complicated situations.

To start the discussion, consider a plane electromagnetic wave penetrating the superconducting half-space $z > 0$. The magnetic field at the boundary of this half-space is specified as follows:

$$\begin{aligned} H_x(0,t) &= H_m[\cos(\omega t + \gamma) + \epsilon f_x(t)], \\ H_y(0,t) &= H_m[\sin(\omega t + \gamma) + \epsilon f_y(t)], \end{aligned} \tag{4.300}$$

where ϵ is some small parameter, while $f_x(t)$ and $f_y(t)$ are given periodic functions of time with the period $\frac{2\pi}{\omega}$.

It is apparent that this plane wave can be construed as a perturbation of the circularly polarized plane wave. By using the Maxwell equations, we find that the distribution of electric field in the half-space $z > 0$ is governed by the following coupled nonlinear partial differential equations:

$$\frac{\partial^2 E_x}{\partial z^2} = \mu_0 \frac{\partial J_x(E_x, E_y)}{\partial t}, \tag{4.301}$$

$$\frac{\partial^2 E_y}{\partial z^2} = \mu_0 \frac{\partial J_y(E_x, E_y)}{\partial t}, \tag{4.302}$$

subject to the boundary conditions

$$\frac{\partial E_x}{\partial z}(0, t) = -\mu_0 H_m [\omega \cos(\omega t + \gamma) + \epsilon f_y'(t)], \tag{4.303}$$

$$\frac{\partial E_y}{\partial z}(0, t) = -\mu_0 H_m [\omega \sin(\omega t + \gamma) - \epsilon f_x'(t)], \tag{4.304}$$

$$E_x(\infty) = E_y(\infty) = 0, \tag{4.305}$$

where functions $J_x(E_x, E_y)$ and $J_y(E_x, E_y)$ are specified by formulas (4.221), (4.222), and (4.223).

Next, we shall look for the periodic solution of the boundary value problem (4.301)–(4.305) in the following form:

$$E_x(z, t) = E_x^0(z, t) + \epsilon e_x(z, t), \tag{4.306}$$

$$E_y(z, t) = E_y^0(z, t) + \epsilon e_y(z, t). \tag{4.307}$$

By substituting expressions (4.306) and (4.307) into Eqs. (4.301) and (4.302) and boundary conditions (4.303)–(4.305) and by equating the terms of like powers of ϵ, we arrive at the following boundary value problems for E_x^0, E_y^0 and e_x, e_y:

$$\frac{\partial^2 E_x^0}{\partial z^2} = \mu_0 \frac{\partial J_x(E_x^0, E_y^0)}{\partial t}, \qquad \frac{\partial^2 E_y^0}{\partial z^2} = \mu_0 \frac{\partial J_y(E_x^0, E_y^0)}{\partial t}, \tag{4.308}$$

$$\frac{\partial E_x^0}{\partial z}(0, t) = -\omega \mu_0 H_m \cos(\omega t + \gamma), \tag{4.309}$$

$$\frac{\partial E^0}{\partial z}(0, t) = -\omega \mu_0 H_m \sin(\omega t + \gamma), \tag{4.310}$$

$$E_x^0(\infty) = E_y^0(\infty) = 0, \tag{4.311}$$

and

$$\frac{\partial^2 e_x}{\partial z^2} = \mu_0 \frac{\partial}{\partial t} \left[\frac{\partial J_x}{\partial E_x}(E_x^0, E_y^0) e_x + \frac{\partial J_x}{\partial E_y}(E_x^0, E_y^0) e_y \right], \tag{4.312}$$

$$\frac{\partial^2 e_y}{\partial z^2} = \mu_0 \frac{\partial}{\partial t}\left[\frac{\partial J_y}{\partial E_x}(E_x^0, E_y^0)e_x + \frac{\partial J_y}{\partial y}(E_x^0, E_y^0)e_y\right], \tag{4.313}$$

$$\frac{\partial e_x}{\partial z}(0,t) = -\mu_0 H_m f_y'(t), \tag{4.314}$$

$$\frac{\partial e_y}{\partial z}(0,t) = \mu_0 H_m f_x'(t). \tag{4.315}$$

$$e_x(\infty, t) = e_y(\infty, t) = 0. \tag{4.316}$$

The boundary value problem (4.308)–(4.311) describes the penetration of circularly polarized plane wave into the superconducting half-space. The solution to this problem has been found in the previous section. For the case when the initial phase, γ, is such that the initial phase of $\mathbf{E}^0$ on the boundary ($z = 0$) is equal to zero, this solution is given by the following expressions:

$$E_x^0(z,t) = E_m\left(1 - \frac{z}{z_0}\right)^{\alpha'} \cos(\omega t + \theta(z)), \tag{4.317}$$

$$E_y^0(z,t) = E_m\left(1 - \frac{z}{z_0}\right)^{\alpha'} \sin(\omega t + \theta(z)), \tag{4.318}$$

$$z_0 = \frac{\sqrt[4]{2n(n+1)(3n+1)^2}}{\sqrt{\omega\mu_0\sigma_m}(n-1)}, \qquad \sigma_m = kE_m^{\frac{1}{n}-1}, \tag{4.319}$$

$$\theta(z) = \alpha'' \ln\left(1 - \frac{z}{z_0}\right), \tag{4.320}$$

$$\alpha' = \frac{2n}{n-1}, \qquad \alpha'' = \frac{\sqrt{2n(n+1)}}{n-1}, \tag{4.321}$$

and E_m can be found from the nonlinear equation:

$$H_m = \frac{|\alpha' + i\alpha''|}{\omega\mu_0 z_0} E_m. \tag{4.322}$$

By substituting (4.317) and (4.318) into Eqs. (4.312) and (4.313) and by using expressions (4.221)–(4.223), after simple but somewhat lengthy transformations (that can be found in Section 2.2) we arrive at the following equations for e_x and e_y:

$$\frac{\partial^2 e_x(z,t)}{\partial z^2} = \mu_0\sigma_m\left(1 - \frac{z}{z_0}\right)^{-2} \frac{\partial}{\partial t}\left[\left\{\frac{1+n}{2n} + \frac{1-n}{2n}\cos(2\omega t + 2\theta(z))\right\}\right.$$
$$\left. \times e_x(z,t) + \frac{1-n}{2n}\sin(2\omega t + 2\theta(z))e_y(z,t)\right], \tag{4.323}$$

$$\frac{\partial^2 e_y(z,t)}{\partial z^2} = \mu_0 \left(1-\frac{z}{z_0}\right)^{-2} \frac{\partial}{\partial t}\left[\frac{1-n}{2n}\sin(2\omega t+2\theta(z))e_x(z,t) + \left\{\frac{1+n}{2n} - \frac{1-n}{2n}\cos(2\omega t+2\theta(z))\right\} e_y(z,t)\right]. \tag{4.324}$$

Equations (4.323) and (4.324) are coupled linear partial differential equations of parabolic type with variable in time and space coefficients. We would like to find the periodic solutions of these equations subject to the boundary conditions (4.314)–(4.316). To this end, we introduce new complex valued state variables:

$$\Phi(z,t) = e_x(z,t) + je_y(z,t), \tag{4.325}$$

$$\psi(z,t) = e_x(z,t) - je_y(z,t). \tag{4.326}$$

By using these state variables, and some simple transformations, we can represent Eqs. (4.323) and (4.324) in the following form:

$$\frac{\partial^2 \Phi}{\partial z^2} = \mu_0\sigma_m \frac{1-n}{2n}\left(1-\frac{z}{z_0}\right)^{-2} \frac{\partial}{\partial t}\left[\frac{1+n}{1-n}\Phi + \left(1-\frac{z}{z_0}\right)^{j2\alpha''} e^{j2\omega t}\psi\right], \tag{4.327}$$

$$\frac{\partial^2 \psi}{\partial z^2} = \mu_0\sigma_m \frac{1-n}{2n}\left(1-\frac{z}{z_0}\right)^{-2} \frac{\partial}{\partial t}\left[\frac{1+n}{1-n}\psi + \left(1-\frac{z}{z_0}\right)^{-j2\alpha''} e^{-j2\omega t}\Phi\right]. \tag{4.328}$$

Assuming that functions $f_x(t)$ and $f_y(t)$ in boundary conditions (4.314) and (4.315) are functions of half-wave symmetry (the case that is usually of most practical interest), we conclude that $e_x(z,t)$ and $e_y(z,t)$ will also be the functions of half-wave symmetry. For this reason, we will use the following Fourier series for $\Phi(z,t)$ and $\psi(z,t)$:

$$\Phi(z,t) = \sum_{k=-\infty}^{\infty} \Phi_{2k+1}(z)e^{j(2k+1)\omega t}, \tag{4.329}$$

$$\psi(z,t) = \sum_{k=-\infty}^{\infty} \psi_{2k+1}(z)e^{j(2k+1)\omega t}. \tag{4.330}$$

It is clear from (4.325), (4.316), (4.329), and (4.330) that

$$\Phi^*_{2k+1}(z) = \psi_{-2k-1}(z), \tag{4.331}$$

$$\psi^*_{2k+1}(z) = \Phi_{-2k-1}(z), \tag{4.332}$$

where the superscript "$*$" means a complex conjugate quantity.

By substituting (4.329) and (4.330) into (4.327) and (4.328) and by equating the terms with the same exponents, after simple transformations we derive

$$\left(1-\frac{z}{z_0}\right)^2 \frac{d^2\Phi_{2k+1}}{dz^2} = j\chi_{2k+1}\left[a\Phi_{2k+1} + (1-\frac{z}{z_0})^{j2\alpha''}\psi_{2k-1}\right], \quad (4.333)$$

$$\left(1-\frac{z}{z_0}\right)^2 \frac{d^2\psi_{2k-1}}{dz^2} = j\chi_{2k-1}\left[a\psi_{2k-1} + \left(1-\frac{z}{z_0}\right)^{-j2\alpha''}\Phi_{2k+1}\right], \quad (4.334)$$

$$(k = 0, \pm 1, \pm 2, \ldots),$$

where we have introduced the following notations:

$$a = \frac{1+n}{1-n}, \qquad \chi_{2k+1} = (2k+1)\omega\mu_0\sigma_m\frac{1-n}{2n}. \quad (4.335)$$

Thus, we have reduced the problem of integration of partial differential equations (4.327)–(4.328) to the solution of an infinite set of ordinary differential equations with respect to Fourier coefficients Φ_{2k+1} and ψ_{2k-1}. The remarkable property of these simultaneous equations is that they are only coupled in pairs. It allows one to solve each pair of these coupled equations separately. After Φ_{2k+1} and ψ_{2k-1} are found, we can compute $\Phi(z,t)$ and $\psi(z,t)$ and then $e_x(z,t)$ and $e_y(z,t)$. Another simplification is that according to (4.331) it suffices to solve coupled equations (4.333) and (4.334) only for nonnegative values of k.

We shall seek a solution of the coupled equations (4.333) and (4.334) in the form:

$$\Phi_{2k+1}(z) = A_{2k+1}\left(1-\frac{z}{z_0}\right)^{\beta}, \quad (4.336)$$

$$\psi_{2k-1} = B_{2k-1}\left(1-\frac{z}{z_0}\right)^{\beta-j2\alpha''}. \quad (4.337)$$

By substituting (4.336) and (4.337) into (4.333) and (4.334), we end up with the following simultaneous homogeneous equations with respect to A_{2k+1} and B_{2k-1}:

$$(\beta^2 - \beta - j\chi_{2k+1}z_0^2 a)A_{2k+1} - j\chi_{2k+1}z_0^2 B_{2k-1} = 0, \quad (4.338)$$

$$-j\chi_{2k-1}z_0^2 A_{2k+1} + [(\beta - j2\alpha'')^2 - (\beta - j2\alpha'') - j\chi_{2k-1}z_0^2 a]B_{2k-1} = 0. \quad (4.339)$$

The above homogeneous equations have nonzero solution for A_{2k+1} and B_{2k-1} if and only if the corresponding determinant is equal to zero. This yields the following characteristic equation for β:

$$\left(\beta^2 - \beta - j\chi_{2k+1} z_0^2 a\right)\left[(\beta - j2\alpha'')^2 - (\beta - j2\alpha'') - j\chi_{2k-1} z_0^2 a\right] + \chi_{2k+1}\chi_{2k-1} z_0^4 = 0. \tag{4.340}$$

From expressions (4.335) and (4.319), we find

$$\chi_{2k+1} z_0^2 a = \frac{(2k+1)(3n+1)(n+1)}{(n-1)^2}\sqrt{\frac{n+1}{2n}}, \tag{4.341}$$

$$\chi_{2k+1}\chi_{2k-1} z_0^4 = \frac{(4k^2-1)(3n+1)^2(n+1)}{(n-1)^2}\sqrt{\frac{n+1}{2n}}. \tag{4.342}$$

From the last two expressions we conclude that the coefficients of the characteristic equation (4.340) depend on the exponent, n, of the power law and k. Consequently, the roots of this equation also depend only on n and k. It can be proven that the above characteristic equation has two roots $\beta_1(n,k)$ and $\beta_2(n,k)$ with positive real parts. After these roots are found, the solution of coupled equations (4.333) and (4.334) can be represented in the form:

$$\Phi_{2k+1}(z) = A_{2k+1}^{(1)}\left(1 - \frac{z}{z_0}\right)^{\beta_1} + A_{2k+1}^{(2)}\left(1 - \frac{z}{z_0}\right)^{\beta_2}, \tag{4.343}$$

$$\psi_{2k-1}(z) = B_{2k-1}^{(1)}\left(1 - \frac{z}{z_0}\right)^{\beta_1 - j2\alpha''} + B_{2k-1}^{(2)}\left(1 - \frac{z}{z_0}\right)^{\beta_2 - j2\alpha''}, \tag{4.344}$$

where for the sake of notational simplicity we have omitted the dependence of β_1 and β_2 on n and k.

From boundary conditions (4.314)–(4.315) and expressions (4.325)–(4.326), we obtain the following equations for $A_{2k+1}^{(1)}, A_{2k+1}^{(2)}, B_{2k-1}^{(1)}$ and $B_{2k-1}^{(2)}$:

$$\beta_1 A_{2k+1}^{(1)} + \beta_2 A_{2k+1}^{(2)} = -z_0\mu_0 H_m[-f'_{y,2k+1} + jf'_{x,2k+1}], \tag{4.345}$$

$$(\beta_1 - j2\alpha'')B_{2k-1}^{(1)} + (\beta_2 - j2\alpha'')B_{2k-1}^{(2)} = -z_0\mu_0 H_m[-f'_{y,2k-1} - jf'_{x,2k-1}], \tag{4.346}$$

$$(\beta_1^2 - \beta_1 - j\chi_{2k+1} z_0^2 a)A_{2k+1}^{(1)} - j\chi_{2k+1} z_0^2 B_{2k-1}^{(1)} = 0, \tag{4.347}$$

$$(\beta_2^2 - \beta_2 - j\chi_{2k+1} z_0^2 a)A_{2k+1}^{(2)} - j\chi_{2k+1} z_0^2 B_{2k-1}^{(2)} = 0, \tag{4.348}$$

where $f'_{x,2k+1}$ and $f'_{y,2k+1}$ are complex Fourier coefficients of f'_x and f'_y.

By solving simultaneous equations (4.345)–(4.348), we can find coefficients $A^{(1)}_{2k-1}, A^{(2)}_{2k+1}, B^{(1)}_{2k-1}$ and $B^{(2)}_{2k-1}$. Then, by using (4.343), (4.344), (4.329)–(4.330) and (4.325)–(4.326), we can determine perturbations $e_x(z,t)$ and $e_y(z,t)$, which in turn can be used in (4.306)–(4.307) to compute the total electric field.

Consider the particular case when

$$f_x(t) = \cos \omega t, \qquad f_y = \sin \omega t. \tag{4.349}$$

This case corresponds to elliptical polarization of the incident field. It is easy to see that in this case the right-hand sides of equations (4.345) and (4.346) are equal to zero for all k except $k = 1$. This means that only first and third harmonics are not equal to zero. We have reached this conclusion because we have considered only first-order perturbations with respect to ϵ. If we consider higher-order perturbations with respect to ϵ, we shall recover higher-order harmonics of electric field.

From the purely mathematical point of view, it is remarkable that the solution of coupled equations (4.323)–(4.324) corresponding to the boundary conditions (4.314)–(4.316), and (4.319) contains only the first and third harmonics. Perhaps this is because the coupled PDEs (4.323)–(4.324) have inherited some symmetry properties from the unperturbed problem corresponding to the circular polarization of the incident wave.

So far, we have dealt with isotropic superconducting media. Now, we proceed to the discussion of nonlinear diffusion in anisotropic media. The first question to be addressed is how the power law that describes gradual resistive transitions can be generalized to the case of anisotropic media. By repeating the same line of reasoning as in Section 2.4, we can reach the conclusion that a reasonable generalization of the power law is given by the following formulas:

$$J_x(E_x, E_y) = (1+\epsilon)kE_x \left(\sqrt{(1+\epsilon)E_x^2 + (1-\epsilon)E_y^2}\right)^{\frac{1}{n}-1}, \tag{4.350}$$

$$J_y(E_x, E_y) = (1-\epsilon)kE_y \left(\sqrt{(1+\epsilon)E_x^2 + (1-\epsilon)E_y^2}\right)^{\frac{1}{n}-1}, \tag{4.351}$$

where ϵ is some relatively small parameter, which accounts for anisotropicity of media. It is clear that the superconductor properties enter into Eqs. (4.350) and (4.351) through parameters n, ϵ, and k.

In the limiting case of $\epsilon = 0$, expressions (4.350) and (4.351) are reduced to

$$J_x^{(0)}(E_x, E_y) = kE_x \left(\sqrt{E_x^2 + E_y^2}\right)^{\frac{1}{n}-1} = kE^{\frac{1}{n}-1}E_x, \tag{4.352}$$

$$J_y^{(0}(E_x, E_y) = kE_y \left(\sqrt{E_x^2 + E_y^2}\right)^{\frac{1}{n}-1} = kE^{\frac{1}{n}-1}E_y, \tag{4.353}$$

which are constitutive relations for isotropic superconducting media with gradual resistive transitions described by the power law.

Thus, the anisotropic media with constitutive relations (4.350) and (4.351) can be mathematically treated as perturbations of isotropic media described by the power law. This suggests that the perturbation technique can be very instrumental in the mathematical analysis of nonlinear diffusion in anisotropic media with constitutive relations (4.350) and (4.351).

Formulas (4.350) and (4.351) lead to power law-type resistive transitions along the x- and y-axis:

$$J_x(E_x) = k_x|E_x|^{\frac{1}{n}} \text{ sign } E_x, \tag{4.354}$$

$$J_y(E_y) = k_y|E_y|^{\frac{1}{n}} \text{ sign } E_y, \tag{4.355}$$

with $k_x = k(1+\epsilon)^{\frac{1+n}{2n}}$ and $k_y = k(1-\epsilon)^{\frac{1+n}{2n}}$.

In the limiting case of $n = \infty$, expressions (4.354) and (4.355) describe ideal ("sharp") resistive transitions with critical currents $J_x^c = (1+\epsilon)k$ and $J_y^c = (1-\epsilon)k$. It is also important to note that the Jacobian matrix for $\mathbf{J}(\mathbf{E})$ defined by Eqs. (4.350) and (4.351) is symmetric. This guarantees the absence of local cyclic (hysteretic type) losses (see Section 2.4).

Now consider a plane circularly polarized electromagnetic wave penetrating the superconducting half-space $z > 0$. The magnetic field on the boundary of this half-space is specified as follows:

$$H_x(0,t) = H_m \cos(\omega t + \gamma), \tag{4.356}$$

$$H_y(0,t) = H_m \sin(\omega t + \gamma). \tag{4.357}$$

By using the Maxwell equation, it is easy to find that the distribution of electric field in the half-space $z > 0$ satisfies the following coupled nonlinear partial differential equations

$$\frac{\partial^2 E_x}{\partial z^2} = \mu_0 \frac{\partial J_x(E_x, E_y)}{\partial t}, \tag{4.358}$$

$$\frac{\partial^2 E_y}{\partial z^2} = \mu_0 \frac{\partial J_y(E_x, E_y)}{\partial t}, \tag{4.359}$$

subject to the boundary conditions:

$$\frac{\partial E_x}{\partial z}(0,t) = -\mu_0\omega H_m \cos(\omega t + \gamma), \qquad \frac{\partial E_y}{\partial z}(0,t) = -\mu_0\omega H_m \sin(\omega t + \gamma), \tag{4.360}$$

$$E_x(\infty) = E_y(\infty) = 0. \tag{4.361}$$

Next, by using the perturbation technique, we shall look for the solution of the boundary value problem (4.358)–(4.361) in the form:

$$E_x(z,t) = E_x^0(z,t) + \epsilon e_x(z,t), \tag{4.362}$$

$$E_y(z,t) = E_y^0(z,t) + \epsilon e_y(z,t). \tag{4.363}$$

We shall also use the following ϵ-expansions for constitutive relations (4.350) and (4.351):

$$J_x(E_x, E_y) = J_x^0(E_x, E_y) + \epsilon J_x^0(E_x, E_y)\left[1 + \frac{1-n}{2n} \cdot \frac{E_x^2 - E_y^2}{E^2}\right] + \dots, \tag{4.364}$$

$$J_y(E_x, E_y) = J_y^0(E_x, E_y) - \epsilon J_y^0(E_x, E_y)\left[1 - \frac{1-n}{2n} \cdot \frac{E_x^2 - E_y^2}{E^2}\right] \dots, \tag{4.365}$$

where $J_x^0(E_x, E_y)$ and $J_y^0(E_x, E_y)$ are defined by expressions (4.352) and (4.353), respectively, while $E = \sqrt{E_x^2 + E_y^2}$.

By substituting expressions (4.362)–(4.363) into Eqs. (4.358)–(4.359) and boundary conditions (4.360)–(4.361), and equating the terms of like powers of ϵ, we end up with the following boundary value problems for E_x^0, E_y^0 and e_x, e_y:

$$\frac{\partial^2 E_x^0}{\partial z^2} = \mu_0 \frac{\partial J_x^0(E_x^0, E_y^0)}{\partial t}, \qquad \frac{\partial^2 E_y^0}{\partial z^2} = \mu_0 \frac{\partial J_y^0(E_x^0, E_y^0)}{\partial t}, \tag{4.366}$$

$$\frac{\partial E_x^0}{\partial z}(0,t) = -\omega\mu_0 H_m \cos(\omega t + \gamma), \qquad \frac{\partial E_y^0}{\partial z} = -\omega\mu_0 H_m \sin(\omega t + \gamma), \tag{4.367}$$

$$E_x^0(\infty) = E_y^0(\infty) = 0, \tag{4.368}$$

and

$$\begin{aligned} &\frac{\partial^2 e_x}{\partial z^2} - \mu_0 \frac{\partial}{\partial t}\left(\frac{\partial J_x^0}{\partial E_x}(E_x^0, E_y^0) e_x + \frac{\partial J_x^0}{\partial Ey}(E_x^0, E_y^0) e_y\right) = \\ &\mu_0 \frac{\partial}{\partial t}\left[J_x^0(E_x^0, E_y^0)\left(1 + \frac{1-n}{2n} \cdot \frac{(E_x^0)^2 - (E_y^0)^2}{(E^0)^2}\right)\right], \end{aligned} \tag{4.369}$$

$$\frac{\partial^2 e_y}{\partial z^2} - \mu_0 \frac{\partial}{\partial t}\left(\frac{\partial J_y^0}{\partial E_x}(E_x^0, E_y^0) e_x + \frac{\partial J_y^0}{\partial E_y}(E_x^0, E_y^0) e_y\right) =$$

$$-\mu_0 \frac{\partial}{\partial t}\left[J_y^0(E_x^0, E_y^0)\left(1 - \frac{1-n}{2n}\cdot\frac{(E_x^0)^2-(E_y^0)^2}{(E^0)^2}\right)\right], \tag{4.370}$$

$$\frac{\partial e_x}{\partial z}(0,t) = \frac{\partial e_y}{\partial z}(0,t) = 0, \qquad e_x(\infty,t) = e_y(\infty,t) = 0. \tag{4.371}$$

The boundary value problem (4.366)–(4.368) describes the diffusion of circularly polarized electromagnetic wave in the isotropic superconducting half-space $z > 0$. The solution to this problem has been found in the previous section. For the case when the initial phase γ in (4.367) is such that the initial phase of $\mathbf{E}^0$ on the boundary ($z = 0$) is equal to zero, this solution is given by formulas (4.317)–(4.322).

By substituting (4.317) and (4.318) into Eqs. (4.369) and (4.370) and by using expressions (4.352) and (4.353), after straightforward but somewhat lengthy transformations we derive the following equations for e_x and e_y:

$$\frac{\partial^2 e_x}{\partial z^2} - \mu_0\sigma_m\left(1-\frac{z}{z_0}\right)^{-2}\frac{\partial}{\partial t}\left[e_x\left(\frac{1+n}{2n}+\frac{1-n}{2n}\cos 2[\omega t+\theta(z)]\right)+ e_y\frac{1-n}{2n}\sin 2\left[\omega t+\theta(z)\right]\right] \tag{4.372}$$

$$= \mu_0\sigma_m E_m\left(1-\frac{z}{z_0}\right)^{\frac{2}{n-1}}\frac{\partial}{\partial t}\left[\frac{3n+1}{4n}\cos\left(\omega t+\theta(z)\right)+ \frac{1-n}{4n}\cos 3\left(\omega t+\theta(z)\right)\right],$$

$$\frac{\partial^2 e_y}{\partial z^2} - \mu_0\sigma_m\left(1-\frac{z}{z_0}\right)^{-2}\frac{\partial}{\partial t}\left[e_x\frac{1-n}{2n}\sin 2[\omega t+\theta(z)] + e_y\left(\frac{1+n}{2n}-\frac{1-n}{2n}\cos 2[\omega t+\theta(z)]\right)\right] \tag{4.373}$$

$$= -\mu_0\sigma_m E_m\left(1-\frac{z}{z_0}\right)^{\frac{2}{n-1}}\frac{\partial}{\partial t}\left[\frac{3n+1}{4n}\sin(\omega t+\theta(z)) - \frac{1-n}{4n}\sin 3(\omega t+\theta(z))\right].$$

To simplify the above equations, we introduced new state variables:

$$\Phi(z,t) = e_x(z,t) + j e_y(z,t), \tag{4.374}$$

$$\psi(z,t) = e_x(z,t) - je_y(z,t). \tag{4.375}$$

By using these state variables, we can transform Eqs. (4.372) and (4.373) as follows:

$$\frac{\partial^2 \Phi}{\partial z^2} - \frac{1-n}{2n}\mu_0\sigma_m\left(1-\frac{z}{z_0}\right)^{-2}\frac{\partial}{\partial t}\left[\frac{1+n}{1-n}\Phi + \left(1-\frac{z}{z_0}\right)^{j2\alpha''} e^{j2\omega t}\psi\right] =$$

$$= \mu_0\sigma_m E_m\left(1-\frac{z}{z_0}\right)^{\frac{2}{n-1}}\frac{\partial}{\partial t}\left[\frac{3n+1}{4n}\left(1-\frac{z}{z_0}\right)^{-j\alpha''} e^{-j\omega t}\right.$$

$$\left. + \frac{1-n}{4n}\left(1-\frac{z}{z_0}\right)^{j3\alpha''} e^{j3\omega t}\right], \tag{4.376}$$

$$\frac{\partial^2 \psi}{\partial z^2} - \frac{1-n}{2n}\mu_0\sigma_m\left(1-\frac{z}{z_0}\right)^{-2}\frac{\partial}{\partial t}\left[\frac{1+n}{1-n}\psi + \left(1-\frac{z}{z_0}\right)^{-j2\alpha''} e^{-j2\omega t}\Phi\right] =$$

$$= \mu_0\sigma_m E_m\left(1-\frac{z}{z_0}\right)^{\frac{2}{n-1}}\frac{\partial}{\partial t}\left[\frac{3n+1}{4n}\left(1-\frac{z}{z_0}\right)^{j\alpha''} e^{j\omega t}\right.$$

$$\left. + \frac{1-n}{4n}\left(1-\frac{z}{z_0}\right)^{-j3\alpha''} e^{-j3\omega t}\right]. \tag{4.377}$$

By looking for the solution of Eqs. (4.376) and (4.377) in terms of Fourier series:

$$\Phi(z,t) = \sum_{k=-\infty}^{\infty} \Phi_{2k+1}(z) e^{j(2k+1)\omega t}, \tag{4.378}$$

$$\psi(z,t) = \sum_{k=-\infty}^{\infty} \psi_{2k+1}(z) e^{j(2k+1)\omega t}, \tag{4.379}$$

it can be shown (see the reasoning in Section 2.4) that only Φ_3, Φ_{-1}, ψ_1, and ψ_{-3} are not equal to zero. For Φ_3 and ψ_1 the following coupled (ordinary differential equations) can be derived:

$$\left(1-\frac{z}{z_0}\right)^2 \frac{d^2\Phi_3}{dz^2} - j\chi_3\left[a\Phi_3 + \left(1-\frac{z}{z_0}\right)^{j2\alpha''}\cdot\psi_1\right] =$$

$$j\zeta_3 E_m\left(1-\frac{z}{z_0}\right)^{\frac{2n}{n-1}+j3\alpha''} \tag{4.380}$$

$$\left(1-\frac{z}{z_0}\right)^2 \frac{d^2\psi_1}{dz^2} - j\chi_1 \left[a\psi_1 + \left(1-\frac{z}{z_0}\right)^{-j2\alpha''} \Phi_3 \right] =$$
$$j\nu_1 E_m \left(1-\frac{z}{z_0}\right)^{\frac{2n}{n-1}+j\alpha''}, \tag{4.381}$$

where

$$\chi_{2k+1} = (2k+1)\omega\mu_0\sigma_m \frac{1-n}{2n}, \quad a = \frac{1+n}{1-n}, \tag{4.382}$$

$$\zeta_3 = 3\omega\mu_0\sigma_m \frac{1-n}{4n}, \quad \nu_1 = \omega\mu_0\sigma_m \frac{3n+1}{4n}. \tag{4.383}$$

The solution of Eqs. (4.380)–(4.381) should be subject to the boundary conditions

$$\frac{d\Phi_3}{dz}(0) = \frac{d\psi_1}{dz}(0) = 0, \qquad \Phi_3(\infty) = \psi_1(\infty) = 0, \tag{4.384}$$

which follow from the boundary conditions (4.371).

Similar ODEs can be derived for φ_{-1} and ψ_{-3}. However, this can be avoided because φ_{-1} and ψ_1 as well as ψ_{-3} and φ_3 are complex conjugate.

The particular solution of ODEs (4.380) and (4.381) has the form:

$$\Phi_3^{(p)}(z) = C_3 \left(1-\frac{z}{z_0}\right)^{\lambda_3}, \qquad \psi_1^{(p)} = C_1 \left(1-\frac{z}{z_0}\right)^{\lambda_1}, \tag{4.385}$$

where

$$\lambda_3 = \frac{2n}{n-1} + j3\alpha'', \qquad \lambda_1 = \frac{2n}{n-1} + j\alpha''. \tag{4.386}$$

Coefficients C_3 and C_1 satisfy the following simultaneous equations:

$$\left[\lambda_3(\lambda_3-1) - j\chi_3 a z_0^2\right] C_3 - j\chi_3 z_0^2 C_1 = j\zeta_3 z_0^2 E_m, \tag{4.387}$$

$$-j\chi_1 z_0^2 C_3 + \left[\lambda_1(\lambda_1-1) - j\chi_1 a z_0^2\right] C_1 = j\nu_1 z_0^2 E_m. \tag{4.388}$$

It is clear from (4.319), (4.382), (4.383), and (4.386) that the coefficients in Eqs. (4.387)–(4.388) depend only on n. This opens the opportunity to compute the ratios C_1/E_m and C_3/E_m as functions of n.

It has been shown before that the solution of homogeneous ODEs corresponding to (4.380)–(4.381) has the form:

$$\Phi_3^{(h)}(z) = A \left(1-\frac{z}{z_0}\right)^{\beta}, \qquad \psi_1^{(h)}(z) = B \left(1-\frac{z}{z_0}\right)^{\beta-j2\alpha''}, \tag{4.389}$$

where β is the solution of the following characteristic equation:

$$(\beta^2-\beta-j\chi_3 a z_0^2)[(\beta-j2\alpha'')^2-(\beta-j2\alpha'')-j\chi_1 a z_0^2]+\chi_3\chi_1 z_0^4 = 0. \tag{4.390}$$

It can be shown that the above characteristic equation has two roots β_1 and β_2 with positive real parts. By using these roots and expressions (4.385) and (4.389), the solution of equations (4.380)–(4.381) can be written as follows:

$$\Phi_3(z) = A_1\left(1 - \frac{z}{z_0}\right)^{\beta_1} + A_2\left(1 - \frac{z}{z_0}\right)^{\beta_2} + C_3\left(1 - \frac{z}{z_0}\right)^{\lambda_3}, \tag{4.391}$$

$$\psi_1(z) = B_1\left(1 - \frac{z}{z_0}\right)^{\beta_1 - j2\alpha''} + B_2\left(1 - \frac{z}{z_0}\right)^{\beta_2 - j2\alpha''} + C_1\left(1 - \frac{z}{z_0}\right)^{\lambda_1}. \tag{4.392}$$

The unknown coefficients A_2, A_2, B_1, and B_2 can be found from the boundary conditions (4.384) at $z = 0$ and from the fact that expressions (4.389) should satisfy homogeneous ODEs corresponding to Eqs. (4.380)–(4.381). This yields the following simultaneous equations for the above coefficients:

$$\beta_1 A_1 + \beta_2 A_2 = -\lambda_3 C_3, \tag{4.393}$$

$$(\beta_1 - j2\alpha'')B_1 + (\beta_2 - j2\alpha'')B_2 = -\lambda_1 C_1, \tag{4.394}$$

$$(\beta_1^2 - \beta_1 - j\chi_3 a z_0^2)A_1 - j\chi_3 z_0^2 B_1 = 0, \tag{4.395}$$

$$(\beta_2^2 - \beta_2 - j\chi_3 a z_0^2)A_2 - j\chi_3 z_0^2 B_2 = 0. \tag{4.396}$$

Again, it is easy to see that the coefficients of characteristic equation (4.390) as well as the coefficients of simultaneous equations (4.393)–(4.396) depend only on n. This allows one to compute the roots β_1 and β_2 as well as the ratios $A_1/E_m, A_2/E_m, B_1/E_m$ and B_2/E_m as functions of n. In the limiting case of $n = \infty$ (ideal resistive transition-critical state model), one can compute specific numerical values of the above quantities. These values are as follows: $\beta_1 = 2 + j\sqrt{2}$, $\beta_2 = 1.921 + j3.699$, $C_1/E_m = \frac{3}{2} - j\frac{9\sqrt{2}}{16}$, $C_3/E_m = j\frac{9\sqrt{2}}{16}$, $A_1/E_m = -0.129 + j0.116$, $A_2/E_m = 0.071 - j0.990$, $B_1/E_m = -0.043 + j0.039$, $B_2/E_m = -1.899 + j0.513$. By using these values all desired quantities can be found. For instance, the magnitudes of the first and third harmonics $\mathbf{e}_1$ and $\mathbf{e}_3$ of the perturbation can be computed as the functions of z. The results of these computations are shown in Fig. 4.33. For gradual resistive transitions (finite n), the roots β_1 and β_2 as well as all the mentioned coefficients have been computed as functions of n. The results of these computations are presented in Figs. 4.34 through 4.37. Finally, Fig. 4.38 shows the dependence on n of $|\mathbf{e}_1|$ and $|\mathbf{e}_3|$ at $z = 0$.

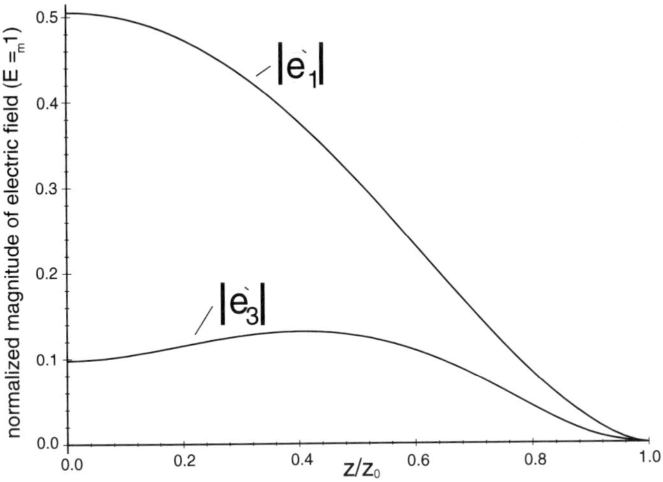

Fig. 4.33

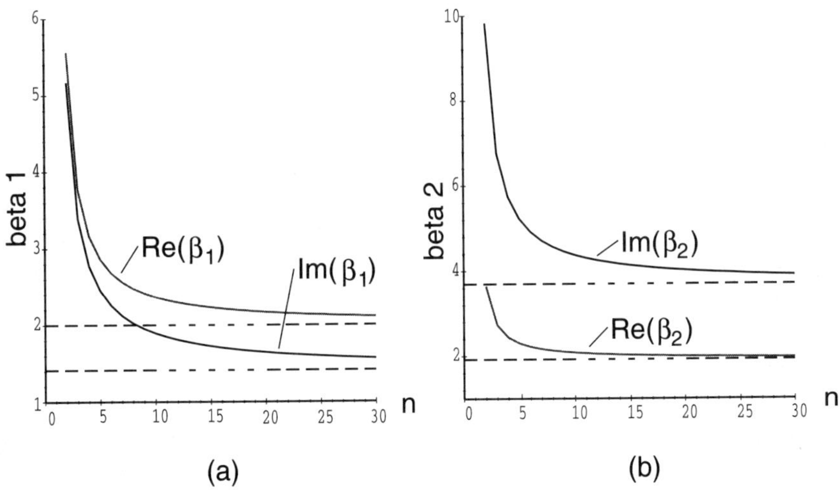

Fig. 4.34

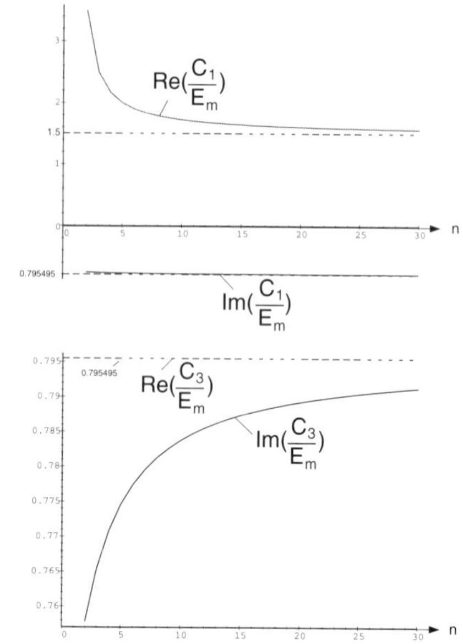

Fig. 4.35

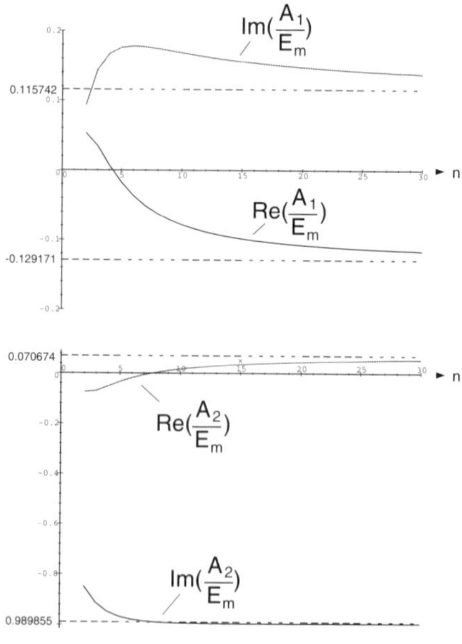

Fig. 4.36

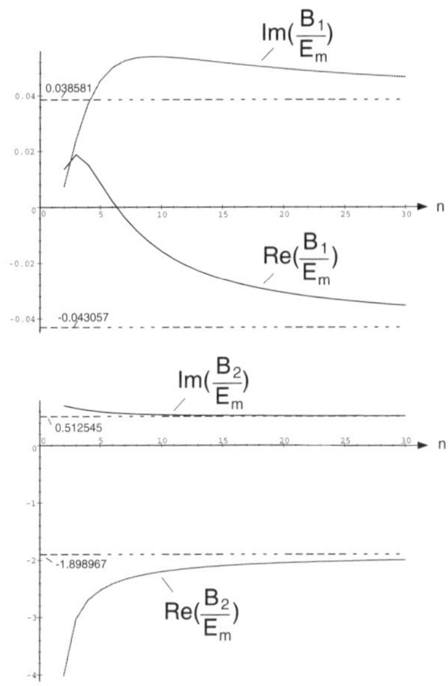

Fig. 4.37

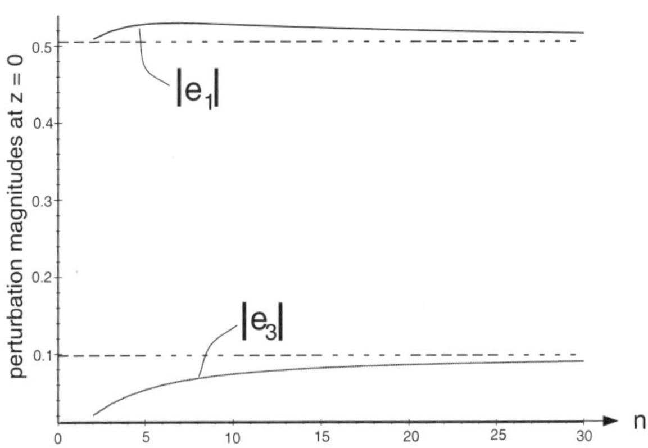

Fig. 4.38

The presented analysis can be extended to the case of nonlinear diffusion of elliptically polarized electromagnetic fields in anisotropic superconducting media described by constitutive relations (4.350)–(4.351). In this case, the perturbation technique with respect to two small parameters can be employed. One small parameter is involved in the constitutive relations, while another is used in the boundary conditions. Mathematical details of this perturbation technique are almost identical to those presented in Section 2.5. For this reason, the discussion of this perturbation technique is omitted here, and the reader is referred to Chapter 2.

REFERENCES

[1] P.W. Anderson and Y.B. Kim, Review of Modern Physics, **36**, pp. 39–43, (1964).
[2] C.P. Bean, Physical Review Letters, **8**, pp. 250–252, (1962).
[3] C.P. Bean, Review of Modern Physics, **36**, pp. 31–39, (1964).
[4] C.P. Bean, Journal of Applied Physics, **41**, pp. 2482–2483, (1970).
[5] M.R. Beasley, R. Labusch, and W.W. Webb, Physical Review, **181**, pp. 682–700, (1969).
[6] E.H. Brandt and A. Gurevich, Physical Review Letters, **76**, pp. 1723–1726, (1996).
[7] E.H. Brandt, Physical Review Letters, **76**, pp. 4030–4033, (1996).
[8] J.W. Elkin, Cryogenics, **2**, p. 603, (1987).
[9] G. Friedman, L. Liu, and J.S. Kouvel, Journal of Applied Physics, **75**, pp. 5683–5685, (1994).
[10] G. Friedman and I.D. Mayergoyz, IEEE Transactions on Magnetics, **28**, pp. 2262–2264, (1992).
[11] C.W. Gardiner, "Handbook of Stochastic Methods," Springer, (1983).
[12] Y.B. Kim, C.F. Hempstead, and A.R. Strnad, Physical Review Letters, **9**, pp. 306–308, (1962).
[13] C.E. Korman and P. Rugkwamsook, IEEE Transactions on Magnetics, **33**, pp. 4176–4178, (1997).
[14] C.E. Korman and I.D. Mayergoyz, IEEE Transactions on Magnetics, **31**, pp. 3545–3547, (1995).
[15] C.E. Korman and I.D. Mayergoyz, IEEE Transactions on Magnetics, **32**, 4204–4209, (1996).
[16] H. London, Phys. Lett. **6**, pp. 162–165, 1963).
[17] I.D. Mayergoyz, "Mathematical models of Hysteresis," Springer, (1991).
[18] I.D. Mayergoyz and T.A. Keim, Journal of Applied Physics, **67**, pp. 5466–5468, (1990).
[19] I.D. Mayergoyz, Journal of Applied Physics, **79**, pp. 6473–6475, (1996).
[20] I.D. Mayergoyz and C.E. Korman, Journal of Applied Physics, **69**, pp. 2128–2134, (1991).
[21] I.D. Mayergoyz and C.E. Korman, Journal of Applied Physics, **75**, pp. 5478–5480, (1994).
[22] I.D. Mayergoyz, Journal of Applied Physics, **76**, pp. 7130–7132, (1994).
[23] I.D. Mayergoyz, Journal of Applied Physics, **75**, pp. 6963–6965, (1994).
[24] I.D. Mayergoyz, Journal of Applied Physics, **76**, pp. 6956–6958, (1994).
[25] I.D. Mayergoyz and M. Neely, Journal of Applied Physics, **79**, pp. 6602–6604, (1996).
[26] I.D. Mayergoyz and M. Neely, Journal of Applied Physics, **81**, pp. 4234–4236, (1997).
[27] C.S. Nikols and D.R. Clarke, Acta Metall., Matter., **39**, p. 995, (1991).

[28] C.J.G. Plummer and J.E. Evetts, IEEE Transactions on Magnetics, **23**, pp. 1179–1182, (1987).
[29] W.M. Vinokur, M.V. Feigelman, and V.B. Geshkenbein, Physical Review Letters, **67**, pp. 915–918, (1991).

CHAPTER 5

Nonlinear Impedance Boundary Conditions and Their Application To the Solution of Eddy Current Problems

5.1 MATHEMATICAL STRUCTURE OF MAXWELL'S EQUATIONS FOR EDDY CURRENT PROBLEMS

In the previous chapters, we extensively discussed the nonlinear diffusion of plane electromagnetic waves in magnetically (or electrically) nonlinear conducting half-space. Special attention has been paid to the calculation of surface impedances. This has been done on purpose. The reason is that, by using these impedances, nonlinear impedance boundary conditions can be formulated. These boundary conditions can then be applied to the computation of eddy currents in conductors of complex shapes provided that the penetration ("skin") depth is small in comparison with geometric dimensions of these conductors. It turns out that the impedance boundary conditions can be most effectively formulated in terms of the magnetic scalar potential. To understand the rationale behind this formulation, it is worthwhile to consider first the mathematical structure of Maxwell's equations for 3-D eddy current problems. This structure is quite peculiar, and it is of interest in its own right.

To start the discussion, consider a conductor V^+ of arbitrary shape subject to a given external ("source") field (see Fig. 5.1). Notations $\mathbf{E}^0$

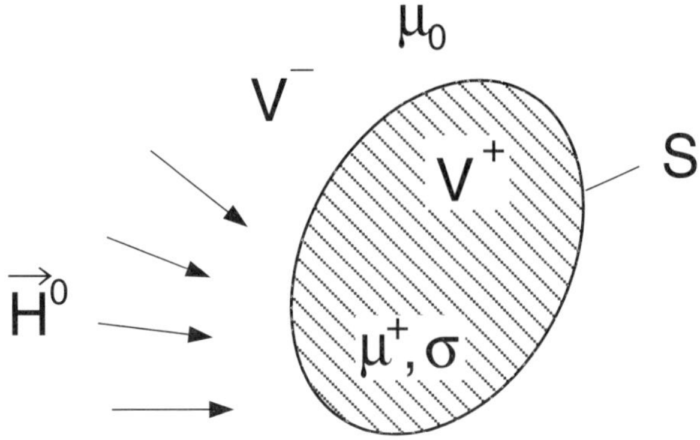

Fig. 5.1

and $\mathbf{H}^0$ will be used for the electric and magnetic vectors, respectively, of the source field. Electric and magnetic vectors of the scattered (reflected) field will be respectively denoted as $\mathbf{E}^-$ and $\mathbf{H}^-$. Within the conductor, the total electromagnetic field vectors, $\mathbf{E}^+$ and $\mathbf{H}^+$, will be employed.

Now, by using Maxwell's equations, the calculation of time harmonic electromagnetic fields can be reduced to the following boundary value problem: find the solution of the equations

$$\text{curl } \mathbf{E}^+ = -j\omega\mu^+\mathbf{H}^+, \tag{5.1}$$

$$\text{curl } \mathbf{H}^+ = \sigma\mathbf{E}^+ + j\omega\epsilon^+\mathbf{E}^+, \tag{5.2}$$

$$\text{curl } \mathbf{E}^- = -j\omega\mu_0\mathbf{H}^-, \tag{5.3}$$

$$\text{curl } \mathbf{H}^- = j\omega\epsilon_0\mathbf{E}^-, \tag{5.4}$$

subject to the boundary conditions

$$\vec{\nu} \times \left(\mathbf{H}^+ - \mathbf{H}^-\right) = \vec{\nu} \times \mathbf{H}^0, \tag{5.5}$$

$$\vec{\nu} \times \left(\mathbf{E}^+ - \mathbf{E}^-\right) = \vec{\nu} \times \mathbf{E}^0. \tag{5.6}$$

It is also tacitly understood that the outgoing Sommerfeld radiation conditions are imposed at infinity on the scattered field vectors $\mathbf{E}^-$ and $\mathbf{H}^-$.

In the above equations, notations μ, ϵ, and σ stand for magnetic permeability, dielectric permittivity, and conductivity, respectively, while $\vec{\nu}$ is a unit vector of outward normal to the conductor boundary, S. It is also clear that the above equations are written for phasor vector quantities.

It can be proven that the boundary value problem (5.1)–(5.6) has a unique solution. The proof can be found in many textbooks on electromagnetics.

It can be remarked that the following equation and integral-type boundary condition

$$\text{div } \mathbf{E}^- = 0, \tag{5.7}$$

$$\oint_S \mathbf{E}^- \cdot \mathbf{d}S = 0 \tag{5.8}$$

can be derived from Eq. (5.4). Indeed, by applying div-operator to both sides of Eq. (5.4), we find

$$0 = \text{div curl } \mathbf{H}^- = j\omega\epsilon_0 \text{ div } \mathbf{E}^-, \tag{5.9}$$

which is tantamount to (5.7). Similarly, by integrating Eq. (5.4) over the boundary S, we obtain

$$\oint_S \text{curl } \mathbf{H}^- \cdot \mathbf{d}S = j\omega\epsilon_0 \oint_S \mathbf{E}^- \cdot \mathbf{d}S. \tag{5.10}$$

By invoking the Stokes theorem, we derive

$$\oint_S \text{curl } \mathbf{H}^- \cdot \mathbf{d}S = 0, \tag{5.11}$$

which, in combination with formula (5.10), is equivalent to boundary condition (5.8).

The boundary value problem (5.1)–(5.6) is typically used for high frequencies (scattering problems). In eddy current problems, it is customary that frequency ω is relatively small. For this reason, displacement currents inside and outside the conductor V^+ can be neglected. This leads to the following degenerate form of the boundary value problem (5.1)–(5.6): find the solution to the equations

$$\text{curl } \mathbf{E}^+ = -j\omega\mu^+\mathbf{H}^+ \tag{5.12}$$

$$\text{curl } \mathbf{H}^+ = \sigma\mathbf{E}^+, \tag{5.13}$$

$$\text{curl } \mathbf{E}^- = -j\omega\mu_0\mathbf{H}^-, \tag{5.14}$$

$$\text{curl } \mathbf{H}^- = 0, \tag{5.15}$$

subject to the boundary conditions

$$\vec{\nu} \times \left(\mathbf{H}^+ - \mathbf{H}^-\right) = \vec{\nu} \times \mathbf{H}^0, \tag{5.16}$$

$$\vec{\nu} \times \left(\mathbf{E}^+ - \mathbf{E}^-\right) = \vec{\nu} \times \mathbf{E}^0. \tag{5.17}$$

The boundary value problem (5.12)–(5.17) does not have a unique solution. Indeed, formula (5.7) does not follow from Eq. (5.15) and, consequently, div$\mathbf{E}^-$ may assume an arbitrary value:

$$\text{div } \mathbf{E}^- \neq 0. \tag{5.18}$$

It is also clear that the boundary condition (5.8) does not follow from Eq. (5.15) either. As a result, integral $\oint_S \mathbf{E}^- \cdot \mathbf{d}S$ may assume an arbitrary value as well:

$$\oint_S \mathbf{E}^- \cdot \mathbf{d}S \neq 0. \tag{5.19}$$

This state of affairs occurs because we have neglected displacement currents and Eq. (5.4) has become homogeneous Eq. (5.15).

Thus, the conclusion can be reached that Eq. (5.7) and boundary condition (5.8) are independent of relations (5.12)–(5.17) and must be incorporated into the mathematical formulation. This leads to the following complete set of differential equations and boundary conditions:

$$\text{curl } \mathbf{E}^+ = -j\omega\mu^+\mathbf{H}^+, \tag{5.20}$$

$$\text{curl } \mathbf{H}^+ = \sigma\mathbf{E}^+, \tag{5.21}$$

$$\text{curl } \mathbf{E}^- = -j\omega\mu_0\mathbf{H}^-, \tag{5.22}$$

$$\text{curl } \mathbf{H}^- = 0, \tag{5.23}$$

$$\text{div } \mathbf{E}^- = 0, \tag{5.24}$$

$$\vec{\nu} \times \left(\mathbf{H}^+ - \mathbf{H}^-\right) = \vec{\nu} \times \mathbf{H}^0, \tag{5.25}$$

$$\vec{\nu} \times \left(\mathbf{E}^+ - \mathbf{E}^-\right) = \vec{\nu} \times \mathbf{E}^0, \tag{5.26}$$

$$\oint_S \mathbf{E}^- \cdot \mathbf{d}S = 0. \tag{5.27}$$

The following result can be proven.

Theorem 1. The boundary value problem (5.20)–(5.27) with appropriate conditions at infinity has a unique solution. ■

The proof of this theorem as well as other theorems from this section can be found in [17].

The fact that the boundary value problem (5.20)–(5.27) has additional Eq. (5.24) and boundary condition (5.27) makes the initial formulation (5.1)–(5.6) look simpler than the formulation (5.20)–(5.27). This fact may also suggest that it would be better to compute eddy currents in conductor V^+ without neglecting displacement currents, that is, by using the initial formulation (5.1)–(5.6). However, the close examination of the situation shows that this is not the case. First, the boundary value problem (5.1)–(5.6) is ill-posed for small frequencies. Indeed, for small frequencies the problem (5.1)–(5.6) is very close to its degenerate form (5.12)–(5.17), which does not have a unique solution. Consequently, discretizations of the boundary value problem (5.1)–(5.6) may result in sets of simultaneous algebraic equations with determinants that are close to zero for small frequencies. Second and more important, the mathematical structure of the boundary value problem (5.20)–(5.27) is simpler than that of the boundary value problem (5.1)–(5.6). Indeed, the magnetic field $\mathbf{H}^-$ is curl-free in V^- (see Eq. (5.23)). For this reason, this field can be expressed in terms of magnetic scalar potential:

$$\mathbf{H}^- = -\nabla\varphi^-. \tag{5.28}$$

In other words, we can use the magnetic scalar potential as a state variable in the region V^- outside the conductor. As far as the conducting region V^+ is concerned, there are two possible choices:

$$\mathbf{H}^+ \text{ is a state variable or } \mathbf{E}^+ \text{ is a state variable,} \tag{5.29}$$

which lead to $\mathbf{H}-\varphi$ and $\mathbf{E}-\varphi$ formulations, respectively.

In $\mathbf{H}-\varphi$ **formulation** one deals with the solution of the equations:

$$\text{curl curl } \mathbf{H}^+ + j\omega\sigma\mu^+\mathbf{H}^+ = 0 \quad \text{in } V^+, \tag{5.30}$$

$$\nabla^2\varphi^- = 0 \quad \text{in } V^-, \tag{5.31}$$

subject to the interface boundary conditions:

$$\vec{\nu}\times\left(\mathbf{H}^+ + \nabla\varphi^-\right) = \vec{\nu}\times\mathbf{H}^0, \tag{5.32}$$

$$\vec{\nu}\cdot\left(\mu^+\mathbf{H}^+ + \mu_0\nabla\varphi^-\right) = \mu_0\vec{\nu}\cdot\mathbf{H}^0, \tag{5.33}$$

and the condition at infinity:

$$\varphi^-(\infty) = 0. \tag{5.34}$$

In the case of $\mathbf{E}-\varphi$ **formulation**, one deals with the solution of the following equations:

$$\text{curl curl } \mathbf{E}^+ + j\omega\sigma\mu^+\mathbf{E}^+ = 0 \quad \text{in } V^+, \tag{5.35}$$

$$\nabla^2 \varphi^- = 0 \qquad \text{in } V^-, \tag{5.36}$$

subject to the interface boundary conditions

$$\vec{\nu} \times \left(\frac{j}{\omega \mu^+} \text{ curl } \mathbf{E}^+ + \nabla \varphi^- \right) = \vec{\nu} \times \mathbf{H}^0, \tag{5.37}$$

$$\vec{\nu} \cdot \left(\frac{j}{\omega} \text{ curl } \mathbf{E}^+ + \mu_0 \nabla \varphi^- \right) = \mu_0 \vec{\nu} \cdot \mathbf{H}^0, \tag{5.38}$$

and the condition at infinity

$$\varphi^-(\infty) = 0. \tag{5.39}$$

It is easy to see that Eqs. (5.30) and (5.35) follow from Eqs. (5.20) and (5.21), while Eq. (5.31) (and (5.36)) can be easily derived from Eq. (5.22), which implies that

$$\text{div } \mathbf{H}^- = 0. \tag{5.40}$$

It is also apparent that the boundary conditions (5.33) and (5.38) express the continuity of the normal component of the magnetic flux density across the conductor boundary S. These boundary conditions can be easily derived from the boundary condition (5.26) and Eqs. (5.20) and (5.22). In other words, the $\mathbf{H} - \varphi$ and $\mathbf{E} - \varphi$ formulations are fully derivable from the boundary value problem (5.20)–(5.27).

The following statement can be proven.

Theorem 2. The boundary value problems (5.30)–(5.34) and (5.35)–(5.39) have unique solutions. ■

In the $\mathbf{H} - \varphi$ and $\mathbf{E} - \varphi$ formulations, vector equations are to be solved only within the conducting region, while the scalar Laplace equation is required to be solved in the region surrounding the conductor. This is a definite advantage over the formulation (5.1)–(5.6), because it leads to the appreciable reduction of the total number of discretized equations to be solved from 3-D eddy current problems. The choice between $\mathbf{H} - \varphi$ and $\mathbf{E} - \varphi$ formulations depends on a particular numerical technique to be used. The $\mathbf{H} - \varphi$ formulation is very suitable for the application of the boundary element technique. By using this approach, the boundary integral equations of minimum order have been derived and extensively used for the calculation of eddy currents [8], [9]. The $\mathbf{E} - \varphi$ formulation, on the other hand, lends itself to the finite element implementation. Furthermore, this formulation is also attractive because it leads directly to the calculation of eddy currents within conductors.

The $\mathbf{H} - \varphi$ and $\mathbf{E} - \varphi$ formulations allow one to compute the magnetic scalar potential φ^- and, consequently, the magnetic field $\mathbf{H}^-$ in the exterior region V^-. To compute the electric field $\mathbf{E}^-$ in the same region requires the solution of the additional boundary value problem, which will be called the **$\mathbf{E}^-$-problem**. In the $\mathbf{E}^-$-problem, one deals with the solution of the following equations:

$$\text{curl } \mathbf{E}^- = j\omega\mu_0 \nabla\varphi^-, \tag{5.41}$$

$$\text{div } \mathbf{E}^- = 0, \tag{5.42}$$

subject to the boundary conditions

$$\vec{\nu} \times \mathbf{E}^- = \vec{\nu} \times (\mathbf{E}^+ - \mathbf{E}^0), \tag{5.43}$$

$$\oint_S \mathbf{E}^- \cdot \mathbf{d}S = 0, \tag{5.44}$$

$$\mathbf{E}^-(\infty) = 0. \tag{5.45}$$

It is apparent that the above formulas are directly derivable from relations (5.22), (5.24), (5.26), (5.27), and (5.28). It is also clear that the solution of the $\mathbf{E}^-$-problem can be attempted only after the $\mathbf{H}-\varphi$ or $\mathbf{E}-\varphi$ formulations have been implemented.

The following statement can be proven.

Theorem 3. The boundary value problem (5.41)–(5.45) has a unique solution. ∎

The proof of the above theorem is instructive because it clearly reveals that the omission of Eq. (5.42) or the boundary condition (5.44) results in nonuniqueness of electric field $\mathbf{E}^-$. We shall not give this proof here, however, we shall rather briefly demonstrate how the boundary condition (5.44) is essential for the uniqueness of $\mathbf{E}^-$-field. To this end, we consider the boundary value problem with all relations (5.41)–(5.45) being homogeneous except the boundary condition (5.44):

$$\text{curl } \mathbf{E}^- = 0 \quad \text{in } V^-, \tag{5.46}$$

$$\text{div } \mathbf{E}^- = 0 \quad \text{in } V^-, \tag{5.47}$$

$$\vec{\nu} \times \mathbf{E}^- = 0 \quad \text{on } S \tag{5.48}$$

$$\oint_S \mathbf{E}^- \cdot \mathbf{d}S = \frac{q}{\varepsilon_0} \neq 0, \tag{5.49}$$

$$\mathbf{E}^-(\infty) = 0. \tag{5.50}$$

It can be easily recognized that the $\mathbf{E}^-$-field described by formulas (5.46)–(5.50) is the "electrostatic" field created by the charged conductor V^+ with total charge q. It is in order to exclude this electrostatic field that the boundary condition (5.45) (or (5.27) for the same matter) is imposed.

By summarizing the discussion presented above, it can be concluded that formulation (5.20)–(5.27) can be decoupled (split) into the $\mathbf{H} - \varphi$ (or $\mathbf{E} - \varphi$) formulation and the $\mathbf{E}^-$-problem, which can be solved sequentially one after another. This decoupling is a peculiar property of 3-D eddy current problems that can be traced back to the neglect of displacement currents. In many engineering applications (such as, for instance, nondestructive testing, induction heating, geological explorations) the solution of the $\mathbf{E}^-$-problem can be avoided. In those applications, all the relevant information (such as distributions of eddy currents and magnetic fluxes) can be directly extracted from $\mathbf{H} - \varphi$ (or $\mathbf{E} - \varphi$) formulation, while the knowledge of the $\mathbf{E}^-$-field is not paramount. Nevertheless, there are some important engineering problems that do require the knowledge of the $\mathbf{E}^-$-field and, consequently, the solution of the $\mathbf{E}^-$-problem. Those problems are related mostly to the area of accelerator technology. For instance, in betatron accelerators, particles are accelerated by electric fields induced in air or in a vacuum. Eddy currents in conducting surroundings of particle rings effect the induced electric fields and, consequently, the acceleration process.

5.2 CALCULATION OF THE SOURCE FIELD $\mathbf{H}^0$

In the previous section, we used the decomposition of the total field $\mathbf{H}^{t-}$ in the exterior region V^- into two components: the source field $\mathbf{H}^0$ and the "reflected" (secondary) field $\mathbf{H}^-$:

$$\mathbf{H}^{t-} = \mathbf{H}^- + \mathbf{H}^0. \tag{5.51}$$

Previously, our attention has been exclusively focused on various differential formulations for 3-D eddy current problems, while the issue of the calculation of the source field $\mathbf{H}^0$ has not been addressed at all. Now, it is this issue that will be the focus of our discussion.

In eddy current problems, a primary source of excitation is usually a coil (or a group of coils) with known (given) distribution of electric currents. This implies that the total field $\mathbf{H}^{t-}$ satisfies the following equations:

$$\text{curl } \mathbf{H}^{t-} = \begin{cases} \mathbf{J}^0 & \text{in } V^0, \\ 0 & \text{in } V^- - V^0, \end{cases} \tag{5.52}$$

$$\text{div } \mathbf{H}^{t-} = 0 \quad \text{in } V^-. \tag{5.53}$$

Here $\mathbf{J}^0$ is the known current density in the coils, while V^0 is the region occupied by the current-carrying coils (see Fig. 5.2).

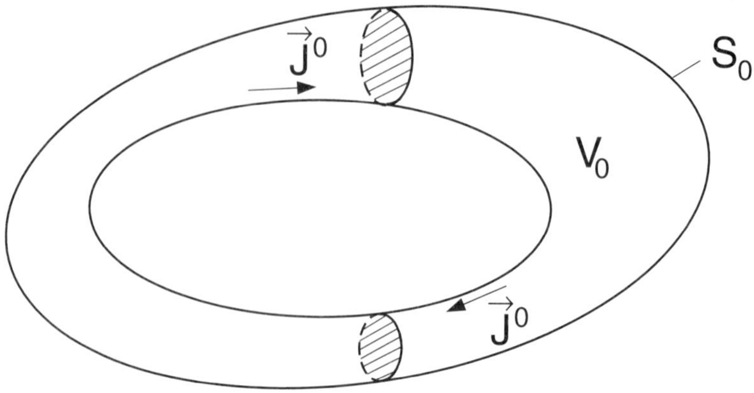

Fig. 5.2

In the previous section, we used the secondary (reflected) field $\mathbf{H}^-$ defined by the equations

$$\text{curl } \mathbf{H}^- = 0 \quad \text{in } V^-, \tag{5.54}$$

$$\text{div } \mathbf{H}^- = 0 \quad \text{in } V^-. \tag{5.55}$$

This prompted the introduction of the magnetic scalar potential

$$\mathbf{H}^- = -\nabla\varphi^-, \tag{5.56}$$

which, according to Eq. (5.55), must satisfy the Laplace equation

$$\nabla^2\varphi^- = 0. \tag{5.57}$$

From relations (5.52)–(5.55) and the decomposition (5.51) we find that the source field $\mathbf{H}^0$ satisfies the equations

$$\text{curl } \mathbf{H}^0 = \begin{cases} \mathbf{J}^0 & \text{in } V^0, \\ 0 & \text{in } V^- - V^0, \end{cases} \tag{5.58}$$

$$\text{div } \mathbf{H}^0 = 0 \quad \text{in } V^0. \tag{5.59}$$

Thus, we can take any solution of Eqs. (5.58)–(5.59) as a source field. One solution of those equations, which is particularly well known, is given by the Biot-Savart law:

$$\boxed{\mathbf{H}^0(Q) = \frac{1}{4\pi}\int_{V^0} \frac{\mathbf{J}^0(M) \times \mathbf{r}_{MQ}}{r_{MQ}^3} dV_M,} \tag{5.60}$$

where Q is a point of observation, M is a point of integration, r_{MQ} is the distance between M and Q, while $\mathbf{r}_{MQ}$ is the vector with magnitude r_{MQ} and direction from M to Q.

The source field given by the formula (5.60) has the transparent physical meaning: this is the field that would exist if the conductor V^+ were not present.

It is easy to see that the source field $\mathbf{H}^0$ given by formula (5.60) satisfies the Eq. (5.59). Indeed, by using simple vector calculus, we can rewrite formula (5.60) as follows:

$$\mathbf{H}^0 = \mathrm{curl}_Q \left(\frac{1}{4\pi} \int_{V^0} \frac{\mathbf{J}^0(M)}{r_{MQ}} dV_M \right), \tag{5.61}$$

which immediately implies the validity of formula (5.59). By using somewhat more complicated vector calculus transformations, it can be shown that the source field (5.60) satisfies the Eq. (5.58) as well.

Numerical implementation of formula (5.60) for the source field encounters two major computational difficulties. First, it requires numerical evaluation of volume integrals. Second, for each new observation point Q, computations have to be performed from scratch. The first difficulty can be somewhat ameliorated in the case when the source current density is curl-free:

$$\mathrm{curl}\ \mathbf{J}^0(M) = 0. \tag{5.62}$$

In this case, the volume integral in formula (5.60) can be reduced to the surface integral. This can be accomplished as follows (see [7]). From the vector calculus, we have

$$\mathrm{curl}_M \left(\frac{\mathbf{J}^0(M)}{r_{MQ}} \right) = \frac{1}{r_{MQ}} \mathrm{curl}\ \mathbf{J}^0(M) - \mathbf{J}^0(M) \times \nabla_M \left(\frac{1}{r_{MQ}} \right). \tag{5.63}$$

From the last expression and formula (5.62), we derive

$$\frac{\mathbf{J}^0(M) \times \mathbf{r}_{MQ}}{r_{MQ}^3} = -\mathrm{curl}_M \left(\frac{\mathbf{J}^0(M)}{r_{MQ}} \right). \tag{5.64}$$

By using the last equation in formula (5.60), we obtain

$$\mathbf{H}^0(Q) = -\frac{1}{4\pi} \int_{V^0} \mathrm{curl}_M \left(\frac{\mathbf{J}_0(M)}{r_{MQ}} \right) dV_M. \tag{5.65}$$

Now, by invoking the following integral relation from the vector calculus (see [5]):

$$\int_V \mathrm{curl}\ \mathbf{a} dV = \oint_S \vec{\nu} \times \mathbf{a} dS, \tag{5.66}$$

we transform formula (5.65) as follows:

$$\mathbf{H}^0(Q) = \frac{1}{4\pi} \oint_{S_0} \frac{\mathbf{J}^0(M) \times \vec{\nu}_M}{r_{MQ}} dS_M, \tag{5.67}$$

where S_0 is the boundary of V_0.

The last formula is simpler than formula (5.60), because it requires the evaluation of surface integrals instead of volume integrals. However, the validity of expression (5.67) is subject to the validity of formula (5.62), which can be rigorously justified only for dc currents. Furthermore, similar to formula (5.60), numerical implementations of formula (5.67) require performing computations from scratch for each new observation point Q. Thus, the conclusion can be easily reached that it is desirable to explore alternative approaches to the calculation of the source field, approaches that are not based on the Biot-Savart law. The main idea of these approaches is to compromise the divergence relation (5.59) in order to achieve simple expressions for the source field. We consider two such approaches. The first approach leads to the exceptionally simple formulas, however, it is applicable only for sufficiently simple geometries of V^0. The second approach is quite general in nature and it is based on the notion of the "Poincaré gauge." It leads to one-dimensional integrals and their evaluation does not start from scratch for each new observation point.

We present the first approach [17] for the case of a coil with rectangular cross-section (see Fig. 5.3). It is assumed that the height of the coil cross-section remains constant along the coil, while its base may change. Under these conditions, the z-component of $\mathbf{J}^0$ is equal to zero and the current density distribution in the coil does not depend on z. Next, we introduce the "extended" source region $\tilde{V}^0$, whose boundary consists of the exterior side surface, S_{sides}, of the coil and two flat surfaces, S_{top} and S_{bottom}. In other words, the extended source region $\tilde{V}^0$ contains the coil itself as well as the region enclosed by the coil and the flat surfaces S_{top} and S_{bottom}. We shall look for the source field, which is confined to the extended source region:

$$\mathbf{H}^0 = 0 \quad \text{in } V^- - \tilde{V}^0. \tag{5.68}$$

Within the extended source region, the source field satisfies the following equations:

$$\text{curl } \mathbf{H}^0 = \begin{cases} \mathbf{J}^0 & \text{in } V^0, \\ 0 & \text{in } \tilde{V}^0 - V^0, \end{cases} \tag{5.69}$$

$$\text{div } \mathbf{H}^0 = 0 \quad \text{in } \tilde{V}^0. \tag{5.70}$$

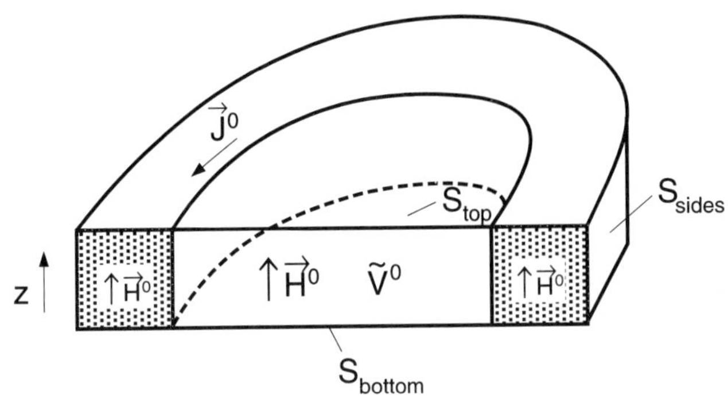

Fig. 5.3

Since the current density distribution does not depend on z, we look for the solution of Eqs. (5.69) and (5.70) in the form:

$$\mathbf{H}^0(x, y) = \mathbf{a}_z H^0(x, y). \tag{5.71}$$

It is apparent that Eq. (5.70) is satisfied, while Eq. (5.69) can be represented as follows:

$$\frac{\partial H^0}{\partial x} = -J_y^0(x, y) \quad \text{in } V^0, \tag{5.72}$$

$$\frac{\partial H^0}{\partial y} = J_x^0(x, y) \quad \text{in } V^0, \tag{5.73}$$

and the above derivatives are equal to zero in the region $\tilde{V}^0 - V^0$.

Now, we introduce the stream function $T(Q)$:

$$T(Q) = \int_{L_{OQ}} J_\nu^0(P) dl_P, \tag{5.74}$$

where L_{OQ} is an arbitrary line in the plane $z = z', (0 \le z' \le h)$, the reference point O belongs to the side surface S_{sides}, and J_ν^0 is the component of the source current density $\mathbf{J}^0$ normal to the line L_{OQ}.

Because

$$\text{div } \mathbf{J}^0 = \frac{\partial J_x^0}{\partial x} + \frac{\partial J_y^0}{\partial y} = 0, \tag{5.75}$$

$$\vec{\nu} \cdot \mathbf{J}^0 \Big|_{\text{sides}} = 0, \tag{5.76}$$

it is clear that the value of the stream function $T(Q)$ does not depend on the particular choice of the reference point 0 on S_{sides} or on the shape of the line L_{OQ}. The value of the stream function depends only on the position of observation point Q in the plane $z = z'$ and it is the same for all z', $(0 \leq z' \leq h)$. It is well known (and it is also easily derivable from the definition (5.74)) that the stream function is related to the current density as follows:

$$\frac{\partial T}{\partial x} = -J_y^0(x, y) \quad \text{in } V^0, \tag{5.77}$$

$$\frac{\partial T}{\partial y} = J_x^0(x, y) \quad \text{in } V^0, \tag{5.78}$$

and the above derivatives are equal to zero in the region $\tilde{V}^0 - V^0$.

Relations (5.72)–(5.73) and (5.77)–(5.78) are identical, so we conclude that

$$\mathbf{H}^0(Q) = \mathbf{a}_z T(Q). \tag{5.79}$$

This is the sought formula for the source field. It is apparent from this formula that

$$\mathbf{H}^0(Q) = 0, \quad \text{if } Q \in S_{\text{sides}}, \tag{5.80}$$

$$\mathbf{H}^0(Q) = \mathbf{a}_z \frac{I}{h} = \text{const}, \quad \text{if } Q \in \tilde{V}^0 - V^0, \tag{5.81}$$

where I is the total current through the coil.

If the coil cross-section as well as the magnitude $|\mathbf{J}|^0$ of the current density are uniform then the source field $\mathbf{H}^0$ varies linearly in normal cross-section of V^0. Otherwise, the calculation of the source field $\mathbf{H}^0$ requires the evaluation of one dimensional integrals (5.74), and in each normal cross-section these integrals are evaluated incrementally for each new observation point Q. In other words, computations of the above integrals do not start from scratch.

Next, we discuss the implications of the above choice of the source field. It is clear from formulas (5.68)–(5.70) that the field $\mathbf{H}^-$ satisfies the Eqs. (5.54)–(5.55). However, the normal components of $\mathbf{H}^-$ are discontinuous across the flat surfaces S_{top} and S_{bottom}:

$$\left[\vec{\nu} \cdot \mathbf{H}^-\right] \Big|_{S_{\text{top}}} = H^0(Q) = T(Q), \tag{5.82}$$

$$\left[\vec{\nu} \cdot \mathbf{H}^-\right] \Bigg|_{S_{\text{bottom}}} = -H^0(Q) = -T(Q), \tag{5.83}$$

where the symbol $[a]|_S$ stands for a discontinuous change in quantity a across S.

Thus, the magnetic scalar potential φ^- must satisfy the Laplace equation in V^-, however, its $z-$ derivatives are discontinuous across S_{top} and S_{bottom}:

$$\left[\frac{\partial \varphi^-}{\partial z}\right]\Bigg|_{S_{\text{top}}} = -H^0(Q) = -T(Q), \tag{5.84}$$

$$\left[\frac{\partial \varphi^-}{\partial z}\right]\Bigg|_{S_{\text{bottom}}} = H^0(Q) = T(Q). \tag{5.85}$$

Formulas (5.84)–(5.85) specify the driving forces for the magnetic scalar potential, and these formulas can be easily incorporated into finite element formulations. These formulas also admit the following physical interpretation: discontinuities (5.84)–(5.85) (as well as (5.82)–(5.85)) can be viewed as caused by fictitious surface magnetic charges distributed over S_{top} and S_{bottom}. The densities of these charges are given by the formulas

$$\sigma_m(Q)\Bigg|_{S_{\text{top}}} = \mu_0 T(Q), \tag{5.86}$$

$$\sigma_m(Q)\Bigg|_{S_{\text{bottom}}} = -\mu_0 T(Q). \tag{5.87}$$

One can easily recognize a close similarity between the modeling of coils with currents by magnetic charges (5.86)–(5.87) and the modeling of ideal permanent magnets by surface magnetic charges. This similarity is not accidental because the ideal permanent magnets can also be modeled by current-carrying coils.

Next, we proceed to the discussion of the second approach, which is applicable to coils of complex geometries. This approach is based on the notion of the Poincaré gauge, which is introduced for magnetic vector potentials [16], [1]. First, we shall briefly describe the basic facts related to this gauge.

It is well known that the magnetic vector potential is introduced by the equation

$$\text{curl } \mathbf{A} = \mathbf{B}, \tag{5.88}$$

where $\mathbf{B}$ is the magnetic flux density.

The last equation does not uniquely define the magnetic vector potential. To make this potential unique, various gauges are introduced additionally to Eq. (5.88). One of the most well-known gauges is the Coulomb gauge:

$$\text{div } \mathbf{A} = 0. \tag{5.89}$$

If we perform the Fourier transform of **A**:

$$\tilde{\mathbf{A}}_{(\mathbf{k})} = \int_{-\infty}^{\infty} \mathbf{A}(\mathbf{r}) e^{j\mathbf{k}\cdot\mathbf{r}} d\mathbf{r}, \tag{5.90}$$

then it is easy to establish that Eq. (5.89) can be written in terms of the transformed potential as follows:

$$\mathbf{k} \cdot \tilde{\mathbf{A}}(\mathbf{k}) = 0. \tag{5.91}$$

In other words, the Coulomb gauge defines the magnetic vector potential, which is orthogonal to the wave vector **k** in the **k**-space. For this reason, the Coulomb gauge is also called the "transverse gauge."

The idea of the Poincaré gauge is to define the vector potential, which is transversed in **r**-space. This means that the Poincaré gauge is mathematically defined by the equation

$$\mathbf{r} \cdot \mathbf{A}(\mathbf{r}) = 0, \tag{5.92}$$

which is dual to Eq. (5.91). The last equation implies that the vector potential in the Poincaré gauge is always tangential to spherical surfaces.

It can be proven that Eq. (5.88) along with the Poincaré gauge (5.92) uniquely defines the magnetic vector potential. Furthermore, there exists quite a simple formula, which expresses this potential in terms of the magnetic flux density:

$$\mathbf{A}(\mathbf{r}) = -\int_{0}^{1} \lambda\mathbf{r} \times \mathbf{B}(\lambda\mathbf{r}) d\lambda. \tag{5.93}$$

Thus, the vector potential can be evaluated by integrating along the ray, which connects the origin with the observation point.

It is this simple formula that makes the Poincaré gauge attractive in quantum mechanics. The reason is that the vector magnetic potential appears in the Schrödinger equation. Thus, if we would like to write the Schrödinger equation for a given external magnetic field, we need the expression for the corresponding magnetic vector potential, and formula (5.93) can be used for this purpose.

The Poincaré gauge can also be very attractive in computational electromagnetics. It enforces the uniqueness of the magnetic vector potential by locally reducing (through Eq. (5.92)) the number of unknowns to two. This can be very advantageous for finite element implementations. Unfortunately, the Poincaré gauge is not well known and is not sufficiently appreciated by the finite element community.

Next, we shall use the Poincaré gauge for a different purpose; namely, we shall apply it to the calculation of the source field $\mathbf{H}^0$. The main equation that the source field must satisfy can be written as follows:

$$\text{curl } \mathbf{H}^0 = \mathbf{J}^0, \tag{5.94}$$

where it is tacitly assumed that $\mathbf{J}^0$ is equal to zero outside V^0. Mathematically, the last equation is identical to formula (5.88), and by itself it does not uniquely define the source field $\mathbf{H}^0$. To make this field unique, we shall impose the Poincaré gauge condition:

$$\mathbf{r} \cdot \mathbf{H}^0(\mathbf{r}) = 0. \tag{5.95}$$

Because the pair of relations (5.94)–(5.95) is mathematically identical to the pair of relations (5.88) and (5.92), we can conclude that the source field $\mathbf{H}^0$ can be expressed in terms of the source current density $\mathbf{J}^0$ in exactly the same way as $\mathbf{A}$ is expressed in terms of $\mathbf{B}$. Thus, we arrive at the formula

$$\boxed{\mathbf{H}^0(\mathbf{r}) = -\int_0^1 \lambda\mathbf{r} \times \mathbf{J}^0(\lambda\mathbf{r})d\lambda.} \tag{5.96}$$

The gauge (5.92) (as well as the gauge (5.95)) is called the Poincaré gauge because formula (5.93) (as well as (5.96)) can be most generally derived from the Poincaré lemma for differential forms [1]. Next, we present a direct derivation of formula (5.96), which is based on vector calculus.

To start the derivation, consider the ray:

$$\mathbf{r}' = \lambda\mathbf{r}, \quad 0 \le \lambda \le 1. \tag{5.97}$$

Next, we write the gauge (5.95) in terms of $\mathbf{r}'$:

$$\mathbf{r}' \cdot \mathbf{H}^0(\mathbf{r}') = 0, \tag{5.98}$$

and take the gradient of both sides of the last equation:

$$\nabla(\mathbf{r}' \cdot \mathbf{H}^0(\mathbf{r}')) = 0. \tag{5.99}$$

By using a well-known formula from vector calculus, we find

$$\begin{aligned} \nabla(\mathbf{r}' \cdot \mathbf{H}^0(\mathbf{r}')) &= \mathbf{H}^0(\mathbf{r}')\nabla \cdot \mathbf{r}' + \mathbf{r}'\nabla \cdot \mathbf{H}^0(\mathbf{r}') \\ &+ \mathbf{H}^0(\mathbf{r}') \times \nabla \times \mathbf{r}' + \mathbf{r}' \times \nabla \times \mathbf{H}^0(\mathbf{r}') = 0. \end{aligned} \tag{5.100}$$

Here

$$\mathbf{H}^0(\mathbf{r}')\nabla \cdot \mathbf{r}' = H_x^0(\mathbf{r}')\frac{\partial \mathbf{r}'}{\partial x'} + H_y^0(\mathbf{r}')\frac{\partial \mathbf{r}'}{\partial y'} + H_z^0(\mathbf{r}')\frac{\partial \mathbf{r}'}{\partial z'}, \tag{5.101}$$

which leads to

$$\mathbf{H}^0(\mathbf{r}')\nabla \cdot \mathbf{r}' = \mathbf{H}^0(\mathbf{r}'). \tag{5.102}$$

Then, we recall that

$$\nabla \times \mathbf{r}' = 0, \qquad \nabla \times \mathbf{H}^0(\mathbf{r}') = \mathbf{J}^0(\mathbf{r}'). \tag{5.103}$$

By substituting expressions (5.102) and (5.103) into formula (5.100), we find

$$\mathbf{H}^0(\mathbf{r}') + \mathbf{r}'\nabla \cdot \mathbf{H}^0(\mathbf{r}') = -\mathbf{r}' \times \mathbf{J}^0(\mathbf{r}'). \tag{5.104}$$

Next, we use formula (5.97) and consider the derivative:

$$\begin{aligned} \frac{d\mathbf{H}^0(\mathbf{r}')}{d\lambda} &= \frac{\partial \mathbf{H}^0(\mathbf{r}')}{\partial x'} \cdot \frac{dx'}{d\lambda} + \frac{\partial \mathbf{H}^0(\mathbf{r}')}{\partial y'} \cdot \frac{dy'}{d\lambda} + \frac{\partial \mathbf{H}^0(\mathbf{r}')}{\partial z'} \cdot \frac{dz'}{d\lambda} \\ &= x\frac{\partial \mathbf{H}^0(\mathbf{r}')}{\partial x'} + y\frac{\partial \mathbf{H}^0(\mathbf{r}')}{\partial y'} + z\frac{\partial \mathbf{H}^0(\mathbf{r}')}{\partial z'} = \frac{1}{\lambda}\left[\mathbf{r}'\nabla \cdot \mathbf{H}^0(\mathbf{r}')\right]. \end{aligned} \tag{5.105}$$

Thus, we have

$$\mathbf{r}'\nabla \cdot \mathbf{H}^0(\mathbf{r}') = \lambda\frac{d\mathbf{H}^0(\lambda\mathbf{r})}{d\lambda}. \tag{5.106}$$

By substituting the last formula into relation (5.104), after simple transformations we arrive at

$$\frac{d}{d\lambda}\left[\lambda\mathbf{H}^0(\lambda\mathbf{r})\right] = -\lambda\mathbf{r} \times \mathbf{J}^0(\lambda\mathbf{r}). \tag{5.107}$$

By integrating the last formula with respect λ from 0 to 1 and taking into account that

$$\int_0^1 \frac{d}{d\lambda}[\lambda\mathbf{H}^0(\lambda\mathbf{r})]d\lambda = \mathbf{H}^0(\mathbf{r}), \tag{5.108}$$

we finally derive the formula (5.96).

Formula (5.96) is valid for arbitrary geometry of the source region V^0, and this is an important advantage of this formula over formula (5.79), which is applicable only to coils of sufficiently simple geometries. In other words, formula (5.96) is as general as the Biot-Savart law (5.60). However, its numerical implementation is appreciably simpler. First, due to the gauge condition (5.95), only two components of the source field should be computed at every observation point. Second, the computation of these components requires only evaluations of one-dimensional integrals along the rays connecting the observation points with the origin. Third, for observation points belonging to the same rays, the above integrals can be evaluated incrementally, rather than being computed from scratch for each new observation point. Finally, the source field is localized. In other words, the source field is equal to zero along the rays, which do not cross the source region V^0. Furthermore, for the rays that do cross V^0, the source field $\mathbf{H}^0$ varies only along those segments of those rays, which are within the source region. As is usually the case, the described computational efficiency comes at some cost. The divergence of the source field $\mathbf{H}^0$ defined by the formula (5.96) is not equal to zero:

$$\text{div } \mathbf{H}^0 \neq 0. \tag{5.109}$$

As a result

$$\text{div } \mathbf{H}^- = - \text{ div } \mathbf{H}^0 \neq 0. \tag{5.110}$$

This means that the magnetic scalar potential φ^- will satisfy the Poisson equation

$$\nabla^2 \varphi^- = \text{div } \mathbf{H}^0, \tag{5.111}$$

rather than the Laplace equation. Finite element discretizations of the above Poisson equations will contain additional source terms of the form:

$$\int_{V_i} \nabla \alpha_i \cdot \mathbf{H}^0 dv, \tag{5.112}$$

where α_i is a finite element function. However, these complications are of rather marginal nature.

5.3 IMPEDANCE BOUNDARY CONDITIONS

It has been demonstrated in Section 5.1 that the analysis of 3-D eddy current problems requires the solution of vector partial differential equations within conductors and scalar (Laplace-Poisson) equations outside the conductors. It is clear that the solution of vector partial differential equations within the conductors constitutes by far the most laborious task of the 3-D eddy current analysis. This task becomes even more daunting when the conductors are magnetically (or electrically) nonlinear. The level of computational difficulties is even further raised when penetration (skin) depths are small in comparison with geometric dimensions of the conductors. This is because, for small skin depths, electromagnetic fields are closely concentrated near conductor boundaries and decay very fast in directions normal to these boundaries. These fast spatial variations of electromagnetic fields are very hard to resolve numerically. Fortunately, the aforementioned numerical difficulties can be completely circumvented by using the idea of impedance boundary conditions. These boundary conditions are based on the notion that for small skin depths electromagnetic fields penetrate locally (i.e., at each boundary point) almost in the same way as plane waves penetrate a conducting half-space. This leads to the conclusion that, at each boundary point, tangential components of electric and magnetic fields are related to one another almost in the same way as in the case of plane waves penetrating the conducting half-space. These relations between the tangential components of electric and magnetic fields can be construed as local impedance boundary conditions. These relations are well known in the case of linear conducting media and they can be expressed mathematically as follows:

$$\vec{\nu} \times \mathbf{E} = \hat{\eta}(\vec{\nu} \times \mathbf{H}), \tag{5.113}$$

where $\vec{\nu}$ is a unit vector of outward normal, while $\hat{\eta}$ is the impedance matrix, which is defined as follows:

$$\hat{\eta} = \sqrt{\frac{\omega\mu}{\sigma}} e^{j\frac{\pi}{4}} \cdot \hat{\Lambda}, \tag{5.114}$$

$$\hat{\Lambda} = \begin{pmatrix} 0 & -1 \\ 1 & 0 \end{pmatrix}. \tag{5.115}$$

In the case of nonlinear conducting media, the general mathematical structure (5.113) of the impedance boundary conditions remains the same, however, the expression for the impedance matrix is different. This matrix can be written as

$$\hat{\eta} = \begin{pmatrix} 0 & -\eta_{12} \\ \eta_{21} & 0 \end{pmatrix}, \tag{5.116}$$

where η_{12} and η_{21} depend on the type of polarization of electromagnetic field at a given point of the conductor boundary. For instance, in the case of circular polarizations we have derived (see Chapter 2) the following expressions:

$$\eta_{12} = \eta_{21} = |\eta| e^{j\varphi}, \tag{5.117}$$

$$|\eta| = \left(\frac{2n}{n+1}\right)^{\frac{1}{4}} \sqrt{\frac{\omega\mu_m}{\sigma}}, \quad \tan\varphi = \sqrt{\frac{n+1}{2n}}. \tag{5.118}$$

Here n is an exponent in power law approximation for magnetization curves, while μ_m is the magnetic permeability at boundary points, which is a nonlinear function of magnetic field magnitude; this makes the impedance boundary conditions (5.113) nonlinear.

In the previous chapters, we have also derived the expressions for η_{12} and η_{21} in the cases of linear and elliptical polarizations. All these expressions can be easily incorporated in the impedance boundary conditions defined by formulas (5.113) and (5.116).

It is important to stress that the impedance boundary conditions are not exact but approximate in nature. They are sufficiently accurate only for small skin depths, and the smaller the skin depths in comparison with geometric dimensions of the conductors the more accurate the impedance boundary conditions. The mathematical justification of the impedance boundary conditions is based on the fact that electromagnetic fields, which are tightly concentrated near the conductor boundaries, can be construed as "boundary layers." This suggests that eddy current problems with small skin depths belong to the class of singularly perturbed problems. For this reason, the impedance boundary conditions have long been justified by using the perturbation analysis. This approach can be traced back to the paper of S.M. Rytov [15]. The impedance boundary conditions have been extensively used in electromagnetic field computations (see, for instance, [3], [17]), and their acceptance has reached the point that they are now presented in textbooks [4].

To understand better the accuracy of the impedance boundary conditions as well as their special importance for **magnetic** conductors, we consider the example problem of a conducting sphere V^+ subject to a uniform time-harmonic magnetic flux density $\mathbf{B}^0$ (see Fig. 5.4). This problem admits the exact analytical solution, which can be compared with the approximate analytical solution obtained by using the impedance boundary conditions.

It will be assumed that the external field $\mathbf{B}^0$ is directed along the z-axis:

$$\mathbf{B}^0 = \mathbf{a}_z B^0. \tag{5.119}$$

Then, it is clear that the posed problem is axially symmetric. As a result, the magnetic vector potential has only the φ-component, A_φ. In the region V^- outside the conducting sphere, this component can be represented as follows:

$$A_\varphi = A_\varphi^- + A_\varphi^0, \tag{5.120}$$

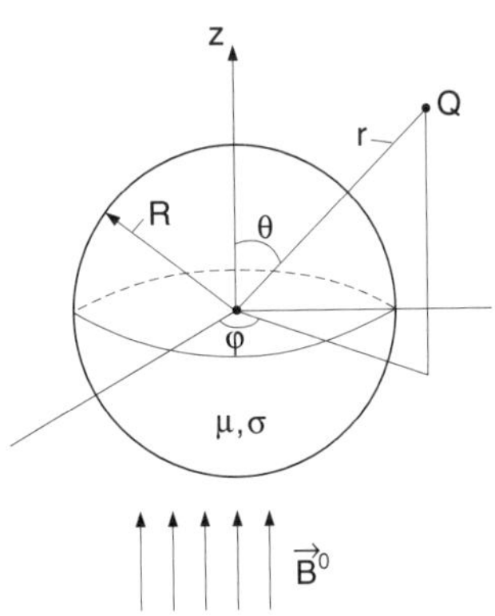

Fig. 5.4

where A_φ^- is the magnetic vector potential of the field in V^- created by the magnetization and eddy currents induced in the magnetic conducting sphere, while A^0 is the magnetic vector potential of the source field $\mathbf{B}^0$.

It is easy to establish that the potential A_φ^0 is related to the source field B^0 by the expression

$$A_\varphi^0 = \frac{1}{2} B^0 r \sin\theta, \tag{5.121}$$

where r, φ, and θ are spherical coordinates with the origin coinciding with the center of the sphere.

Indeed, consider a circular path $L_{r,\theta}$ defined by equations r =const and θ =const. The source flux, which links this path, can be evaluated by using the line integral of A_φ^0 along $L_{r,\theta}$ as well as the surface integral of B_0 over the circular region $S_{r,\theta}$ enclosed by $L_{r,\theta}$. This leads to

$$\oint_{L_{r,\theta}} A_\varphi^0 dl = A_\varphi^0 2\pi r \sin\theta = \int_{S_{r,\theta}} B^0 ds = B^0 \pi r^2 \sin^2\theta, \tag{5.122}$$

which yields formula (5.121).

The potential, A_φ^-, satisfies in the region V^- the following equation:

$$\frac{1}{r^2}\frac{\partial}{\partial r}\left(r^2 \frac{\partial A_\varphi^-}{\partial r}\right) + \frac{1}{r^2 \sin\theta}\frac{\partial}{\partial\theta}\left(\sin\theta \frac{\partial A_\varphi^-}{\partial\theta}\right) - \frac{A_\varphi^-}{r^2\sin^2\theta} = 0, \quad \text{for } R \le r < \infty, \tag{5.123}$$

where R is the radius of the conducting sphere.

It is assumed that the frequency of the source field is large enough that the skin depth in the conducting sphere is quite small in comparison with R. Under these conditions, we can use the impedance boundary conditions on the sphere boundary S_R. We shall first reformulate these boundary conditions in terms of A_φ. To this end, we shall invoke the relations:

$$E_\varphi = -j\omega A_\varphi, \tag{5.124}$$

$$H_\theta = -\frac{1}{\mu_0 r}\frac{\partial}{\partial r}(rA_\varphi). \tag{5.125}$$

By assuming the conducting medium of the sphere to be linear and by using formulas (5.124) and (5.125), we represent the impedance boundary conditions (5.113)–(5.115) in the form:

$$\left[A_\varphi - \frac{1-j}{2}\cdot\frac{\mu}{\mu_0}\cdot\frac{\delta}{r}\frac{\partial}{\partial r}(rA_\varphi)\right]\Bigg|_{r=R} = 0, \tag{5.126}$$

where δ is the skin depth:

$$\delta = \sqrt{\frac{2}{\omega\sigma\mu}}. \tag{5.127}$$

In order to satisfy the boundary condition (5.126), we look for the solution of Eq. (5.123) in the form that has the same dependence on θ as A_φ^0. It is easy to see that this solution is given by formula

$$A_\varphi^- = \frac{a_1 B^0}{2r^2}\sin\theta, \tag{5.128}$$

where coefficient a_1 should be determined from the boundary condition (5.126).

By substituting relations (5.128) and (5.121) into formula (5.120) and then into the boundary condition (5.126), after straightforward transformations we derive the following expression for a_1:

$$a_1 = \frac{2 - (1+j)\frac{\mu_0}{\mu}\cdot\frac{R}{\delta}}{1 + (1+j)\frac{\mu_0}{\mu}\cdot\frac{R}{\delta}} R^3. \tag{5.129}$$

The posed problem can also be solved exactly, i.e., without invoking impedance boundary conditions. For the exact analytical solution, potential A_φ^- has also the form given by formula (5.128), however, the coefficient a_1 is determined by the equation

$$a_1^{ex} = \frac{(\mu_0 + 2\mu)(\beta R \mathrm{cth}\beta R - 1) - \mu_0(\beta R)^2}{\mu_0(\beta R)^2 - (\mu_0 - \mu)(\beta R \mathrm{cth}\beta R - 1)} R^3, \tag{5.130}$$

where

$$\beta = \sqrt{j\omega\sigma\mu} = \frac{1+j}{\delta}. \tag{5.131}$$

In the case when the skin depth is much smaller than the radius of the sphere:

$$\frac{R}{\delta} \gg 1, \tag{5.132}$$

we have

$$|\beta R| \gg 1, \tag{5.133}$$

$$\mathrm{cth}\beta R \approx 1. \tag{5.134}$$

By using the last two formulas in Eq. (5.130), we easily obtain the following asymptotic expression for a_1^{ex} :

$$a_1^{ex} \approx \frac{2\mu\beta R - \mu_0(\beta R)^2}{\mu\beta R + \mu_0(\beta R)^2} R^3. \tag{5.135}$$

By dividing the numerator and denominator of the fraction in (5.135) by $\mu\beta R$ and taking into account (5.131), we derive

$$a_1^{ex} \approx \frac{2-(1+j)\frac{\mu_0}{\mu}\cdot\frac{R}{\delta}}{1+(1+j)\frac{\mu_0}{\mu}\cdot\frac{R}{\delta}}R^3, \tag{5.136}$$

which coincides with (5.129). Thus, as far as the external field is concerned, the impedance boundary conditions lead to the solution, which is asymptotically (for small skin depth) the same as the exact solution.

As an example, consider the conducting sphere with $R = 0.1m$, $\mu = 10^2\mu_0$, $\sigma = 0.833.10^7 \frac{1}{\Omega\cdot m}$ and assume that the frequency of the external field is $f = 60\text{sec}^{-1}$. For these data, we have

$$\delta = 0.00225m, \tag{5.137}$$

and the condition (5.132) is met. By substituting these data into formula (5.136) (or (5.129)), we find

$$a_1 = (0.89 - j0.59)R^3. \tag{5.138}$$

Now, let us compare the impedance boundary conditions with other approximate boundary conditions, which have long been used for the solution of electromagnetic problems. In the case of nonmagnetic conductor and small skin depths, the boundary condition of perfect (ideal) conductor is customarily used. This boundary condition can be written as follows:

$$\vec{\nu} \times \mathbf{E} = 0, \tag{5.139}$$

which reflects that tangential components of electric fields on the boundary of good conductors is almost equal to zero.

From the above boundary condition, we find

$$\vec{\nu} \cdot \mathbf{H} = 0, \tag{5.140}$$

which means that eddy currents, induced in good conductors, shield these conductors from the magnetic fields and force the magnetic field lines to go around the conductors.

Consider the solution of the above problem in the case of boundary condition (5.139). By invoking the relation (5.124), the above boundary condition can be written as follows:

$$A_\varphi = 0. \tag{5.141}$$

Now, by using the last equation as well as formulas (5.120), (5.121), and (5.128), we easily derive

$$a_1 = -R^3. \tag{5.142}$$

According to formula (5.129), the impedance boundary conditions, applied to the nonmagnetic ($\mu = \mu_0$) conducting sphere, lead to the result

$$a_1 = \frac{2 - (1+j)\frac{R}{\delta}}{1 + (1+j)\frac{R}{\delta}} R^3. \tag{5.143}$$

By comparing formulas (5.142) and (5.143) and taking inequality (5.132) into account, we conclude that, in the case of nonmagnetic conductors and small skin depth, the perfect conductor boundary conditions and impedance boundary conditions lead to asymptotically equivalent results. In other words, as far as the external fields are concerned, the perfect conductor boundary conditions give fairly accurate results (first approximations), whereas the impedance boundary conditions only slightly improve them. The situation is fundamentally different in the case of magnetic ($\mu \gg \mu_0$) conductors and small skin depths. It is evident from comparison of formulas (5.138) and (5.142) that the perfect conductor boundary conditions do not lead to asymptotically accurate results, whereas the impedance boundary conditions do (we have already proved this). The physical reason for this difference is not difficult to see. The point is that the perfect conductor boundary conditions enforce zero normal component of the magnetic field at the conductor boundary (see formula (5.140)), whereas for magnetic conductors this component of the magnetic field can be quite large due to the large magnetic permeability of the conductor.

It may be conjectured that, in the case of magnetic ($\mu \gg \mu_0$) conductors, accurate results can be achieved by using the "ideal magnetic boundary conditions":

$$\vec{\nu} \times \mathbf{H} = 0. \tag{5.144}$$

These boundary conditions are exact in the case when $\mu = \infty$ and no current is enclosed by "ideal" magnetic objects. For real magnetic objects, this boundary conditions can be regarded as approximate ones.

Now, consider the solution of the problem shown in Fig. 5.4 in the case of boundary condition (5.144). By invoking the relation (5.125), this boundary condition can be written as follows:

$$\left. \frac{\partial}{\partial r}(rA_\varphi) \right|_{r=R} = 0. \tag{5.145}$$

Now, by using the last equation as well as formulas (5.120), (5.121), and (5.128), we easily derive

$$a_1 = 2R^3. \tag{5.146}$$

By comparing formulas (5.138) and (5.146), the conclusion can be easily reached that the ideal magnetic boundary conditions do not lead to asymptotically accurate results either, i.e., the results that are accurate for small skin depths. The reason is not difficult to find. The thing is that, in magnetic ($\mu \gg \mu_0$) conductors with small skin depths $\left(\frac{R}{\delta} \gg 1\right)$, there is always some trade-off between electric and magnetic shielding effects. The degree of this trade-off is determined by the delicate interplay of such parameters as permeability μ, conductivity σ and frequency f. The "ideal" boundary conditions (5.139) or (5.144) reflect only electric or magnetic shielding effects separately, however, they do not account for possible trade-offs between these shielding effects. The impedance boundary conditions, on the other hand, do account for these trade-offs, and they lead to asymptotically accurate results for any values of μ, σ, and f, provided that the condition (5.132) is satisfied. This clearly reveals the special significance of impedance boundary conditions for **magnetic** conductors. Although our conclusion has been motivated by the solution of the model problem of the conducting sphere subject to uniform magnetic fields, it is apparent that this conclusion is of general nature. It is also important to stress that although the applicability of impedance boundary conditions is limited to the case of small skin depths, nevertheless, for the magnetic conductors this case is quite typical. The reason is that due to high values of magnetic permeability of these conductors the skin depth is quite small even for low frequencies. For magnetically nonlinear conductors, there is further decrease of the skin depth, because the magnetic permeability inside the conductors is increased due to the attenuation of magnetic field.

Until now, we have discussed the solution of the problem of the conducting sphere for the linear case when the magnetic permeability of the conducting sphere has been assumed to be constant and independent of magnetic field. It is instructive to solve the same problem in nonlinear formulation by using the nonlinear impedance boundary conditions. Due to the axial symmetry of the posed problem, it is clear that there is only a θ-component of the magnetic field tangential to the sphere. This suggests that at each boundary point the electromagnetic field penetrates the sphere as a linearly polarized plane ware. Thus, we can use the nonlinear impedance computed in Section 1.5 in the boundary condition (5.113). This, together with formulas (5.124) and (5.125), leads to the following nonlinear impedance boundary conditions:

$$\left[-j\omega A_\varphi - \frac{|\eta(|H_\theta|)|e^{j\varphi}}{\mu_0 r}\frac{\partial}{\partial r}(rA_\varphi)\right]\Bigg|_{r=R} = 0, \tag{5.147}$$

where

$$|\eta| = 1.32\sqrt{\frac{\omega\mu_m(|H_\theta|)}{\sigma}}, \quad \tan\varphi = 0.6. \tag{5.148}$$

The previous expressions are taken from Figs. 1.18 a and 1.18 b for the "squareness" parameter $\chi \approx 0.8$.

By using the method of separation of variables, a general solution to Eq. (5.123) can be represented in the form:

$$A_{\varphi}^{-}(r,\theta) \approx \sum_{k=0}^{N} a_{2k+1} r^{(-2k+2)} P_{2k+1}^{(1)}(\cos\theta), \tag{5.149}$$

where $P_{2k+1}^{(1)}(\cos\theta)$ are associated Legendre functions.

Then, the θ-component of the total magnetic field is given by

$$H_{\theta}(r,\theta) = \frac{1}{\mu_0} \sum_{k=0}^{N} (2k+1) a_{2k+1} r^{-(2k+3)} P_{2k+1}^{(1)}(\cos\theta) + H^0 \sin\theta. \tag{5.150}$$

Now the algorithm of the solution of the problem can be stated as the sequence of the following steps: (1) first, we assume μ_m in (5.148) to be constant and analytically solve the problem in exactly the same way as we did before; (2) by using the found solution, we can evaluate H_θ and $\mu_m(|H_\theta|)$ in formula (5.148); (3) by substituting formulas (5.121) and (5.149) into the boundary condition (5.147) and requiring that this boundary condition is satisfied in the following "collocation" points

$$\theta_i = \frac{i}{N+2} \cdot \frac{\pi}{2}, \qquad (i = 1, 2, \ldots N+1), \tag{5.151}$$

we obtain and solve simultaneous equations for coefficients a_{2k+1}; (4) by using the found coefficients and formula (5.150), we can evaluate H_θ and correct the previous values of $\mu_m(H_\theta)$; (5) then, we sequentially repeat steps 3 and 4 until the convergence is reached.

The described calculations have been performed for the following sample values: $R = 0.25m$, $H^0 = 4.10^3 \frac{A}{m}$, $f = 60\ \mathrm{sec}^{-1}$, and $\sigma = 0.833 \cdot 10^7 \frac{1}{\Omega \cdot m}$. Three terms ($N = 2$) have been used in the expansion (5.149) (calculations for larger number N of terms resulted practically in the same values of H_θ)). Our computations produced the following magnetic field:

$$\begin{aligned} H_\theta(R,\theta) = &- [13.33 + j12.08 - (2.934 + j2.555)\cos\theta \\ &+ (0.269 - j0.086)\cos 4\theta] \sin\theta. \end{aligned} \tag{5.152}$$

To compare the solutions obtained for linear and nonlinear formulations, the following function is introduced

$$\mathcal{F}(\theta) = \frac{|H(R,\theta)|}{\max_\theta |H(R,\theta)|}, \tag{5.153}$$

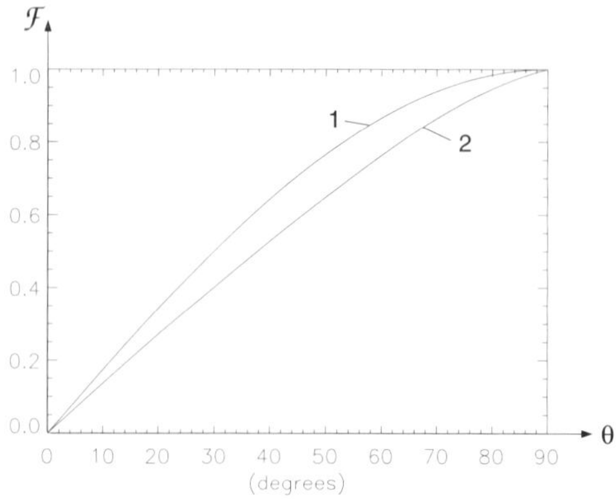

Fig. 5.5

and its values are plotted in Fig. 5.5, with curve 1 corresponding to the linear formulation and curve 2 to "nonlinear" solution (5.152). It is apparent from the figure that $|H(\theta)|$ is increased faster in the linear case. This can be explained by the fact that saturation gradually builds up as θ is increased. As a result, $\mu_m(|H_\theta|)$ is gradually reduced, while the skin depth is increased. This, in turn, causes more uniform (with respect to θ) pulling of magnetic field lines inside the conductor than in the linear case. As a consequence, there is a more gradual increase of $|H_\theta|$.

The impedance boundary conditions (5.113) are formulated in terms of **E** and **H** vectors, while the magnetic fields outside the conductors can be characterized in terms of magnetic scalar potential φ^-. For this reason, it is very desirable to formulate the impedance boundary conditions in terms of the same potential as well. If this is accomplished, then 3-D eddy current problems can be reduced to the calculation of magnetic scalar potential, and vectorial computations can be completely avoided. Furthermore, it turns out that the scalar potential formulation of the impedance boundary conditions is very convenient for finite element implementations. These implementations result in two (volume and **surface**) stiffness matrices, which makes the assembly process of finite element equations rather simple. All these issues are discussed in the next section.

5.4 FINITE ELEMENT IMPLEMENTATION OF IMPEDANCE BOUNDARY CONDITIONS

To arrive at the scalar potential formulation of the impedance boundary conditions, we introduce orthogonal curvilinear coordinates τ_1, τ_2, and ν. These coordinates are such that τ_1- and τ_2-coordinate lines lie on the conductor boundary S, while ν-coordinate lines are normal to S. Let

$\mathbf{a}_1, \mathbf{a}_2$, and $\mathbf{a}_3$ be the triad of orthogonal unit vectors that are tangent to the τ_1-, τ_2-, and ν-coordinate lines, respectively, and that form a right-handed set in this order. Also, let h_1, h_2, and h_3 be the corresponding metric coefficients. The introduced curvilinear coordinates will be extensively used in subsequent derivation. However, the final finite element formulation will contain no traces of these coordinates; in this sense, it will be coordinate invariant.

We begin our derivation by invoking the Maxwell equation:

$$\operatorname{curl} \mathbf{E} = -j\omega\mu_0 \mathbf{H}, \tag{5.154}$$

and the following expression for the curl-operator in the curvilinear coordinates:

$$\operatorname{curl} \mathbf{E} = \frac{1}{h_1 h_2 h_3} \begin{vmatrix} h_1\mathbf{a}_1 & h_2\mathbf{a}_2 & h_3\mathbf{a}_3 \\ \frac{\partial}{\partial \tau_1} & \frac{\partial}{\partial \tau_2} & \frac{\partial}{\partial \nu} \\ h_1 E_{\tau_1} & h_2 E_{\tau_2} & h_3 E_\nu \end{vmatrix}. \tag{5.155}$$

From the last two equations, we find

$$-j\omega\mu_0 H_\nu = \frac{1}{h_1 h_2}\left[\frac{\partial}{\partial \tau_1}(h_2 E_{\tau_2}) - \frac{\partial}{\partial \tau_2}(h_1 E_{\tau_1})\right]. \tag{5.156}$$

Next, we shall write the impedance boundary conditions (5.113) and (5.116) in terms of the introduced curvilinear coordinates:

$$E_{\tau_1} = -\eta_{12} H_{\tau_2}, \tag{5.157}$$

$$E_{\tau_2} = \eta_{21} H_{\tau_1}. \tag{5.158}$$

By substituting the last two formulas into Eq. (5.156), we end up with the following **H**-formulation of the impedance boundary conditions:

$$-j\omega\mu_0 H_\nu = \frac{1}{h_1 h_2}\left[\frac{\partial}{\partial \tau_1}(h_2 \eta_{21} H_{\tau_1}) + \frac{\partial}{\partial_{\tau_2}}(h_1 \eta_{12} H_{\tau_2})\right]. \tag{5.159}$$

It is convenient to introduce vector $\mathbf{c}_\tau$:

$$\mathbf{c}_\tau = \hat{\zeta}\mathbf{H}_\tau, \tag{5.160}$$

where

$$\mathbf{c}_\tau = \begin{pmatrix} c_{\tau_1} \\ c_{\tau_2} \end{pmatrix}, \quad \mathbf{H}_\tau = \begin{pmatrix} H_{\tau_1} \\ H_{\tau_2} \end{pmatrix}, \tag{5.161}$$

and

$$\hat{\zeta} = \begin{pmatrix} \eta_{21} & 0 \\ 0 & \eta_{12} \end{pmatrix}. \tag{5.162}$$

Then the boundary condition (5.159) can be written in the form:

$$-j\omega\mu_0 H_\nu = \frac{1}{h_1 h_2}\left[\frac{\partial}{\partial\tau_1}\left(h_2 c_{\tau_1}\right) + \frac{\partial}{\partial\tau_2}\left(h_1 c_{\tau_2}\right)\right]. \tag{5.163}$$

To understand the meaning of the right-hand side of formula (5.163) and how it can be cast in finite element terms, we shall need some machinery of the vector field theory on curvilinear surfaces (manifolds). This theory is of general interest and it has already been used successfully for eddy current analysis [6] [14]. To make our exposition self-contained and easily readable, we shall present next some selected and relevant facts from this theory. These facts will be very instrumental in the development of finite element formulations discussed in this and subsequent sections.

Consider some vector field **c** that is tangential to curvilinear surface S. We shall next introduce the notion of divergence of vector field **c** on S. To this end, consider an arbitrary point P on S and a closed path L on S that encloses the point P (see Fig. 5.6). Let **n** be a unit vector that is tangential to S and normal to L. Then, the divergence of **c** on S (i.e., $\mathrm{div}_S\mathbf{c}$) is defined as follows:

$$\mathrm{div}_S\mathbf{c} = \lim_{\Delta S\to 0}\frac{\oint_L c_n dl}{\Delta S}, \tag{5.164}$$

where ΔS is the area enclosed by L. We shall next derive the expression for $\mathrm{div}_S\mathbf{c}$ in terms of curvilinear coordinates τ_1 and τ_2. Consider an integration path L formed by infinitesimally small portions of τ_1- and τ_2-coordinate lines (see Fig. 5.7). These coordinate lines are specified by coordinates τ_2', $\tau_2' + d\tau_2, \tau_1'$, and $\tau_1' + d\tau_1$, respectively. By using Fig. 5.7, the path integral in formula (5.164) can be evaluated as follows:

$$\begin{aligned}\oint_L c_n d\ell \approx\; & c_{\tau_1} h_2 d\tau_2\Big|_{\tau_1'+d\tau_1} - c_{\tau_1} h_2 d\tau_2\Big|_{\tau_1'} + \\ & c_{\tau_2} h_1 d\tau_1\Big|_{\tau_2'+d\tau_2} - c_{\tau_2} h_1 d\tau_1\Big|_{\tau_2'}.\end{aligned} \tag{5.165}$$

By taking into account that $d\tau_1$ and $d\tau_2$ are small, the last expression can be simplified as follows:

$$\oint_L c_n d\ell \approx \frac{\partial}{\partial\tau_1}\left(h_2 c_{\tau_1}\right) d\tau_1 d\tau_2 + \frac{\partial}{\partial\tau_2}\left(h_1 c_{\tau_2}\right) d\tau_1 d\tau_2. \tag{5.166}$$

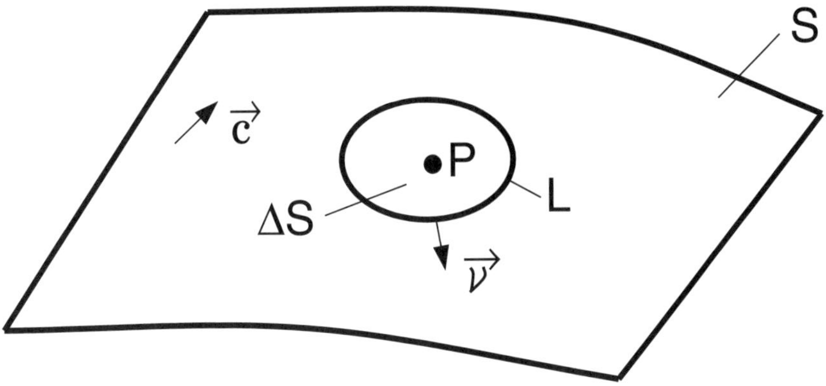

Fig. 5.6

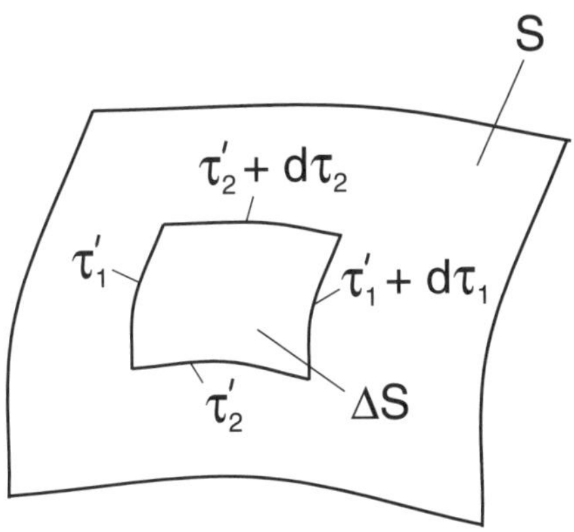

Fig. 5.7

On the other hand, the area enclosed by the path L can be evaluated as

$$\Delta S \approx h_1 h_2 d\tau_1 d\tau_2. \tag{5.167}$$

By substituting formulas (5.166) and (5.167) into the definition (5.164) and taking into account that expressions (5.166) and (5.167) become more and more accurate as ΔS goes to zero, we finally derive

$$\mathrm{div}_S \mathbf{c} = \frac{1}{h_1 h_2}\left[\frac{\partial}{\partial \tau_1}\left(h_2 c_{\tau_1}\right) + \frac{\partial}{\partial \tau_2}\left(h_1 c_{\tau_1}\right)\right]. \tag{5.168}$$

By comparing formulas (5.168) and (5.163) and taking into account relation (5.160), we can represent the impedance boundary condition (5.163) in the following compact form:

$$\boxed{H_\nu = \frac{j}{\omega \mu_0}\mathrm{div}_S\left(\hat{\zeta}\mathbf{H}_\tau\right),} \tag{5.169}$$

where the impedance matrix $\hat{\zeta}$ is defined by (5.162).

The above form of the impedance boundary conditions will be used in the development of the finite element formulation. Before proceeding to this formulation, we shall establish some additional useful facts from the vector field theory on curvilinear surfaces. First, we shall prove the "divergence theorem":

$$\oint_S \operatorname{div}_S \mathbf{c} dS = 0. \tag{5.170}$$

To this end, let us consider some smooth closed surface S and let us partition it into small pieces ΔS_k (see Fig. 5.8). Then, we can write

$$\oint_S \operatorname{div}_S \mathbf{c} dS = \lim_{N\to\infty} \sum_{k=1}^{N} \operatorname{div}_S \mathbf{c}\Big|_k \cdot \Delta S_k, \tag{5.171}$$

where symbol "$\Big|_k$" means that divergence is evaluated at some interior point of ΔS_k. According to the definition (5.164) of divergence, we have

$$\operatorname{div}_S \mathbf{c}\Big|_k \approx \frac{\oint_{L_k} c_n d\ell}{\Delta S_k}, \tag{5.172}$$

where L_k is the boundary of ΔS_k. By substituting formula (5.172) into expression (5.171), we obtain

$$\oint_S \operatorname{div}_S \mathbf{c} dS = \lim_{N\to\infty} \sum_{k=1}^{N} \oint_{L_k} c_n d\ell. \tag{5.173}$$

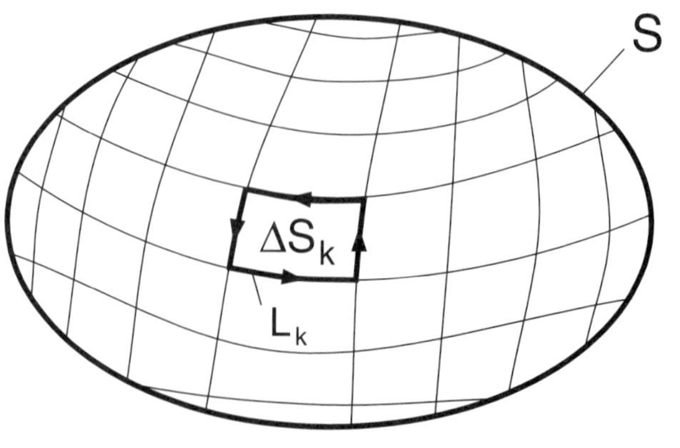

Fig. 5.8

Because each boundary L_k of piece ΔS_k has common parts with the boundaries of adjacent pieces and because normal $\mathbf{n}$ has opposite directions for adjacent pieces, the complete cancellation occurs in the sum in formula (5.173) and we conclude that

$$\sum_{k=1}^{N} \oint_{L_k} c_n d\ell = 0. \tag{5.174}$$

From the last equation and formula (5.173), we arrive at the divergence theorem (5.170).

It is easy to see that in the case when the surface S is not closed but bounded by some curve L, the cancellation in the sum in formula (5.173) is not complete. Indeed, the integrals over those parts of L_k, which are shared with the boundary L of S, are not cancelled. As a result, we have

$$\sum_{k=1}^{N} \oint_{L_k} c_n d\ell = \oint_{L} c_n d\ell. \tag{5.175}$$

Thus, the divergence theorem takes the form:

$$\int_S \operatorname{div}_S \mathbf{c} dS = \oint_L c_n d\ell. \tag{5.176}$$

Next, we derive the expression for $\operatorname{div}_S(\psi \mathbf{c})$, which will be instrumental in our subsequent discussion. By using formula (5.168), we have

$$\operatorname{div}_S(\psi \mathbf{c}) = \frac{1}{h_1 h_2} \left[\frac{\partial}{\partial \tau_1} \left(h_2 \psi c_{\tau_1}) + \frac{\partial}{\partial \tau_2} (h_1 \psi c_{\tau_2} \right) \right]. \tag{5.177}$$

By using the "product rule," we find

$$\begin{aligned} \operatorname{div}_S(\psi \mathbf{c}) =& \psi \frac{1}{h_1 h_2} \left[\frac{\partial}{\partial \tau_1} (h_2 c_{\tau_1}) + \frac{\partial}{\partial \tau_2} (h_1 c_{\tau_2}) \right] \\ &+ c_{\tau_1} \cdot \frac{1}{h_1} \frac{\partial \psi}{\partial \tau_1} + c_{\tau_2} \cdot \frac{1}{h_2} \frac{\partial \psi}{\partial \tau_2}. \end{aligned} \tag{5.178}$$

By using again formula (5.168) and the following expression for the gradient of scalar function ψ along S:

$$\operatorname{grad}_S \psi = \mathbf{a}_1 \frac{1}{h_1} \frac{\partial \psi}{\partial \tau_1} + \mathbf{a}_2 \frac{1}{h_2} \frac{\partial \psi}{\partial \tau_2}, \tag{5.179}$$

we derive

$$\mathrm{div}_S(\psi\mathbf{c}) = \psi\mathrm{div}_S\mathbf{c} + \mathbf{c}\cdot\mathrm{grad}_S\psi. \tag{5.180}$$

Now we apply the divergence theorem to $\mathrm{div}_S(\psi\mathbf{c})$:

$$0 = \oint_S \mathrm{div}(\psi\mathbf{c})dS = \oint_S \psi\mathrm{div}_S\mathbf{c}dS + \oint_S \mathbf{c}\cdot\mathrm{grad}\ \psi dS. \tag{5.181}$$

Thus, we arrive at the following formula:

$$\oint_S \psi\mathrm{div}_S\mathbf{c}dS = -\oint_S \mathbf{c}\cdot\mathrm{grad}_S\psi dS, \tag{5.182}$$

which will be used in the derivation of finite element formulations.

Next, we proceed to the formulation of a generic boundary value problem for 3-D eddy current analysis by using the impedance boundary condition (5.169) and the magnetic scalar potential. As before, we shall use the following decomposition for the magnetic field outside the conductor:

$$\mathbf{H} = -\nabla\varphi^- + \mathbf{H}^0, \tag{5.183}$$

where $\mathbf{H}^0$ is the source field, which can be computed by using one of the techniques discussed in Section 5.2. If $\mathbf{H}^0$ is computed by using the Biot-Savart law, then:

$$\mathrm{div}\ \mathbf{H}^0 = 0. \tag{5.184}$$

By using this fact, from relation (5.183) we find that the magnetic scalar potential satisfies the Laplace equation

$$\nabla^2\varphi^- = 0 \qquad \text{in } V^-. \tag{5.185}$$

By using the relation (5.183), we represent the impedance boundary condition (5.169) in the form:

$$\frac{\partial\varphi^-}{\partial\nu} = \frac{j}{\omega\mu_0}\mathrm{div}_S\left[\hat{\zeta}\left(\nabla_S\varphi^- - \mathbf{H}^0_\tau\right)\right] + H^0_\nu. \tag{5.186}$$

In addition, we impose the following zero conduction at infinity:

$$\varphi(\infty) = 0. \tag{5.187}$$

Formulas (5.185)–(5.187) constitute the boundary value problem for the magnetic scalar potential.

Now we shall reduce this boundary value problem to the weak Galerkin form. To this end, we shall use the following Green formula:

$$\int_{V^-} (\psi \nabla^2 \varphi + \nabla \psi \cdot \nabla \varphi) \mathrm{dv} = - \oint_S \psi \frac{\partial \varphi}{\partial \nu} dS, \tag{5.188}$$

where ψ and φ are arbitrary and sufficiently smooth functions, and $\frac{\partial \varphi}{\partial \nu}$ is the derivative with respect to the inward (to V^-) normal.

By choosing

$$\varphi = \varphi^-, \tag{5.189}$$

and taking into account the Eq. (5.185) and the boundary condition (5.186), we transform the Green formula (5.188) as follows:

$$\begin{aligned} &\int_{V^-} \nabla \psi \cdot \nabla \varphi^- dv + \frac{\hat{j}}{\omega \mu_0} \oint_S \psi \ \mathrm{div}_S (\hat{\zeta} \ \mathrm{grad}_S \varphi^-) dS = \\ &\frac{j}{\omega \mu_0} \oint_S \psi \ \mathrm{div}_S \left(\hat{\zeta} \mathbf{H}_\tau^0 \right) dS - \oint_S \psi H_\nu^0 dS. \end{aligned} \tag{5.190}$$

To proceed further, we shall use the expression (5.182) to transform the second and third integrals in formula (5.190):

$$\oint_S \psi \ \mathrm{div}_S (\hat{\zeta} \ \mathrm{grad}_S \varphi^-) dS = - \oint_S \mathrm{grad}_S \psi \cdot \hat{\zeta} \ \mathrm{grad}_S \varphi^- dS, \tag{5.191}$$

$$\oint_S \psi \ \mathrm{div}_S \left(\hat{\zeta} \mathbf{H}_\tau^0 \right) dS = - \oint_S \mathrm{grad}_S \psi \cdot \hat{\zeta} \mathbf{H}_\tau^0 dS. \tag{5.192}$$

By substituting formulas (5.191) and (5.192) into the relation (5.190), we arrive at the following weak Galerkin form:

$$\boxed{\begin{aligned} &\int_{V^-} \nabla \psi \cdot \nabla \varphi^- dv - \frac{j}{\omega \mu_0} \oint_S \nabla_S \psi \cdot \hat{\zeta} \nabla_S \varphi^- dS = \\ &- \frac{j}{\omega \mu_0} \oint_S \nabla_S \psi \cdot \hat{\zeta} \mathbf{H}_\tau^0 dS - \oint_S \psi H_\nu^0 dS. \end{aligned}} \tag{5.193}$$

Thus, we have proved that if some function φ^- satisfies the boundary value problem (5.185)–(5.187) it will also satisfy the weak Galerkin form (5.193). This form is called "weak" because it contains only first-order derivatives of φ^-, while the Laplace Eq. (5.185) requires φ^- to be twice differentiable.

Thus, regularity conditions on φ^- in the form (5.193) are weaker than those in the boundary value problem (5.185)–(5.187). The form (5.193) is called the "Galerkin form" for historical reasons. The point is that forms like (5.193) were first proposed by the Russian engineer B. Galerkin as alternatives to variational formulations. It can be shown that if the Galerkin form (5.193) is satisfied for any (sufficiently smooth) function ψ and function φ^- is twice differentiable, then φ^- is the solution of the boundary value problem (5.185)–(5.187). In this sense, the boundary value problem (5.185)–(5.187) and the weak Galerkin form (5.193) are equivalent.

The weak Galerkin form is convenient for the construction of approximate finite element solution as well as for the proof of various statements concerning the boundary value problem (5.185)–(5.187). We demonstrate the latter, by proving the uniqueness of the solution to the boundary value problem (5.185)–(5.187) for the linear case. For this case, the Galerkin form (5.193) can be written as follows:

$$\begin{aligned}&\int_{V^-} \nabla\psi \cdot \nabla\varphi^- dv - \frac{j\eta}{\omega\mu_0}\oint_S \nabla_S\psi \cdot \nabla_S\varphi^- dS = \\ &-\frac{j\eta}{\omega\mu_0}\oint_S \nabla_S\psi \cdot \mathbf{H}_\tau^0 dS - \oint_S \psi \mathbf{H}_\tau^0 dS,\end{aligned} \tag{5.194}$$

where η is the surface impedance defined as $\sqrt{\frac{\omega\mu}{\sigma}}e^{j\frac{\pi}{4}}$.

If there are two solutions φ_1^- and φ_2^- to the boundary value problem (5.185)–(5.187), then their difference:

$$\tilde{\varphi} = \varphi_1^- - \varphi_2^- \tag{5.195}$$

will satisfy the following homogeneous Galerkin form:

$$\begin{aligned}&\int_{V^-} \nabla\psi \cdot \nabla\tilde{\varphi} dv + \frac{\mathrm{Im}(\eta)}{\omega\mu_0}\oint_S \nabla_S\psi \cdot \nabla_S\tilde{\varphi} dS \\ &-j\frac{\mathrm{Re}(\eta)}{\omega\mu_0}\oint_S \nabla_S\psi \cdot \nabla_S\tilde{\varphi} dS = 0.\end{aligned} \tag{5.196}$$

The last Galerkin form holds for any function ψ, so it should hold for

$$\psi = \tilde{\varphi}^*, \tag{5.197}$$

where $\tilde{\varphi}^*$ is a complex conjugate to $\tilde{\varphi}$.

By substituting relation (5.197) into the form (5.196), we obtain

$$\int_{V^-} |\nabla\tilde{\varphi}|^2 dv + \frac{\text{Im}(\eta)}{\omega\mu_0} \oint_S |\nabla_S\tilde{\varphi}|^2 dS - \\ j\frac{\text{Re}(\eta)}{\omega\mu_0} \oint_S |\nabla_S\tilde{\varphi}|^2 dS = 0. \tag{5.198}$$

By equating the real and imaginary parts of the left-hand side of formula (5.198) to zero, we find

$$\oint_S |\nabla_S\tilde{\varphi}|^2 dS = 0, \tag{5.199}$$

$$\int_{V^-} |\nabla\tilde{\varphi}|^2 dv = 0. \tag{5.200}$$

From the last two formulas, we derive

$$\tilde{\varphi} = \text{const} \quad \text{in } V. \tag{5.201}$$

By taking into account that $\tilde{\varphi}$ should be equal to zero at infinity, we conclude that

$$\tilde{\varphi} \equiv 0 \quad \text{in } V^-. \tag{5.202}$$

This, according to (5.195), proves the uniqueness of the solution of the boundary value problem (5.185)–(5.187).

Next, we shall use the weak Galerkin form for the development of the finite element technique. To this end, we shall mesh the region V^- and introduce the following finite element approximation for φ^-:

$$\varphi^- \approx \sum_{n=1}^{N} \varphi_n^- \alpha_n, \tag{5.203}$$

where φ_n^- are unknown "node" values of φ^-, while α_n are local support finite element functions. The term "local support" means that functions α_n are equal to zero outside some small regions (elements) centered around mesh points. This eventually leads to the sparse structure of finite element equations.

By substituting expression (5.203) into the Galerkin form (5.198) and by choosing sequentially:

$$\psi = \alpha_i, \qquad (i = 1, 2, \dots N),$$

we end up with the following finite element equations:

$$\sum_{n=1}^{N} \gamma_{ni}\varphi_n^- = f_i, \qquad (i = 1, 2, \dots N), \tag{5.204}$$

where

$$\gamma_{ni} = \int_{V^-} \nabla\alpha_i \cdot \nabla\alpha_n dv - \frac{j\eta}{\omega\mu_0} \oint_S \nabla_S\alpha_i \cdot \nabla_S\alpha_n dS, \tag{5.205}$$

$$f_i = -\frac{j\eta}{\omega\mu_0} \oint_S \nabla_S\alpha_i \cdot \mathbf{H}_\tau^0 dS - \oint_S \alpha_i H_\nu^0 dS. \tag{5.206}$$

A remarkable feature of these equations is that the contribution of surface integrals

$$\oint_S \nabla_S\alpha_i \cdot \nabla_S\alpha_n dS \tag{5.207}$$

is very similar in form to the contribution of volume integrals

$$\int_{V^-} \nabla\alpha_i \cdot \nabla\alpha_n dv. \tag{5.208}$$

For this reason, the submatrix formed by the surface integrals (5.207) can be called the "surface stiffness matrix." Its assembly is quite similar to the assembly of the volume stiffness matrix (5.208). This suggests that the numerical implementation of the discussed technique may require very minor new software development.

In our previous discussion of finite element formulation, it has been assumed that the source field is computed by using the Biot-Savart law. However, this formulation can be easily extended to accommodate other techniques for the calculation of the source field. We demonstrate this for the case when the source field is created by a coil with rectangular cross-section (see Fig. 5.3). It has been shown in Section 5.2 that in this case the source field $\mathbf{H}^0$ can be confined to the extended source region $\tilde{V}^0$ and it is given by formula (5.79). As a result, the relation (5.183) takes the form:

$$\mathbf{H} = \begin{cases} -\nabla\varphi^- & \text{in } V^- - \tilde{V}^0 \\ -\nabla\varphi^- + \mathbf{a}_z T(Q) & \text{in } \tilde{V}^0. \end{cases} \tag{5.209}$$

By using the last formula as well as formulas (5.84) and (5.85), we end up with the following boundary value problem: find the solution to the Laplace equation

$$\nabla^2\varphi^- = 0 \quad \text{in } V^-, \tag{5.210}$$

subject to the boundary conditions

$$\frac{\partial \varphi^-}{\partial \nu} = \frac{j}{\omega \mu_0} \mathrm{div}_S \left[\hat{\zeta} \mathrm{grad}_S \varphi \right] \text{ on } S, \tag{5.211}$$

$$\left[\frac{\partial \varphi^-}{\partial \nu}(Q) \right] \Bigg|_{S_{\text{top}}} = -T(Q), \tag{5.212}$$

$$\left[\frac{\partial \varphi^-}{\partial \nu}(Q) \right] \Bigg|_{S_{\text{bottom}}} = T(Q), \tag{5.213}$$

and the zero condition at infinity

$$\varphi^-(\infty) = 0. \tag{5.214}$$

By using the Green formula (5.188) and the same line of reasoning as before, the boundary value problem (5.210)–(5.214) can be reduced to the following weak Galerkin form:

$$\boxed{\begin{aligned} &\int_{V^-} \nabla \psi \cdot \nabla \varphi^- dv - \frac{j}{\omega \mu_0} \oint_S \nabla_S \psi \cdot \hat{\zeta} \nabla_S \varphi^- dS = \\ &\int_{S_{\text{top}}} \psi T dS - \int_{S_{\text{bottom}}} \psi T dS. \end{aligned}} \tag{5.215}$$

By using the last Galerkin form and the following finite element approximation for the potential:

$$\varphi^- \approx \sum_{n=1}^{N} \varphi_n^- \alpha_n, \tag{5.216}$$

we easily derive the finite element equations:

$$\sum_{n=1}^{N} \gamma_{ni} \varphi_n^- = f_i, \qquad (i = 1, 2, \ldots N), \tag{5.217}$$

where

$$\gamma_{ni} = \int_{V^-} \nabla \alpha_i \cdot \nabla \alpha_n dv - \frac{j}{\omega \mu_0} \oint_S \alpha_i \cdot \hat{\zeta} \nabla_S \alpha_n dS, \tag{5.218}$$

$$f_i = \int_{S_{\text{top}}} \alpha_i T dS - \int_{S_{\text{bottom}}} \alpha_i T dS. \tag{5.219}$$

It is clear that the stream function $T(Q)$ of the source currents is the "driving force" in the above finite element formulation.

In the case of nonlinear impedance boundary conditions, the matrix elements of $\hat{\zeta}$ depend on $|\nabla_S \varphi^-|$, which makes the finite element Eqs. (5.217)–(5.219) nonlinear. For this reason, iterative techniques must be employed to solve the above equations. One of the simplest iterative techniques can be described by the following formulas:

$$\sum_{n=1}^{N} \gamma_{ni}^{(k)} \varphi_n^{(k+1)} = f_i, \tag{5.220}$$

$$\gamma_{ni}^{(k)} = \int_{V^-} \nabla \alpha_i \cdot \nabla \alpha_n dv - \frac{j}{\omega \mu_0} \oint_S \nabla \alpha_i \cdot \hat{\zeta}^{(k)} \nabla_S \alpha_n dS, \tag{5.221}$$

where the matrix elements of $\hat{\zeta}^{(k)}$ are computed by using $|\nabla_S \varphi^{(k)}|$, which is the magnetic scalar potential computed on the previous step of iterations.

5.5 IMPEDANCE BOUNDARY CONDITIONS FOR THIN MAGNETIC CONDUCTING SHELLS AND THEIR FINITE ELEMENT IMPLEMENTATION

The mathematical machinery developed in the previous section can also be used for the calculation of eddy currents in thin magnetic conducting shells. To start the discussion, let us consider the problem of computing 3-D electromagnetic field created by given (known) time-harmonic "source" currents $\mathbf{J}^0$ in coil V^0 in the presence of thin magnetic conducting shell V^s (see Fig. 5.9). The shell region, V^s, is bounded by surfaces S_1 and S_2. The posed problem is a simple one. Nevertheless, the finite element formulation developed for this problem can be easily extended to more complex and realistic situations.

Within the conducting shell, electric $\mathbf{E}$ and magnetic $\mathbf{H}$ fields satisfy the following equations:

$$\text{curl } \mathbf{H} = \sigma \mathbf{E}, \tag{5.222}$$

$$\text{curl } \mathbf{E} = -j\omega \mathbf{B}(\mathbf{H}). \tag{5.223}$$

The last two equations can be written in the following symmetric form:

$$\text{curl } \mathbf{H} = \mathbf{J}_e, \tag{5.224}$$

$$\text{curl } \mathbf{E} = \mathbf{J}_m, \tag{5.225}$$

where

$$\mathbf{J}_e = \sigma \mathbf{E} \tag{5.226}$$

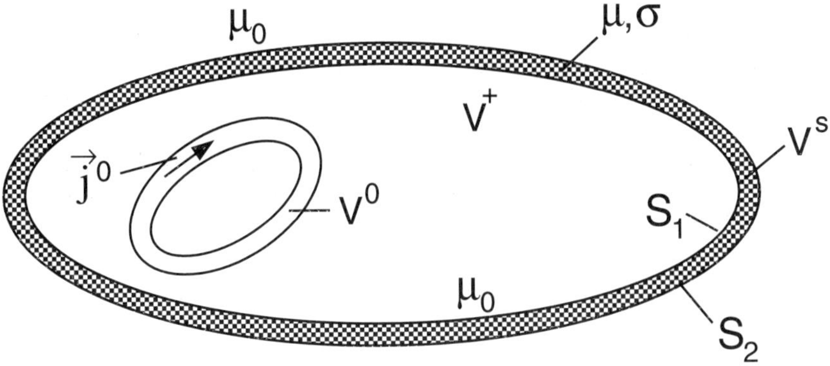

Fig. 5.9

is an electric current density, while

$$\mathbf{J}_m = -j\omega\mathbf{B}(\mathbf{H}) \tag{5.227}$$

can be construed as a magnetic current density.

Next, we shall replace the actual magnetic conducting shell by its "middle" surface S and the actual distributions of electric and magnetic currents by the equivalent distributions of surface electric $\mathbf{i}_e$ and magnetic $\mathbf{i}_m$ currents on S (see Fig. 5.10). These equivalent surface currents are related to the actual currents by the expressions:

$$\mathbf{i}_e = \int_\Delta \mathbf{J}_e d\ell, \tag{5.228}$$

$$\mathbf{i}_m = \int_\Delta \mathbf{J}_m d\ell, \tag{5.229}$$

where the integration is performed over the thickness Δ of the shell.

It is clear that the tangential components of magnetic and electric fields on S are related to the surface currents $\mathbf{i}_e$ and $\mathbf{i}_m$ by the formulas

$$\vec{\nu} \times (\mathbf{H}^+ - \mathbf{H}^-) = \vec{\nu} \times (\vec{\nu} \times \mathbf{i}_e), \tag{5.230}$$

$$\vec{\nu} \times (\mathbf{E}^+ - \mathbf{E}^-) = \vec{\nu} \times (\vec{\nu} \times \mathbf{i}_m), \tag{5.231}$$

where the superscripts "+" and "–" indicate the values of physical quantities inside and outside S, respectively.

In the case of thin shells, it can be assumed with high accuracy that **locally** (at each point) electromagnetic fields penetrate these shells in the same way as plane waves penetrate thin conducting plates. This is the

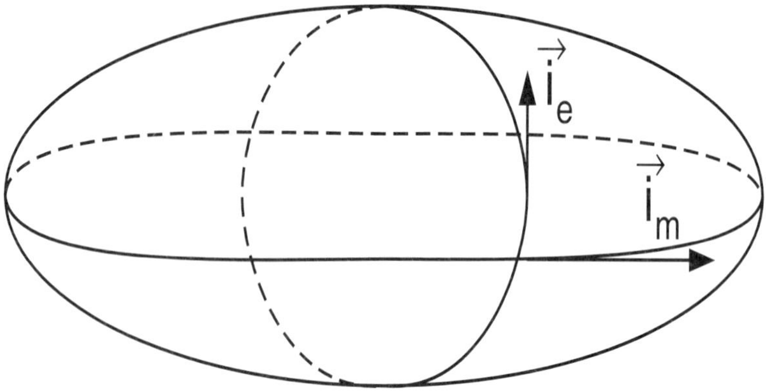

Fig. 5.10

"plane wave" approximation. By using this approximation and formulas (5.228) and (5.229), we can easily derive the following expressions for $\mathbf{i}_e$ and $\mathbf{i}_m$ in terms of the boundary values of electric and magnetic fields:

$$\vec{\nu} \times \mathbf{i}_e = \frac{1}{\zeta} \vec{\nu} \times (\mathbf{E}^+ + \mathbf{E}^-), \tag{5.232}$$

$$\vec{\nu} \times \mathbf{i}_m = -\eta \vec{\nu} \times (\mathbf{H}^+ + \mathbf{H}^-), \tag{5.233}$$

where the factors ζ and η have the dimension of surface impedance and are given by the formulas

$$\zeta = \frac{\beta}{\sigma} \coth \frac{\beta\Delta}{2}, \tag{5.234}$$

$$\eta = \frac{j\omega\mu}{\beta} \tanh \frac{\beta\Delta}{2}, \tag{5.235}$$

and, as before,

$$\beta = \sqrt{j\omega\sigma\mu}. \tag{5.236}$$

By substituting relations (5.232) and (5.233) into formula (5.230) and (5.231), we arrive at the following impedance boundary conditions:

$$\vec{\nu} \times (\mathbf{E}^+ + \mathbf{E}^-) = \zeta(\vec{\nu} \times (\mathbf{H}^+ - \mathbf{H}^-)) \times \vec{\nu}, \tag{5.237}$$

$$\vec{\nu} \times (\mathbf{E}^+ - \mathbf{E}^-) = \eta(\vec{\nu} \times (\mathbf{H}^+ + \mathbf{H}^-)) \times \vec{\nu}. \tag{5.238}$$

These impedance boundary conditions allow one to avoid the calculation of electromagnetic fields within the conducting shell. In other words, the analysis of eddy currents in the magnetic conducting shell is reduced to the solution of the field equations in the regions exterior to the shell subject to the impedance boundary conditions (5.237) and (5.238).

In the practically important case when the "skin" depth is large in comparison with the shell thickness, we have

$$\tanh \frac{\beta\Delta}{2} \approx \frac{\beta\Delta}{2}. \tag{5.239}$$

This leads to the following simplified expressions for ζ and η:

$$\zeta = \frac{2}{\sigma\Delta}, \tag{5.240}$$

$$\eta = \frac{j\omega\mu\Delta}{2}. \tag{5.241}$$

In this case, the impedance boundary conditions (5.237) and (5.238) can be actually derived without using the "plane wave" approximation but instead evaluating integrals in formulas (5.228) and (5.229) on the basis of simple averaging of electric and magnetic current densities on both sides of the conducting shell. It is clear from this remark that the "plane wave" approximation is needed to justify the impedance boundary conditions (5.237) and (5.238) only in the case when the skin depth is small in comparison with the thickness of the shell.

We shall next formulate the impedance boundary conditions in terms of the magnetic scalar potential. To this end, we introduce the same curvilinear coordinates τ_1, τ_2, and ν as in the previous section. By using these curvilinear coordinates, we can represent the boundary conditions (5.237) and (5.238) in the form:

$$E^+_{\tau_1} + E^-_{\tau_1} = -\zeta\left(H^+_{\tau_2} - H^-_{\tau_2}\right), \tag{5.242}$$

$$E^+_{\tau_2} + E^-_{\tau_2} = \zeta\left(H^+_{\tau_1} - H^-_{\tau_1}\right), \tag{5.243}$$

$$E^+_{\tau_1} - E^-_{\tau_1} = -\eta\left(H^+_{\tau_2} + H^-_{\tau_2}\right), \tag{5.244}$$

$$E^+_{\tau_2} - E^-_{\tau_2} = \eta\left(H^+_{\tau_1} + H^-_{\tau_1}\right). \tag{5.245}$$

Next, we invoke the Maxwell equation:

$$\text{curl}\,\mathbf{E}^{\pm} = -j\omega\mu_0\mathbf{H}^{\pm}, \tag{5.246}$$

whose ν-component can be written as follows:

$$-j\omega\mu_0 H^{\pm}_{\nu} = \frac{1}{h_1 h_2}\left[\frac{\partial}{\partial\tau_1}\left(h_2 E^{\pm}_{\tau_2}\right) - \frac{\partial}{\partial\tau_2}\left(h_1 E^{\pm}_{\tau_1}\right)\right]. \tag{5.247}$$

By using impedance boundary conditions (5.242)–(5.245) in Eq. (5.247), we derive

$$-j\omega\mu_0\left(H_\nu^+ + H_\nu^-\right) = \frac{1}{h_1 h_2}\left\{\frac{\partial}{\partial\tau_1}\left[h_2\zeta\left(H_{\tau_1}^+ - H_{\tau_1}^-\right)\right] + \frac{\partial}{\partial\tau_2}\left[h_1\zeta\left(H_{\tau_2}^+ - H_{\tau_2}^-\right)\right]\right\}, \tag{5.248}$$

$$-j\omega\mu_0\left(H_\nu^+ - H_\nu^-\right) = \frac{1}{h_1 h_2}\left\{\frac{\partial}{\partial\tau_1}\left[h_2\eta\left(H_{\tau_1}^+ + H_{\tau_1}^-\right)\right] + \frac{\partial}{\partial\tau_2}\left[h_1\eta\left(H_{\tau_2}^+ + H_{\tau_2}^-\right)\right]\right\}. \tag{5.249}$$

Now, by recalling the formula (5.168), we represent the last two relations in the following forms:

$$H_\nu^+ + H_\nu^- = \frac{j}{\omega\mu_0}\operatorname{div}_S\left[\zeta\left(\mathbf{H}_\tau^+ - \mathbf{H}_\tau^-\right)\right], \tag{5.250}$$

$$H_\nu^+ - H_\nu^- = \frac{j}{\omega\mu_0}\operatorname{div}_S\left[\eta\left(\mathbf{H}_\tau^+ + \mathbf{H}_\tau^-\right)\right], \tag{5.251}$$

where, as before, we used the notation

$$\mathbf{H}_\tau^\pm = \mathbf{a}_{\tau_1} H_{\tau_1}^\pm + \mathbf{a}_{\tau_2} H_{\tau_2}^\pm. \tag{5.252}$$

Formulas (5.250) and (5.251) constitute the magnetic field formulation of impedance boundary conditions (5.237)-(5.238). By adding and subtracting formulas (5.250) and (5.251), the above impedance boundary conditions can also be written as follows:

$$H_\nu^+ = \frac{j}{2\omega\mu_0}\operatorname{div}_S\left[(\zeta+\eta)\mathbf{H}_\tau^+ - (\zeta-\eta)\mathbf{H}_\tau^-\right], \tag{5.253}$$

$$H_\nu^- = \frac{j}{2\omega\mu_0}\operatorname{div}_S\left[(\zeta-\eta)\mathbf{H}_\tau^+ - (\zeta+\eta)\mathbf{H}_\tau^-\right]. \tag{5.254}$$

Next, we proceed to the formulation of the boundary value problem with respect to the magnetic scalar potential. As before, we shall use the following decompositions of the magnetic field in the regions V^+ and V^-, respectively:

$$\mathbf{H}^+ = \mathbf{H}^0 - \nabla\varphi^+, \tag{5.255}$$

$$\mathbf{H}^- = -\nabla\varphi^-, \tag{5.256}$$

where $\mathbf{H}^0$ is the source field.

For the sake of simplicity, it will be assumed that the coil V^0 has a rectangular cross-section (see Fig. 5.3). In this case, the source field is confined to the extended source region $\tilde{V}^0$ and it is given by the formula

$$\mathbf{H}^0 = \begin{cases} \mathbf{a}_z T(Q) & \text{in } \tilde{V}^0, \\ 0 & \text{in } V - \tilde{V}^0. \end{cases} \tag{5.257}$$

By using the last five formulas, we arrive at the following boundary value problem: find the solution to the Laplace equations

$$\nabla^2 \varphi^+ = 0 \quad \text{in } V^+, \tag{5.258}$$

$$\nabla^2 \varphi^- = 0 \quad \text{in } V^-, \tag{5.259}$$

subject to the boundary conditions

$$\frac{\partial \varphi^+}{\partial \nu} = \frac{j}{2\omega\mu_0} \operatorname{div}_S \left[(\zeta + \eta) \nabla_S \varphi^+ - (\zeta - \eta) \nabla_S \varphi^- \right] \text{ on } S, \tag{5.260}$$

$$\frac{\partial \varphi^-}{\partial \nu} = \frac{j}{2\omega\mu_0} \operatorname{div}_S \left[(\zeta - \eta) \nabla_S \varphi^+ - (\zeta + \eta) \nabla_S \varphi^- \right] \text{ on } S, \tag{5.261}$$

$$\left. \left[\frac{\partial \varphi^+}{\partial \nu} \right] \right|_{S_{\text{top}}} = -T(Q), \tag{5.262}$$

$$\left. \left[\frac{\partial \varphi^+}{\partial \nu} \right] \right|_{S_{\text{bottom}}} = T(Q), \tag{5.263}$$

and the following condition at infinity:

$$\varphi^-(\infty) = 0. \tag{5.264}$$

To reduce the above boundary value problem to weak Galerkin forms, we invoke the Green formula (5.188). By applying this formula to regions V^+ and V^- and by taking into account Eqs. (5.258) and (5.259) and the boundary conditions (5.262) and (5.263), we obtain the following relations:

$$\int_{V^-} \nabla\psi \cdot \nabla\varphi^- dv + \oint_S \psi \frac{\partial \varphi^-}{\partial \nu} dS = 0, \tag{5.265}$$

$$\int_{V^+} \nabla\psi \cdot \nabla\varphi^+ dv - \oint_S \psi \frac{\partial \varphi^+}{\partial \nu} dS = \int_{S_{\text{top}}} \psi T dS - \int_{S_{\text{bottom}}} \psi T dS. \tag{5.266}$$

Now, by using the boundary conditions (5.260) and (5.261) in the formulas (5.265) and (5.266), we obtain

$$\int_{V^-} \nabla\psi \cdot \nabla\varphi^- dv + \frac{j}{2\omega\mu_0} \oint_S \psi \,\mathrm{div}_S \left[(\zeta - \eta)\nabla_S \varphi^+ - (\zeta + \eta)\nabla_S \varphi^- \right] dS = 0, \tag{5.267}$$

$$\int_{V^+} \nabla\psi \cdot \nabla\varphi^+ dv - \frac{j}{2\omega\mu_0} \oint_S \psi \mathrm{div}_S \left[(\zeta + \eta)\nabla_S \varphi^+ - (\zeta - \eta)\nabla_S \varphi^- \right] dS = \int_{S_{\mathrm{top}}} \psi T dS - \int_{S_{\mathrm{bottom}}} \psi T dS. \tag{5.268}$$

To proceed further, we shall use generic formula (5.182) in order to transform the surface integrals in expressions (5.267) and (5.268). As a result, we arrive at the following coupled Galerkin's forms:

$$\int_{V^-} \nabla\psi \cdot \nabla\varphi^- dv + \frac{j}{2\omega\mu_0} \left[\oint_S (\zeta + \eta)\nabla_S \psi \cdot \nabla_S \varphi^- dS - \oint_S (\zeta - \eta)\nabla_S \psi \cdot \nabla_S \varphi^+ dS \right] = 0, \tag{5.269}$$

$$\int_{V^+} \nabla\psi \cdot \nabla\varphi^+ dv + \frac{j}{2\omega\mu_0} \left[\oint_S (\zeta + \eta)\nabla_S \psi \cdot \nabla_S \varphi^+ dS - \oint_S (\zeta - \eta)\nabla_S \psi \cdot \nabla_S \varphi^- dS \right] = \int_{S_{\mathrm{top}}} \psi T dS - \int_{S_{\mathrm{bottom}}} \psi T dS. \tag{5.270}$$

The weak Galerkin's forms (5.269) and (5.270) are the natural starting point for finite element discretizations. Indeed, by using the following finite element approximations for φ^+ and φ^-, respectively:

$$\varphi^+ \approx \sum_{n=1}^{N^+} \varphi_n^+ \alpha_n^+, \tag{5.271}$$

$$\varphi^- \approx \sum_{n=1}^{N^-} \varphi_n^- \alpha_n^-, \tag{5.272}$$

and by substituting them into Galerkin's forms (5.269) and (5.270), we naturally end up with the finite element equations:

$$\sum_{n=1}^{N^-} \gamma_{ni}^- \varphi_n^- + \sum_{n=1}^{N^+} \lambda_{ni}^+ \varphi_n^+ = 0, \quad (i = 1, 2, \dots N^-), \tag{5.273}$$

$$\sum_{n=1}^{N^+} \gamma_{ni}^+ \varphi_n^+ + \sum_{n=1}^{N^-} \lambda_{ni}^- \varphi_n^- = f_i, \quad (i = 1, 2, \dots N^+). \tag{5.274}$$

The matrix coefficients and the right-hand sides of these finite element equations are given by the following formulas:

$$\gamma_{ni}^+ = \int_{V^+} \nabla \alpha_i^+ \cdot \nabla \alpha_n^+ dv + \frac{j}{2\omega\mu_0} \oint_S (\zeta + \eta) \nabla_S \alpha_i^+ \cdot \nabla_S \alpha_n^+ dS \tag{5.275}$$

$$\lambda_{ni}^- = \frac{-j}{2\omega\mu_0} \oint_S (\zeta - \eta) \nabla_S \alpha_i^+ \cdot \nabla_S \alpha_n^- dS, \tag{5.276}$$

$$\gamma_{ni}^- = \int_{V^-} \nabla \alpha_i^- \cdot \nabla \alpha_n^- dv + \frac{j}{2\omega\mu_0} \oint_S (\zeta + \eta) \nabla_S \alpha_i^- \cdot \nabla_S \alpha_n^- dS, \tag{5.277}$$

$$\lambda_{ni}^+ = \frac{-j}{2\omega\mu_0} \oint_S (\zeta - \eta) \nabla_S \alpha_i^- \cdot \nabla_S \alpha_n^+ dS, \tag{5.278}$$

$$f_i = \int_{S_{\text{top}}} \alpha_i^+ T ds - \int_{S_{\text{bottom}}} \alpha_i^+ T dS. \tag{5.279}$$

As before, we note that a remarkable feature of the above finite element equations is that the contribution of the surface integrals into the overall equation matrix closely resembles the contribution of the volume integrals. For this reason, the submatrices formed by the above surface integrals can be called "surface stiffness matrices." Their assembly is quite similar to the assembly of volume stiffness matrices, which is a standard procedure in any finite element program. This suggests that the numerical implementation of the above finite element equations may require very little additional software development with respect to already existing ones.

Finally, we shall establish the relationship between the stream function $T_S(M)$ for eddy currents $\mathbf{i}_e$ in the thin conducting shell and the values of the magnetic scalar potential φ^+ and φ^- on S. First, we shall define the stream function $T_S(M)$. Consider two points M and Q on S (see Fig. 5.11). Let L_{QM} be an arbitrary path that lies on S and connects points M and Q. Then, the stream function can be defined as follows:

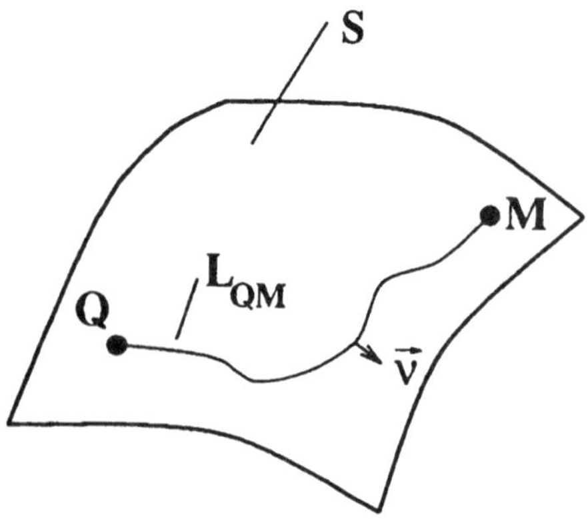

Fig. 5.11

$$T_S(M) - T_S(Q) = \int_{L_{QM}} \mathbf{i}_e(P) \cdot \mathbf{n}_P d\ell_P, \tag{5.280}$$

where $\mathbf{n}_P$ is a unit vector that is tangent to S and normal to the path L_{QM}.

If the point Q is chosen to be a reference point:

$$T_S(Q) = 0, \tag{5.281}$$

then

$$T_S(M) = \int_{L_{QM}} \mathbf{i}_e(P) \cdot \mathbf{n}_P d\ell_P. \tag{5.282}$$

It is apparent that the value of the steam function $T_S(M)$ does not depend on the choice of the integration path between the points M and Q. It is also clear from (5.282) that the "level" lines of $T_S(M)$ (lines of equal values of $T_S(M)$) coincide at every instant of time with the lines of eddy current $\mathbf{i}_e(t)$. Thus, if the level lines of $T_S(M)$ are somehow found, they can be used for the visualization of distributions of eddy currents $\mathbf{i}_e(t)$ on S. This, in turn, may benefit the post-processing of the finite element solution.

By using formula (5.230) in the definition (5.282) of the stream function $T_S(M)$, we find

$$T_S(M) = \int_{L_{QM}} \mathbf{H}^-(P) \cdot d\mathbf{l}_P - \int_{L_{QM}} \mathbf{H}^+(P) \cdot d\mathbf{l}_P. \tag{5.283}$$

By recalling formulas (5.255)–(5.257), the last expression can be transformed as follows:

$$T_S(M) = \left[\varphi^+(M) - \varphi^-(M)\right] - \left[\varphi^+(Q) - \varphi^-(Q)\right]. \tag{5.284}$$

Because the stream function is defined up to an arbitrary constant, the last (Q-dependent) term in the relation (5.284) can be omitted, and we arrive at the following result:

$$T_S(M) = \varphi^+(M) - \varphi^-(M). \tag{5.285}$$

Thus, the stream function is equal to the "step" change in the magnetic scalar potential across the conducting shell. By using this fact and the finite element solution for the magnetic scalar potential, the lines of "equipotential differences" ($\varphi^+(M,t) - \varphi^-(M,t) = \text{const}$) can be determined at any instant of time t. These lines will coincide with the lines of eddy currents at the same instant of time t. For this reason, these lines can be used for the visualization of eddy current distributions.

5.6 CALCULATION OF EDDY CURRENTS IN THIN NONMAGNETIC CONDUCTING SHELLS

The finite element technique discussed in the previous section requires the solution of finite element equations in the regions around magnetic conducting shells. In this sense, this is a "volume" solution technique. It turns out that in the case of thin **nonmagnetic** conducting shells the "boundary" solution technique can be developed. This technique requires only discretizations of shell surfaces. The development of this boundary technique utilizes the same mathematical machinery of vector calculus on curvilinear manifolds that has been developed and extensively used in the previous two sections. The development of this boundary technique also involves additional mathematical tools from functional analysis. These tools bring new and interesting insights into the nature of eddy current problems. The value of these insights may extend far beyond the utility of the technique itself. For instance, by using these mathematical tools, efficient (i.e., easily computable) estimates for eddy current losses are derived. These estimates allow one to obtain useful information concerning eddy current losses without resorting to laborious eddy current calculations.

To start the discussion, consider a thin nonmagnetic conducting shell shown in Fig. 5.12. The electric field within this conducting shell satisfies the equation

$$\text{curl } \mathbf{E} = -j\omega\mathbf{B}. \tag{5.286}$$

By introducing the magnetic vector potential

$$\mathbf{B} = \text{curl } \mathbf{A}, \tag{5.287}$$

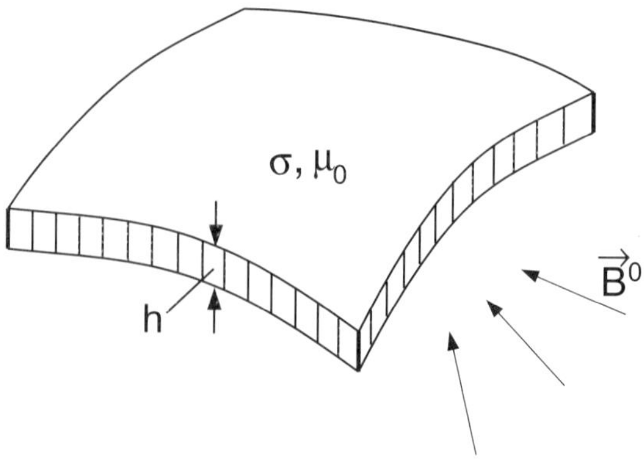

Fig. 5.12

from the last two formulas we derive

$$\mathbf{E} = -j\omega\mathbf{A} - \nabla U, \tag{5.288}$$

where U can be interpreted as an electric scalar potential.

Next, by assuming that distribution of eddy currents over the shell thickness is more or less uniform, we replace the actual conducting shell by its "middle" surface S and the actual distribution of eddy currents within the shell by the equivalent distribution of surface currents $\mathbf{i}$ (see Fig. 5.13):

$$\mathbf{i} = \sigma h \mathbf{E}_\tau, \quad (\mathbf{E}_\tau = \mathbf{a}_1 E_{\tau_1} + \mathbf{a}_2 E_{\tau_2}). \tag{5.289}$$

Here, we use the same curvilinear coordinates τ_1 and τ_2 on the surface S that are described in Section 5.4.

By substituting formula (5.288) into the last expression, we find:

$$\mathbf{i} = -j\omega\sigma h \mathbf{A}_\tau - \sigma h \nabla_S U, \tag{5.290}$$

where, as before, $\nabla_S U$ means a gradient along the surface S.

The magnetic vector potential $\mathbf{A}_\tau$ has two distinct components: $\mathbf{A}^i$, which is due to the distribution of eddy currents $\mathbf{i}$ over S, and $\mathbf{A}^0$, which is due to the source field. Thus, we can write:

$$\mathbf{A} = \mathbf{A}^i + \mathbf{A}^0. \tag{5.291}$$

By substituting the last formula into (5.290), we obtain

$$\mathbf{i} = -j\omega\sigma h(\mathbf{A}^i_\tau + \mathbf{A}^0_\tau) - \sigma h \nabla_S U. \tag{5.292}$$

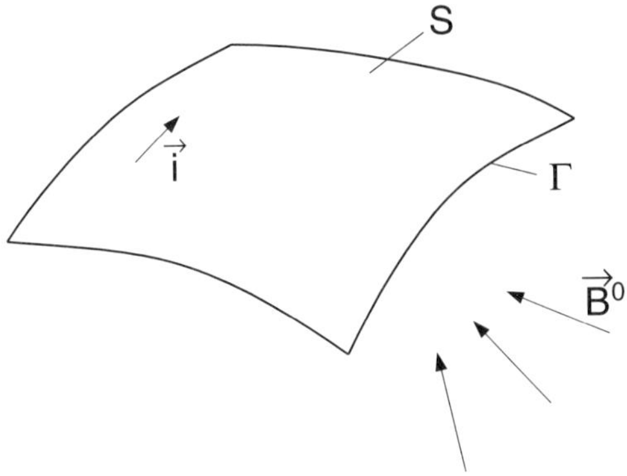

Fig. 5.13

The following expression can be used for $\mathbf{A}^i$:

$$\mathbf{A}^i(Q) = \frac{\mu_0}{4\pi} \int_S \frac{\mathbf{i}(M)}{r_{MQ}} dS_M, \tag{5.293}$$

where, as before, Q and M are observation and integration points, respectively, while r_{MQ} is the distance between Q and M.

By substituting the last formula into Eq. (5.292) and then projecting this equation along the directions of τ_1- and τ_2-coordinate lines, we derive the following integral equations:

$$\begin{aligned} i_{\tau_1}(Q) + \frac{j\omega\mu_0\sigma h}{4\pi} \Bigg[\int_S i_{\tau_1}(M) \frac{\mathbf{a}_1(Q) \cdot \mathbf{a}_1(M)}{r_{MQ}} dS_M \\ + \int_S i_{\tau_2}(M) \frac{\mathbf{a}_1(Q) \cdot \mathbf{a}_2(M)}{r_{MQ}} dS_M \Bigg] \\ + \sigma h \mathbf{a}_1(Q) \cdot \nabla_S U = -j\omega\sigma h A^0_{\tau_1}(Q), \end{aligned} \tag{5.294}$$

$$\begin{aligned} i_{\tau_2}(Q) + \frac{j\omega\mu_0 h}{4\pi} \Bigg[\int_S i_{\tau_1}(M) \frac{\mathbf{a}_2(Q) \cdot \mathbf{a}_1(M)}{r_{MQ}} dS_M \\ + \int_S i_{\tau_2}(M) \frac{\mathbf{a}_2(Q) \cdot \mathbf{a}_2(M)}{r_{MQ}} dS_M \Bigg] \\ + \sigma h \mathbf{a}_2(Q) \cdot \nabla_S U = -j\omega\sigma h A^0_{\tau_2}(Q). \end{aligned} \tag{5.295}$$

The coupled surface integral Eqs. (5.294)–(5.295) are not complete, because we have not yet specified how the electric potential U must be computed. This potential is chosen in such a way that the following equation and boundary condition are satisfied:

$$\mathrm{div}_S \mathbf{i} = 0 \quad \text{on } S, \tag{5.296}$$

$$\mathbf{i} \cdot \mathbf{n} = 0 \quad \text{on } \Gamma, \tag{5.297}$$

where $\mathbf{n}$ is a unit vector tangential to S and normal to the boundary Γ of S. (By the way, this boundary may consist of several closed lines if the conducting shell has holes. Our subsequent discussion is directly applicable to this more general case.)

By substituting the relation (5.292) into formulas (5.296) and (5.297), we arrive at the following Neumann boundary value problem for U on curvilinear surface S:

$$\text{div}_S \, \text{grad}_S U = -j\omega \, \text{div}_S(\mathbf{A}_\tau^i + \mathbf{A}_\tau^0) \quad \text{on } S, \tag{5.298}$$

$$\mathbf{n} \cdot \nabla_S U = -j\omega(\mathbf{A}_\tau^i + \mathbf{A}_\tau^0) \cdot \mathbf{n} \quad \text{on } \Gamma. \tag{5.299}$$

The boundary value problem (5.298)–(5.299) is coupled with surface integral Eqs. (5.294)–(5.295) through formula (5.293). This boundary value problem along with the above integral equations constitute a complete "boundary" formulation for the calculation of eddy currents $\mathbf{i}$. This means that these integral equations and the boundary value problem can be jointly discretized and simultaneously solved. The finite element discretization of the boundary value problem (5.298)–(5.299) is based on its reduction to the weak Galerkin's form. To arrive at this form, we shall first multiply equation (5.298) by an arbitrary and sufficiently smooth function ψ and integrate over S:

$$\int_S \psi \, \text{div}_S \, \text{grad}_S U dS = -j\omega \int_S \psi \text{div}_S(\mathbf{A}_\tau^i + \mathbf{A}_\tau^0) dS. \tag{5.300}$$

Next, we shall use generic formula (5.180), which leads to the following equalities:

$$\psi \text{div}_S \, \text{grad}_S U = \text{div}_S(\psi \text{grad}_S U) - \nabla_S \psi \cdot \nabla_S U, \tag{5.301}$$

$$\psi \text{div}_S(\mathbf{A}_\tau^i + \mathbf{A}_\tau^0) = \text{div}_S[\psi(\mathbf{A}_\tau^i + \mathbf{A}_\tau^0)] - \nabla_S \psi'(\mathbf{A}^i + \mathbf{A}^0). \tag{5.302}$$

By substituting these equations into formula (5.300), and then using the divergence theorem in the form (5.176), we derive

$$\int_S \nabla_S \psi \cdot \nabla_S U dS - \oint_\Gamma \psi \mathbf{n} \cdot \nabla_S U dl = -j\omega \int_S \nabla_S \psi \cdot (\mathbf{A}^i + \mathbf{A}^0) dS$$
$$+ j\omega \oint_\Gamma \psi(\mathbf{A}_\tau^i + \mathbf{A}_\tau^0) \cdot \mathbf{n} dl. \tag{5.303}$$

Finally, by using the boundary condition (5.299), we arrive at the following Galerkin's form:

$$\int_S \nabla_S \psi \cdot \nabla_S U dS + j\omega \int_S \nabla_S \psi \cdot \mathbf{A}^i dS = -j\omega \int_S \nabla_S \psi \cdot \mathbf{A}^0 dS, \qquad (5.304)$$

which can be discretized along with integral equations (5.294)–(5.295).

The finite element discretization of the Galerkin's form (5.304) is straightforward. The first term of this form leads to a sparse "stiffness" matrix, which we have already encountered in the two previous sections. On the other hand, the discretization of the surface integral Eq. (5.294)–(5.295) leads to fully populated matrices for eddy currents **i**. These fully populated matrices are the main impediment in the numerical solution of the overall set of discretized equations. This numerical difficulty can be ameliorated by using the following iterative technique:

$$\int_S \nabla_S \psi \cdot \nabla_S U^{(k+1)} dS = -j\omega \int_S \nabla_S \psi \cdot (\mathbf{A}^{i^{(k)}} + \mathbf{A}_0) dS, \qquad (5.305)$$

$$\begin{aligned} i_{\tau_1}^{(k+1)}(Q) =& (1-\alpha) i_{\tau_1}^{(k)}(Q) - \alpha \frac{j\omega\mu_0\sigma h}{4\pi} \left[\int_S i_{\tau_1}^{(k)}(M) \frac{\mathbf{a}_1(Q)\cdot \mathbf{a}_2(M)}{r_{MQ}} dS_M \right. \\ & \left. + \int_S i_{\tau_2}^{(k)}(M) \frac{\mathbf{a}_1(Q)\cdot \mathbf{a}_2(M)}{r_{MQ}} dS_M \right] \\ & - \alpha\sigma h \mathbf{a}_1(Q) \cdot \nabla_S U^{(k+1)}(Q) - \alpha j\omega\sigma h A_{\tau_1}^0(Q), \end{aligned} \qquad (5.306)$$

$$\begin{aligned} i_{\tau_2}^{(k+1)}(Q) =& (1-\alpha) i_{\tau_2}^{(k)}(Q) - \alpha \frac{j\omega\mu_0\sigma h}{4\pi} \left[\int_S i_{\tau_1}^{(k)}(M) \frac{\mathbf{a}_2(Q)\cdot \mathbf{a}_1(M)}{r_{MQ}} dS_M \right. \\ & \left. + \int_S i_{\tau_2}^{(k)}(M) \frac{\mathbf{a}_2(Q)\cdot \mathbf{a}_2(M)}{r_{MQ}} dS_M \right] \\ & - \alpha\sigma h \mathbf{a}_2(Q) \cdot \nabla_S U^{(k+1)}(Q) - \alpha j\omega\sigma h A_{\tau_2}^0(Q), \end{aligned} \qquad (5.307)$$

where α is some parameter that can be chosen to guarantee the convergence of iterations (5.305)–(5.307).

The above iterative technique can be summarized as follows. Suppose that the kth iteration is found. Then, by using a finite element discretization of the Galerkin's form, we can find $U^{(k+1)}$. By using this electric potential and **explicit** formulas (5.306) and (5.307) we can compute the iteration, $\mathbf{i}^{(k+1)}$, for eddy currents. Thus, at each step of iterations, we have to solve simultaneous equations with a sparse "surface stiffness" matrix, while the direct solution of algebraic equations with fully populated matrices is replaced by explicit iterations (5.306)–(5.307). The evaluation of integrals in formulas (5.306)–(5.307) can be run in parallel for various observation points Q, which is another advantage of the above iterative technique.

Theorem. Iterations (5.305)–(5.307) converge globally (i.e., for any choice of initial guess) if parameter α satisfies the inequalities:

$$\boxed{0 < \alpha < \frac{2}{1 + (\omega\mu_0\sigma h\mathcal{D})^2},} \tag{5.308}$$

where $\mathcal{D}$ is the diameter of S (that is the largest distance between two points on S).

Proof. The proof is somewhat lengthy and uses some standard facts from the functional analysis [2]. However, some results developed in the course of this proof are of interest in their own right.

The central point of the proof is the interpretation of the "boundary" formulation (5.294)–(5.295) and (5.298)–(5.299) as a special operator equation. This interpretation is achieved by treating the subtraction of $\sigma h \nabla_S U$ in formula (5.292) as an operation of orthogonal projection. In order to give the exact meaning to the last statement, we have to introduce some Hilbert spaces. First, consider the Hilbert space $\mathcal{H}$ of vector valued functions on S with the inner product:

$$< \mathbf{b}(M), \mathbf{c}(M) >= \int_S \mathbf{b}(M) \cdot \mathbf{c}^*(M) dS_M, \tag{5.309}$$

where $\mathbf{c}^*(M)$ is a vector whose Cartesian components are complex conjugate to the Cartesian components of the vector $\mathbf{c}(M)$. In the space $\mathcal{H}$ consider a linear subset of continuously differentiable vector functions, which satisfy the following two conditions:

$$\text{div}_S \mathbf{b}(M) = 0 \quad \text{on } S, \tag{5.310}$$

$$\mathbf{b}(M) \cdot \mathbf{n}_M = 0 \quad \text{on } \Gamma. \tag{5.311}$$

The closure of the above subset (in the norm $||\mathbf{b}|| = \sqrt{< \mathbf{b}, \mathbf{b} >}$ of the space $\mathcal{H}$) is a subspace of $\mathcal{H}$. The notation $\mathcal{H}_0$ will be used for the notation of this subspace. It is clear from formulas (5.296)–(5.297) that the surface eddy current density $\mathbf{i}(M)$ belongs to the subspace $\mathcal{H}_0$.

Next, we shall prove that for any $\mathbf{b}(M) \in \mathcal{H}_0$ and an arbitrary function ψ we have

$$< \mathbf{b}(M), \nabla_S \psi >= 0. \tag{5.312}$$

Indeed, we derive

$$< \mathbf{b}(M), \nabla_S \psi >= \int_S \mathbf{b}(M) \cdot \nabla_S \psi dS_M = \int_S \text{div}_S(\psi^* \mathbf{b}(M)) dS_M$$
$$- \int_S \psi^* \text{div}_S \mathbf{b}(M) dS_M = \oint_\Gamma \psi^* \mathbf{b}(M) \cdot \mathbf{n} dl_M - \int_S \psi^* \ \text{div}_S \mathbf{b}(M) dS_M. \tag{5.313}$$

Now, by recalling formulas (5.310) and (5.311), we arrive at the identity (5.312). Strictly speaking, this identity has been proved for sufficiently smooth $\mathbf{b}(M)$ from $\mathcal{H}_0$. However, by using the continuity argument, this results can be extended to any $\mathbf{b}(M)$ from $\mathcal{H}_0$. This type of subtlety is tacitly assumed in the subsequent discussion.

Proceeding further, we introduce the vector

$$\mathbf{c}(Q) = -j\omega(\mathbf{A}_\tau^i(Q) + \mathbf{A}_\tau^0(Q)) \tag{5.314}$$

and rewrite formula (5.292) as follows:

$$\mathbf{i}(Q) = \sigma h(\mathbf{c}(Q) - \nabla_S U). \tag{5.315}$$

Because the potential U is chosen to guarantee that $\mathbf{i}(Q)\epsilon\mathcal{H}_0$, according to (5.312), we have

$$< \mathbf{i}(Q), \nabla_S U >= 0. \tag{5.316}$$

By using formula (5.315), the last equation can be written as follows:

$$< \mathbf{c}(Q) - \nabla_S U, \nabla_S U >= 0. \tag{5.317}$$

This means that the operation of subtraction of $\nabla_S U$ is the operation of orthogonal projection on $\mathcal{H}_0$. Thus, we can symbolically write:

$$\mathbf{c} - \nabla_S U = \hat{P}\mathbf{c}, \tag{5.318}$$

where $\hat{P}$ is the linear operator of orthogonal projection on $\mathcal{H}_0$. It is known that this operator is self-adjoint (Hermitian) and that

$$||\hat{P}|| = 1, \tag{5.319}$$

where $|| \cdot ||$ in (5.319) stands for the norm of the operator.

From formulas (5.314), (5.315), and (5.318), we find

$$\mathbf{i} = -j\omega\sigma h(\hat{P}\mathbf{A}_\tau^i + \hat{P}\mathbf{A}_\tau^0). \tag{5.320}$$

Next, we introduce the matrix integral operator:

$$\hat{K} = \begin{pmatrix} \hat{K}_{11} & \hat{K}_{12} \\ \hat{K}_{21} & \hat{K}_{22} \end{pmatrix}, \tag{5.321}$$

with

$$\hat{K}_{nn}\varphi = \int_S \varphi(M) \frac{\mathbf{a}_n(Q) \cdot \mathbf{a}_n(M)}{r_{MQ}} dS_M, \tag{5.322}$$

$$\hat{K}_{mm}\varphi = \int_S \varphi(M) \frac{\mathbf{a}_m(Q) \cdot \mathbf{a}_n(M)}{r_{MQ}} dS_M, \tag{5.323}$$

$$(m = 1, 2; \quad n = 1, 2).$$

From formulas (5.320)–(5.323) we find that surface integral Eqs. (5.294)–(5.295) can be represented in the following operator form:

$$\mathbf{i} + j\frac{\omega\mu_0\sigma h}{4\pi}\hat{P}\hat{K}\mathbf{i} = -j\omega\sigma h\hat{P}\mathbf{A}_\tau^0. \tag{5.324}$$

It is easy to see that operator $\hat{K}$ is self-adjoint:

$$< \hat{K}\mathbf{b}, \mathbf{c} >=< \mathbf{b}, \hat{K}\mathbf{c} > . \tag{5.325}$$

This can be checked by using the definition (5.309) of the inner product and formulas (5.321)–(5.323).

Now, we are going to prove a very important fact that operator $\hat{P}\hat{K}$ is self-adjoint on the subspace $\mathcal{H}_0$. Let

$$\mathbf{b}(M) \in \mathcal{H}_0 \quad \text{and} \quad \mathbf{c}(M) \in \mathcal{H}_0, \tag{5.326}$$

which means that

$$\hat{P}\mathbf{b} = \mathbf{b}, \qquad \hat{P}\mathbf{c} = \mathbf{c}. \tag{5.327}$$

Then, by using the last formulas and the fact that $\hat{P}$ and $\hat{K}$ are self-adjoint operators, we derive

$$\begin{aligned} < \hat{P}\hat{K}\mathbf{b}, \mathbf{c} > &=< \hat{K}\mathbf{b}, \hat{P}\mathbf{c} >=< \hat{K}\mathbf{b}, \mathbf{c} >=< \mathbf{b}, \hat{K}\mathbf{c} > \\ &=< \hat{P}\mathbf{b}, \hat{K}\mathbf{c} >=< \mathbf{b}, \hat{P}\hat{K}\mathbf{c} >, \end{aligned} \tag{5.328}$$

which means that operator $\hat{P}\hat{K}$ is self-adjoint on $\mathcal{H}_0$.

Next, we shall rewrite Eq. (5.324) in the following concise form:

$$\mathbf{i} + j\hat{T}\mathbf{i} = \mathbf{f}, \tag{5.329}$$

where

$$\hat{T} = \frac{\omega\mu_0\sigma h}{4\pi}\hat{P}\hat{K}, \tag{5.330}$$

$$\mathbf{f} = -j\omega\sigma h\hat{P}\mathbf{A}_\tau^0. \tag{5.331}$$

It is apparent that operator $\hat{T}$ is self-adjoint.

By using parameter α, we make the following equivalent transformation of operator Eq. (5.329):

$$\mathbf{i} = (1-\alpha)\mathbf{i} - j\alpha\hat{T}\mathbf{i} + \alpha\mathbf{f}. \tag{5.332}$$

By introducing the operator

$$\hat{C}_\alpha = (1-\alpha)\hat{I} - j\alpha\hat{T}, \tag{5.333}$$

we rewrite Eq. (5.332) as follows:

$$\mathbf{i} = \hat{C}_\alpha\mathbf{i} + \alpha\mathbf{f}. \tag{5.334}$$

Now we shall establish the fact that if α satisfies inequalities (5.308), then the operator $\hat{C}_\alpha$ is a contraction:

$$||\hat{C}_\alpha|| < 1, \tag{5.335}$$

and iterations

$$\mathbf{i}^{(k+1)} = \hat{C}_\alpha\mathbf{i}^{(k)} + \alpha\mathbf{f} \tag{5.336}$$

globally converge.

To prove inequality (5.335), we use the chain of the following identities:

$$\begin{aligned} ||\hat{C}_\alpha\mathbf{b}||^2 &= < (1-\alpha)\mathbf{b} - j\alpha\hat{T}\mathbf{b}, (1-\alpha)\mathbf{b} - j\alpha\hat{T}\mathbf{b} > \\ &= (1-\alpha)^2 < \mathbf{b}, \mathbf{b} > + j\alpha(1-\alpha) < \mathbf{b}, \hat{T}\mathbf{b} > \\ &\quad - j\alpha(1-\alpha) < \hat{T}\mathbf{b}, \mathbf{b} > + \alpha^2 < \hat{T}\mathbf{b}, \hat{T}\mathbf{b} > . \end{aligned} \tag{5.337}$$

Operator $\hat{T}$ is self-adjoint, consequently the second and third terms before the last one in formula (5.337) are cancelled out. As a result, we obtain

$$||\hat{C}_\alpha\mathbf{b}||^2 = (1-\alpha)^2||\mathbf{b}||^2 + \alpha^2||\hat{T}\mathbf{b}||^2. \tag{5.338}$$

By using inequality

$$||\hat{T}\mathbf{b}||^2 \le ||\hat{T}||^2||\mathbf{b}||^2, \tag{5.339}$$

from formula (5.338) we find

$$||\hat{C}_\alpha\mathbf{b}||^2 \le \left[(1-\alpha)^2 + \alpha^2||\hat{T}||^2\right]||\mathbf{b}||^2, \tag{5.340}$$

which means that

$$||\hat{C}_\alpha||^2 \leq (1-\alpha)^2 + \alpha^2 ||\hat{T}||^2. \tag{5.341}$$

We easily find that

$$(1-\alpha)^2 + \alpha^2 ||\hat{T}||^2 < 1 \tag{5.342}$$

and inequality (5.335) is valid, if

$$0 < \alpha < \frac{2}{1 + ||\hat{T}||^2}. \tag{5.343}$$

Next, we shall estimate the norm of operator $\hat{T}$. According to formula (5.330), we have

$$||\hat{T}||^2 \leq \left(\frac{\omega\mu_0\sigma h}{4\pi}\right)^2 ||\hat{P}||^2 ||\hat{K}||^2. \tag{5.344}$$

The matrix integral operator $\hat{K}$ is an operator with weak singularity. According to [13], the following estimate is valid for the norm of $\hat{K}$:

$$||\hat{K}|| \leq 4\pi\mathcal{D}. \tag{5.345}$$

By using formulas (5.319) and (5.345) in the inequality (5.344), we derive

$$||\hat{T}||^2 \leq (\omega\mu_0\sigma h\mathcal{D})^2. \tag{5.346}$$

Inequalities (5.343) are guaranteed if we replace $||\hat{T}||^2$ in (5.343) by its estimate (5.346). This leads to the inequality (5.308) for α. Thus, it is established that iterations (5.336) are globally convergent if formula (5.308) is valid. By using formulas (5.332), (5.331), (5.330), as well as the definition of operation $\hat{K}$, the iterations (5.336) can be represented in the form:

$$\mathbf{i}^{(k+1)} = (1-\alpha)\mathbf{i}^{(k)} - \alpha j\omega\sigma h\hat{P}(\mathbf{A}_\tau^{i(k)} + \mathbf{A}_\tau^0). \tag{5.347}$$

By recalling the definition of projection operator $\hat{P}$, from the last formula we find

$$\mathbf{i}^{(k+1)} = (1-\alpha)\mathbf{i}^{(k)} - \alpha j\omega\sigma h(\mathbf{A}_\tau^{i^{(k)}} + \mathbf{A}_\tau^0) - \alpha\sigma h\nabla_S U^{(k)}, \tag{5.348}$$

where $U^{(k)}$ is the solution of the boundary value problem (5.298)–(5.299) with $\mathbf{A}_\tau^i$ replaced by $\mathbf{A}_\tau^{i^{(k)}}$. This boundary value problem is, in turn, tantamount to the weak Galerkin's form (5.305). Finally, it is clear that iterations (5.348) are equivalent to the iterations (5.306)–(5.307). ∎

Any value of α that satisfies inequalities (5.308) can be used in iterations (5.305)–(5.307). However, the best estimate for the convergence rate of these iterations can be obtained for

$$\alpha = \frac{1}{1 + (\omega\mu_0\sigma h\mathcal{D})^2} \tag{5.349}$$

For this value of α, from formulas (5.341) and (5.346) we find

$$||\hat{C}_\alpha|| \leq \frac{\omega\mu_0\sigma h\mathcal{D}}{\sqrt{1 + (\omega\mu_0\sigma h\mathcal{D})^2}}. \tag{5.350}$$

This means that iterations (5.305)–(5.307) converge faster than a geometric series with the ratio equal to the right-hand side of formula (5.350). In fact, this convergence may be appreciably faster.

Now, by using the mathematical machinery developed in the proof of the above theorem, we shall derive easily computable estimates for eddy current losses in conducting shells.

First, the eddy current losses can be expressed as follows:

$$L_e = \frac{\sigma h}{2}\int_S |\mathbf{E}|^2 dS = \frac{1}{2\sigma h}\int_S |\mathbf{i}|^2 dS = \frac{1}{2\sigma h}||\mathbf{i}||^2. \tag{5.351}$$

Second, we invoke the operator Eq. (5.329) to estimate $||\mathbf{i}||^2$. According to the above equation, we find

$$\begin{aligned} ||\mathbf{f}||^2 &= <\mathbf{f},\mathbf{f}> = <\mathbf{i} + j\hat{T}\mathbf{i}, \mathbf{i} + j\hat{T}\mathbf{i}> = ||\mathbf{i}||^2 - j <\mathbf{i}, \hat{T}\mathbf{i}> \\ &\quad + j <\hat{T}\mathbf{i}, \mathbf{i}> + ||\hat{T}\mathbf{i}||^2. \end{aligned} \tag{5.352}$$

The operator $\hat{T}$ is self-adjoint, so the second and the third terms before the last one in formula (5.352) are cancelled out. As a result, we have

$$||\mathbf{i}||^2 + ||\hat{T}\mathbf{i}||^2 = ||\mathbf{f}||^2. \tag{5.353}$$

Now it is obvious that

$$||\mathbf{i}||^2 \leq ||\mathbf{f}||^2. \tag{5.354}$$

By using formulas (5.351) and (5.331), from the last inequality we obtain

$$L_e \leq \frac{\omega^2\sigma h}{2}||\hat{P}\mathbf{A}_\tau^0||^2. \tag{5.355}$$

It is clear from the above derivation that the estimate (5.355) is obtained by neglecting the term $\hat{T}\mathbf{i}$. This term describes the part of the surface eddy current density, which is induced by the electromagnetic field created by eddy currents themselves. For this reason, this part can be called the **eddy current reaction**. Thus, we reach the conclusion: **the actual eddy current losses are always smaller than those computed by neglecting the eddy current reaction**. In other words, **the eddy current reaction reduces eddy current losses.**

To proceed further, we use the inequality:

$$||\hat{P}\mathbf{A}_\tau^0|| \leq ||\hat{P}|| \cdot ||\mathbf{A}_\tau^0||. \tag{5.356}$$

By recalling formula (5.319) from (5.355) and (5.356), we derive

$$L_e \leq \frac{\omega^2 \sigma h}{2} ||\mathbf{A}_\tau^0||^2, \tag{5.357}$$

which means that

$$L_e \leq \frac{\omega^2 \sigma h}{2} \int_S |\mathbf{A}_\tau^0|^2 dS. \tag{5.358}$$

In the last step of our derivation, we have replaced $||\hat{P}\mathbf{A}_\tau^0||$ by $||\mathbf{A}_\tau^0||$. In doing so, we have neglected "shape effects." Thus, it can be concluded that **"shape effects" reduce eddy current losses as well.**

It is clear that the estimate (5.363) is easily computable. And this is the main attractive feature of the above estimate.

The inequality (5.358) has been derived for the case of time harmonic source fields. However, by using this inequality and the Fourier transform, we can derive the easily computable estimates for eddy current losses in the case of arbitrary time varying source fields. The derivation proceeds as follows:

$$L_e = \frac{1}{\sigma h} \int_S \left(\int_{-\infty}^{\infty} |\mathbf{i}(M,t)|^2 dt \right) dS_M. \tag{5.359}$$

By using the Fourier transform and Parseval's relation, we obtain

$$L_e = \frac{1}{\sigma h} \int_{-\infty}^{\infty} \left(\int_S |\mathbf{i}(M,\omega)|^2 dS_M \right) d\omega. \tag{5.360}$$

According to (5.358), for each frequency ω we have the inequality:

$$\frac{1}{\sigma h} \int_S |\mathbf{i}(M,\omega)|^2 dS_M \leq \omega^2 \sigma h \int_S |\mathbf{A}_\tau^0(M,\omega)|^2 dS_M. \tag{5.361}$$

By substituting the last inequality into formula (5.360), we find

$$L_e \leq \sigma h \int_S \left(\int_{-\infty}^{\infty} |\omega \mathbf{A}_\tau^0(M,\omega)|^2 d\omega \right)^2 dS_M. \tag{5.362}$$

By using Parseval's relation again, we finally derive

$$L_e \leq \sigma h \int_S \left(\int_{-\infty}^{\infty} |\frac{\partial \mathbf{A}_\tau^0(M,t)}{\partial t}|^2 dt \right) dS_M. \tag{5.363}$$

All the results obtained in this section can also be derived for eddy currents in "bulk" nonmagnetic conductors [14]. In the case of these conductors, the following relation is valid for the volume eddy current density

$$\mathbf{J} = -j\omega\sigma(\mathbf{A}^J + \mathbf{A}^0) - \sigma\nabla U \tag{5.364}$$

where $\mathbf{A}^0$ is the magnetic vector potential of the source field, $\mathbf{A}^J$ is the magnetic vector potential of the field created by the eddy currents and this potential is given by the expression:

$$\mathbf{A}^J(Q) = \frac{\mu_0}{4\pi} \int_{V^+} \frac{\mathbf{J}(M)}{r_{MQ}} dv_M, \tag{5.365}$$

while the electric potential U is chosen to guarantee the following conditions for the eddy current density:

$$\text{div } \mathbf{J} = 0 \quad \text{in } V^+, \tag{5.366}$$

$$\mathbf{J} \cdot \vec{\nu} = 0 \quad \text{on } S, \tag{5.367}$$

with V^+ and S being the region occupied by the conductor and its boundary, respectively.

By substituting formula (5.365) into Eq. (5.364), we arrive at the following integral equation for the eddy current density.

$$\mathbf{J}(Q) + \frac{j\omega\mu_0\sigma}{4\pi} \int_{V^+} \frac{\mathbf{J}(M)}{r_{MQ}} dv_M + \sigma\nabla U = -j\omega\sigma\mathbf{A}^0(Q). \tag{5.368}$$

This integral equation is not complete and should be supplemented by the following interior Newmann boundary value problem for U:

$$\nabla^2 U = 0 \quad \text{in} V^+, \tag{5.369}$$

$$\frac{\partial U}{\partial \nu} = -j\omega(\mathbf{A}^J + \mathbf{A}^0) \cdot \vec{\nu} \quad \text{on } S. \tag{5.370}$$

The last two relations are easily derivable from formulas (5.364), (5.365), (5.368) and Coulomb gauge conditions for the magnetic vector potentials $\mathbf{A}^J$ and $\mathbf{A}^0$. By using the same line of reasoning as before, it can be demonstrated [14] that the operation of subtraction of $\sigma\nabla U$ in formula

(5.364) can be interpreted as the operation of orthogonal projection on subspace $\mathcal{H}_0(V^+)$ defined by constraints (5.366)–(5.367). On the basis of this fact, the formulation (5.368)–(5.370) can be reduced to the operator equation of the form (5.324) with the self-adjoint operator $\hat{P}\hat{K}$ on $\mathcal{H}_0(V^+)$. By employing this operator equation and the same reasoning as before, it can be proved that the following iterations:

$$\nabla^2 U^{(k+1)} = 0, \tag{5.371}$$

$$\frac{\partial U^{(k+1)}}{\partial \nu} = -j\omega(\mathbf{A}^{J^{(k)}} + \mathbf{A}^0) \cdot \vec{\nu}, \tag{5.372}$$

$$\begin{aligned} \mathbf{J}^{(k+1)}(Q) =& (1-\alpha)\mathbf{J}^{(k)}(Q) - j\alpha\frac{\omega\mu_0\sigma}{4\pi}\int_{V^+}\frac{\mathbf{J}^{(k)}(M)}{r_{MQ}}dv_M \\ &- \alpha\sigma\nabla U^{(k+1)}(Q) - j\alpha\omega\sigma\mathbf{A}^0(Q) \end{aligned} \tag{5.373}$$

globally converge if parameter α is chosen to satisfy inequalities:

$$0 < \alpha < \frac{2}{1 + \frac{4}{3}\left(\frac{R}{\delta}\right)^4}, \tag{5.374}$$

where R is the radius of the smallest ball that contains V^+, while δ is the skin depth.

The best estimate for the convergence rate is obtained for the following value of α:

$$\alpha = \frac{1}{1 + \frac{4}{3}\left(\frac{R}{\delta}\right)^4}. \tag{5.375}$$

For this value of α, the iterations (5.371)–(5.373) converge at least as fast as the geometric series with the ratio:

$$q = \frac{\frac{2}{\sqrt{3}}\left(\frac{R}{\delta}\right)^2}{\sqrt{1 + \frac{4}{3}\left(\frac{R}{\delta}\right)^4}}. \tag{5.376}$$

By repeating the same reasoning as before, we can also derive the following easily computable estimates for eddy current losses:

$$L_e \leq \frac{\omega^2\sigma}{2}\int_{V^+}|\mathbf{A}^0(M)|^2 dv_M, \tag{5.377}$$

$$L_e \leq \sigma\int_{V^+}\left(\int_{-\infty}^{\infty}|\frac{\partial \mathbf{A}^0}{\partial t}(M,t)|^2 dt\right) dv_M. \tag{5.378}$$

Next, we shall illustrate our previous discussion with some examples.

Example 1. Consider a thin spherical conducting shell subject to uniform time-harmonic source magnetic field $\mathbf{H}^0$. By using formula (5.121) for $\mathbf{A}^0$ and estimate (5.358), we find

$$L_e \leq \tilde{L}_e = \frac{\pi}{3}\omega^2 \mu_0^2 \sigma h H_0^2 R^4, \tag{5.379}$$

where R is the radius of the shell.

This problem can also be solved analytically by using the method of separation of variables. The analytical solution leads to the following expression for eddy current losses:

$$L_e = \frac{\pi\omega^2 \mu_0^2 \sigma h H_0^2 R^4}{3(1 + \frac{\omega^2 \mu_0^2 \sigma^2 h^2 R^2}{9})}, \tag{5.380}$$

which is exact up to small terms of order $\frac{h}{R}$.

From formulas (5.379) and (5.380), we derive

$$\eta = \frac{\tilde{L}_e}{L_e} = 1 + \frac{(\lambda R)^2}{9}, \tag{5.381}$$

where $\lambda = \omega\mu_0\sigma h$.

The function $\eta = f(\lambda R)$ is shown in Fig. 5.14. It is apparent from this figure that the estimate (5.358) is fairly accurate for $0 \leq \lambda R \leq 1.5$. ■

Example 2. Consider a nonmagnetic conducting ball of radius R subject to uniform time-harmonic source field $\mathbf{H}^0$. By using the estimate, (5.377), we find

$$L_e \leq \tilde{L}_e = \frac{2\pi\omega\mu_0 H^2 R^2}{15} \cdot \left(\frac{R}{\delta}\right)^2. \tag{5.382}$$

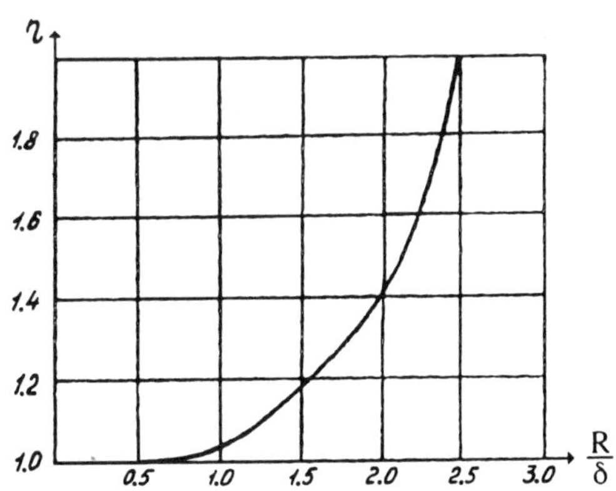

Fig. 5.14

The exact analytical solution of this problem results in the following exact expression for eddy current losses:

$$L_e = \frac{3}{2}\pi\omega\mu_0 H_0^2 R^2 \left[\frac{\mathrm{sh}\left(\frac{2R}{\delta}\right) + \sin\left(\frac{2R}{\delta}\right)}{\mathrm{ch}\left(\frac{2R}{\delta}\right) - \cos\left(\frac{2R}{\delta}\right)} - \frac{R}{\delta} \right]. \tag{5.383}$$

From the last two formulas we find

$$\eta = \frac{\tilde{L}_e}{L_e} = \frac{4}{45} \cdot \frac{\left(\frac{R}{\delta}\right)^2}{\frac{\mathrm{sh}\left(\frac{2R}{\delta}\right)+\sin\left(\frac{2R}{\delta}\right)}{\mathrm{ch}\left(\frac{2R}{\delta}\right)-\cos\left(\frac{2R}{\delta}\right)}}. \tag{5.384}$$

The function $\eta = f\left(\frac{R}{\delta}\right)$ is shown in Fig. 5.15. It is clear from this figure that the estimate (5.377) is fairly accurate for $0 \le \frac{R}{\delta} \le 1.5$. ■

Example 3 (see [14]). Consider a slotted copper cube with 1 cm long edges and $\sigma = 5.8 \times 10^7 \frac{1}{\Omega m}$, subject to source magnetic field created by a

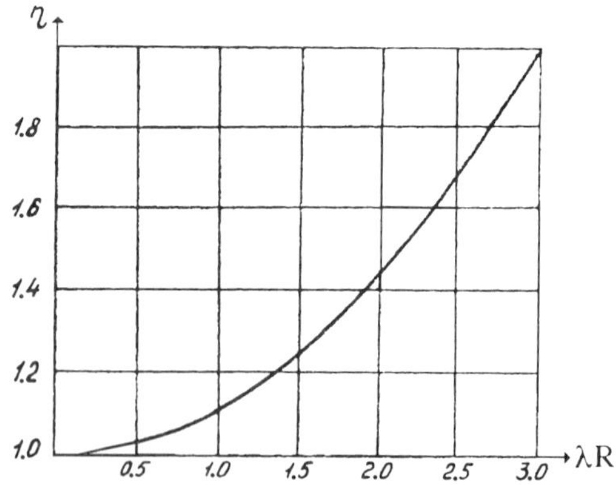

Fig. 5.15

Fig. 5.16

current through the "rectangular frame" conductor around the cube (see Fig. 5.16). This problem was solved for different frequencies varying from 1 Hz to 500 Hz by using the iteration (5.371)–(5.373). The convergence criterion used was one part per million. The finite element mesh used in calculations contained 30,282 nodes and 32,513 (hexahedron and prism) elements. Some sample results of calculations are presented in the table below. In this table, L_{est} stands for the right-hand side of the estimate (5.377), while L_e are actual eddy current losses computed by using the above iterations. It is clear from the table below that the actual numbers of iterations in many cases are substantially smaller than the numbers of iterations estimated on the basis of the q-value given by formula (5.376). Thus, as expected, the actual convergence is faster than the theoretical one predicted by the estimate (5.376). It is also worthwhile to note that the estimate (5.376) for eddy current losses is quite realistic even for fairly small skin depths (see the last column of the table).

f[Hz]	δ[cm]	$\frac{R}{\delta}$	α	q	Iterations **Actual**	Iterations **Estimated**	$\frac{L_{\text{est}}}{L_e}$
1	6.6	0.13	0.99961	0.01983	4	4	1.01986
5	3.0	0.29	0.99023	0.09867	5	6	1.01992
10	2.1	0.41	0.96217	0.19451	6	6	1.02011
20	1.5	0.59	0.86409	0.36866	9	14	1.02089
30	1.2	0.72	0.73861	0.51127	12	21	1.02220
40	1.0	0.83	0.61382	0.62144	16	30	1.02402
50	0.93	0.93	0.50427	0.70408	21	40	1.02638
60	0.85	1.02	0.41398	0.76552	27	52	1.02924
100	0.66	1.31	0.20275	0.89289	62	122	1.04588
500	0.295	2.93	0.010070	0.99495	1526	2731	1.67280

■

5.7 ANALYSIS OF THIN MAGNETIC SHELLS SUBJECT TO STATIC MAGNETIC FIELDS

This analysis is important for at least two reasons. First, this analysis is of interest in its own right because thin magnetic shells are widely used for shielding of static magnetic fields in many enginering applications. Second, the static analysis can also be of interest as an approximate one in situations when magnetic conducting shells are employed for shielding of time-harmonic fields. The stronger the magnetic shielding effects in comparison with the eddy current shielding effects the more accurate this approximation. In other words, this approximation is justifiable when distributions of magnetic flux density are fairly uniform over the shell thickness. In that case, the static analysis can be performed first to find the distribution of magnetic field and then eddy current losses can be approximately evaluated by using appropriate formulas from Section 2.6.

Below, the analysis of thin magnetic shells subject to static magnetic fields will be based on boundary conditions that are mathematically similar to those used for thin magnetic conducting shells in Section 5.5. For this reason, the same mathematical machinery will be utilized to carry out finite element formulations of this analysis. However, we shall also develop additional mathematical tools to rigorously study the convergence of finite element discretizations as well as the global convergence of special iterative techniques for the solution of nonlinear finite element equations.

We begin our discussion with the derivation of boundary conditions for thin magnetic shells. To this end, consider the infinitesimal volume as shown in Fig. 5.17. The boundary of this volume consists of three pieces: ΔS_{top}, ΔS_{bottom}, and S_{sides}. Pieces ΔS_{top} and ΔS_{bottom} lie on the exterior and interior boundaries of the shell, respectively, and have almost the same area ΔS, while piece S_{sides} is entirely within the shell. By using the principle of continuity of the magnetic flux, we find

$$\int_{S_{\text{top}}} \mathbf{B} \cdot \mathbf{dS} + \int_{S_{\text{bottom}}} \mathbf{B} \cdot \mathbf{dS} + \int_{S_{\text{sides}}} \mathbf{B} \cdot \mathbf{dS} = 0. \tag{5.385}$$

Each term of the left-hand side of the formula (5.390) can be evaluated as follows:

$$\int_{S_{\text{top}}} \mathbf{B} \cdot \mathbf{dS} \approx \left(\mathbf{B}^- \cdot \vec{\nu}\right) \Delta S, \tag{5.386}$$

$$\int_{S_{\text{bottom}}} \mathbf{B} \cdot \mathbf{dS} \approx -\left(\mathbf{B}^+ \cdot \vec{\nu}\right) \Delta S, \tag{5.387}$$

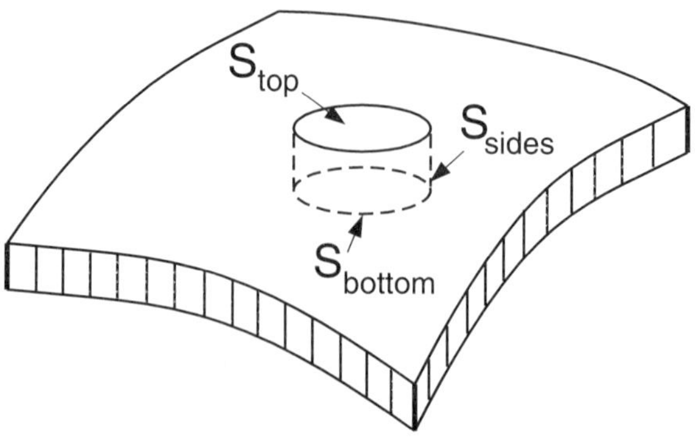

Fig. 5.17

$$\int_{S_{\text{sides}}} \mathbf{B} \cdot \mathbf{dS} \approx h \oint_L \mathbf{B} \cdot \mathbf{n} dl. \tag{5.388}$$

Here $\vec{\nu}$ is the unit outward normal to the middle surface S of the shell, h is the thickness of the shell, L is the closed line of intersection between S_{sides} and S, while $\vec{n}$ is a unit vector tangential to S and normal to L.

By substituting formulas (5.386)–(5.388) into Eq. (5.385), we obtain

$$B_\nu^- \Delta S - B_\nu^+ \Delta S + h \oint_L B_n dl \approx 0, \tag{5.389}$$

which can be rewritten as follows:

$$H_\nu^+ - H_\nu^- \approx \frac{h}{\mu_0} \frac{\oint_L B_n dl}{\Delta S}. \tag{5.390}$$

The smaller ΔS, the more accurate the last equality. Thus, in the limit we obtain the exact relation:

$$H_\nu^+ - H_\nu^- = \frac{h}{\mu_0} \lim_{\Delta S \to 0} \frac{\oint_L B_m dl}{\Delta S}. \tag{5.391}$$

By recalling the definition (5.164) of the divergence of a vector field on a curvilinear surface, from the last formula we find

$$\boxed{H_\nu^+ - H_\nu^- = \frac{h}{\mu_0} \operatorname{div}_S \mathbf{B}.} \tag{5.392}$$

Now we replace the actual thin magnetic shell by its middle surface S and we use the boundary condition (5.392) on this surface. It is tacitly assumed that $\mathbf{B}$ in the above boundary condition is the total magnetic flux density, which is tangential to S. This assumption is justified by the fact that within the actual thin magnetic shells the normal components of $\mathbf{B}$ are very small and can be neglected.

In addition to the boundary condition (5.392), it is assumed that tangential components of the magnetic field $\mathbf{H}$ are continuous across S:

$$\vec{\nu} \times \mathbf{H}^+ = \vec{\nu} \times \mathbf{H}^-. \tag{5.393}$$

This assumption is justified by the small thickness of the shell as well as by the continuity of the tangential components of the magnetic field across the exterior and interior boundaries of the actual shell.

Next, we shall introduce the magnetic scalar potential φ and formulate the boundary value problem for this potential. For the sake of simplicity, we assume that the shell is closed and that the source field is created by a coil with a rectangular cross-section. In this case, the source field can be confined to the extended source region $\tilde{V}^0$ and it is given by the formula (see Section 5.2):

$$\mathbf{H}^0(Q) = \begin{cases} \mathbf{a}_z T(Q) & \text{in } \tilde{V}^0, \\ 0 & \text{outside } \tilde{V}^0. \end{cases} \tag{5.394}$$

To be specific, we shall also assume that the coil and the extended source region $\tilde{V}^0$ are located within the shell. In this case, the magnetic scalar potential can be introduced as follows:

$$\mathbf{H}^+ = \mathbf{H}^0 - \nabla\varphi^+, \tag{5.395}$$

$$\mathbf{H}^- = -\nabla\varphi^-. \tag{5.396}$$

By using formulas (5.392)–(5.396), we find that the magnetic scalar potential φ satisfies the equations:

$$\nabla^2\varphi^+ = 0 \quad \text{in } V^+, \tag{5.397}$$

$$\nabla^2\varphi^- = 0 \quad \text{in } V^-, \tag{5.398}$$

and the following boundary conditions:

$$\varphi^+ = \varphi^- \text{ on } S, \tag{5.399}$$

$$\frac{\partial\varphi^-}{\partial\nu} - \frac{\partial\varphi^+}{\partial\nu} = \frac{h}{\mu_0}\mathrm{div}_S\mathbf{B}(-\nabla_S\varphi), \tag{5.400}$$

$$\left[\frac{\partial\varphi^+}{\partial\nu}\right]\Bigg|_{S_{\mathrm{top}}} = -T(Q), \tag{5.401}$$

$$\left[\frac{\partial\varphi^+}{\partial\nu}\right]\Bigg|_{S_{\mathrm{bottom}}} = T(Q), \tag{5.402}$$

$$\varphi^-(\infty) = 0. \tag{5.403}$$

To arrive at the finite element formulation, we shall first reduce the boundary value problem (5.397)–(5.403) to the appropriate weak Galerkin's form.

To this end, we shall recall the scalar Green formulas, which can be written in the form:

$$\int_{V^+} \nabla\psi \cdot \nabla\varphi^+ dv - \oint_S \psi \frac{\partial \varphi^+}{\partial \nu} dS = \int_{S_{\text{top}}} \psi T dS - \int_{S_{\text{bottom}}} \psi T dS, \quad (5.404)$$

$$\int_{V^-} \nabla\psi \cdot \nabla\varphi^- dv + \oint_S \psi \frac{\partial \varphi^-}{\partial \nu} dS = 0. \quad (5.405)$$

Here we have already taken into account Eqs. (5.397)–(5.398), discontinuity conditions (5.401)–(5.402) and condition (5.403) at infinity. By adding formulas (5.404) and (5.405) and taking into account the boundary condition (5.400), we obtain

$$\begin{aligned} &\int_V \nabla\psi \cdot \nabla\varphi dv + \frac{h}{\mu_0} \oint_S \psi \; \text{div}_S \mathbf{B}(-\nabla_S \varphi) dS = \\ &\int_{S_{\text{top}}} \psi T dS - \int_{S_{\text{bottom}}} \psi T dS, \end{aligned} \quad (5.406)$$

where $V = V^+ + V^-$.

Now, by applying identity (5.182) to the surface integral in (5.406), we arrive at the following Galerkin form:

$$\int_V \nabla\psi \cdot \nabla\varphi dv - \frac{h}{\mu_0} \oint_S \nabla_S \psi \cdot \mathbf{B}(-\nabla_S \varphi) dS = \int_{S_{\text{top}}} \psi T dS - \int_{S_{\text{bottom}}} \psi T dS. \quad (5.407)$$

This Galerkin form is considered on the set of continuous functions. In this way, the boundary condition (5.399) is taken into account. Thus, we have proved that if the magnetic scalar potential φ is a solution of the boundary value problem (5.397)–(5.403), then this potential satisfies the weak Galerkin form (5.407) for any sufficiently regular function ψ. The inverse statement can be established as well. Namely, it can be proved that if the potential φ is twice differentiable in V^+ and V^- and satisfies the weak Galerkin form (5.407) for any function ψ then this potential is a solution of the boundary value problem (5.397)–(5.403). The main idea of this proof is to exploit the arbitrariness of function ψ in the Galerkin form (5.407).

It is clear from the above discussion that the boundary value problem (5.397)–(5.403) is equivalent to the Galerkin form (5.407). However, the above Galerkin form can be preferable for the following reasons. First, this Galerkin form is a more concise statement of the problem; this statement is expressed in one formula (5.407). Second, the Galerkin form can be directly used for finite element discretizations. Finally, the Galerkin form is more convenient for derivations of various integral estimates that are indispensable in proving various mathematical facts. We shall first illustrate the last statement by proving the uniqueness theorem for the boundary value problem (5.397)–(5.403). Before directly proceeding to the proof of the uniqueness theorem, we recall that constitutive relations $\mathbf{B}(\mathbf{H})$ for unhysteretic media satisfy the following constraints (see Section 2.4): (a) Jacobian matrices for constitutive relations are symmetric; (b) the following inequalities are valid:

$$c|\Delta\mathbf{H}|^2 \leq \Delta\mathbf{B}\cdot\Delta\mathbf{H} \leq C|\Delta\mathbf{H}|^2. \tag{5.408}$$

Here $\Delta\mathbf{H}$ is an arbitrary increment of the magnetic field and $\Delta\mathbf{B}$ is the corresponding increment of the magnetic flux density.

The left inequality in (5.408) expresses the fact that the magnetic medium is passive, while the right inequality in (5.408) reflects saturation phenomena of magnetic media.

Inequalities (5.408) can also be written in the following equivalent form:

$$c|\vec{\xi}|^2 \leq \vec{\xi}\cdot\hat{J}(\mathbf{H})\vec{\xi} \leq C|\vec{\xi}|^2, \tag{5.409}$$

where $\hat{J}(\mathbf{H})$ is a three-dimensional symmetric Jacobian matrix with matrix elements evaluated at the field $\mathbf{H}$ and $\vec{\xi}$ is an arbitrary three-dimensional vector. The values of constants c and C depend on particular forms of constitutive relations. In the case of nonlinear isotropic media with $\mathbf{B}(\mathbf{H}) = \mu(H)\mathbf{H}$, we have established (see (2.315)) that

$$c = \min_H \mu_d(H), \quad C = \max_H \mu(H), \tag{5.410}$$

where minimum and maximum are taken over the relevant range of variation of H.

Now we are ready to prove the validity of the following statement.

Uniqueness Theorem. For unhysteretic magnetic media with constitutive relations that satisfy inequalities (5.408) (or (5.409)), the boundary value problem (5.397)–(5.403) has the unique solution.

Proof. Suppose that there are two solutions $\varphi^{(1)}$ and $\varphi^{(2)}$ of the boundary value problem (5.397)–(5.403). Then, these solutions must satisfy the

Galerkin form (5.407):

$$\int_V \nabla\psi\cdot\nabla\varphi^{(1)}dv - \frac{h}{\mu_0}\oint_S \nabla_S\psi\cdot\mathbf{B}\left(-\nabla_S\varphi^{(1)}\right)dS = \int_{S_{\text{top}}}\psi T dS - \int_{S_{\text{bottom}}}\psi T dS, \tag{5.411}$$

$$\int_V \nabla\psi\cdot\nabla\varphi^{(2)}dv - \frac{h}{\mu_0}\oint_S \nabla_S\psi\cdot\mathbf{B}\left(-\nabla_S\varphi^{(2)}\right)dS = \int_{S_{\text{top}}}\psi T dS - \int_{S_{\text{bottom}}}\psi T dS, \tag{5.412}$$

By subtracting formula (5.412) from formula (5.411), by introducing the notation

$$\tilde{\varphi} = \varphi^{(1)} - \varphi^{(2)}, \tag{5.413}$$

and by taking $\psi = \tilde{\varphi}$, we derive

$$\int_V |\nabla\tilde{\varphi}|^2 dv - \frac{h}{\mu_0}\oint_S \nabla_S\tilde{\varphi}\cdot\left[\mathbf{B}\left(-\nabla_S\varphi^{(1)}\right) - \mathbf{B}\left(-\nabla_S\varphi^{(2)}\right)\right]dS = 0. \tag{5.414}$$

Consider the increment of the magnetic field:

$$\Delta\mathbf{H} = -\nabla_S\left(\varphi^{(1)} - \varphi^{(2)}\right) = -\nabla_S\tilde{\varphi}. \tag{5.415}$$

The corresponding increment of the magnetic flux density is

$$\nabla\mathbf{B} = \mathbf{B}\left(-\nabla_S\varphi^{(1)}\right) - \mathbf{B}\left(-\nabla_S\varphi^{(2)}\right). \tag{5.416}$$

According to the left inequality of (5.408), we have

$$-\nabla_S\tilde{\varphi}\cdot\left[\mathbf{B}\left(-\nabla_S\varphi^{(1)}\right) - \mathbf{B}\left(-\nabla_S\varphi^{(2)}\right) \geq c|\nabla_S\tilde{\varphi}|^2. \tag{5.417}$$

By using the last inequality in the formula (5.414), we arrive at the contradiction

$$0 = \int_V |\nabla\tilde{\varphi}|^2 dv - \frac{h}{\mu_0}\oint_S \nabla_S\tilde{\varphi}\cdot\left[\mathbf{B}\left(-\nabla_S\varphi^{(1)}\right) - \mathbf{B}\left(-\nabla_S\varphi^{(2)}\right)\right]dS \geq \int_V |\nabla\tilde{\varphi}|^2 dv + \frac{hc}{\mu_0}\oint_S |\nabla_S\tilde{\varphi}|^2 dS > 0. \tag{5.418}$$

This contradiction is removed only when

$$\nabla\tilde{\varphi} \equiv 0. \tag{5.419}$$

Because $\tilde{\varphi}(\infty) = 0$, from the last equation we find

$$\tilde{\varphi} \equiv 0, \tag{5.420}$$

and

$$\varphi^{(1)} = \varphi^{(2)}. \tag{5.421}$$

The uniqueness is proved. ∎

Next, we proceed to the finite element discretization of the Galerkin form (5.407). We shall look for the approximate solution in the form:

$$\varphi \approx \varphi_N = \sum_{n=1}^{N} \varphi_n \alpha_n, \tag{5.422}$$

where, as before, α_n are local support finite element functions, while φ_n are unknown node values of φ.

By substituting the expression (5.422) into the Galerkin form (5.407) and by choosing sequentially

$$\psi = \alpha_i, \quad (i = 1, 2, \dots N), \tag{5.423}$$

we end up with the following nonlinear algebraic equations:

$$\begin{aligned}&\sum_{n=1}^{N} \varphi_n \int_V \nabla\alpha_i \cdot \nabla\alpha_n dv - \frac{h}{\mu_0} \oint_S \nabla_S \alpha_i \cdot \mathbf{B}\left(-\sum_{n=1}^{N} \varphi_n \nabla_S \alpha_n\right) dS \\ &= \int_{S_{\text{top}}} \alpha_i T dS - \int_{S_{\text{bottom}}} \alpha_i T dS, \quad (i = 1, 2, \dots N).\end{aligned} \tag{5.424}$$

By solving these equations with respect to unknowns φ_n, we can find the approximate solution φ_N.

Three important questions can be posed concerning nonlinear finite element Eq. (5.424). Is the solution to these equations unique? Does the finite element solution φ_N converge to the exact solution φ? Is it possible to design iterative techniques for the solution of nonlinear finite element equations (5.424) that are **globally** convergent? Next we shall present affirmative answers to all these questions.

To establish the uniqueness of the solution of the finite element equations (5.424), we shall rewrite them in the following equivalent form:

$$\int_V \nabla\alpha_i \cdot \nabla\varphi_N dv - \frac{h}{\mu_0}\oint_S \nabla_S\alpha_i \cdot \mathbf{B}\left(-\nabla_S\varphi_N\right) dS$$
$$= \int_{S_{\text{top}}} \alpha_i T dS - \int_{S_{\text{bottom}}} \alpha_i T dS, \quad (i = 1, 2, \ldots N). \tag{5.425}$$

By multiplying each equation in (5.425) by an arbitrary coefficient d_i and by adding them together, we obtain

$$\int_V \nabla\psi_N \cdot \nabla\varphi_N dv - \frac{h}{\mu_0}\oint_S \nabla_S\psi_N \cdot \mathbf{B}(-\nabla\varphi_N) dS$$
$$= \int_{S_{\text{top}}} \psi_N T dS - \int_{S_{\text{bottom}}} \psi_N T dS, \tag{5.426}$$

where we introduce the notation

$$\psi_N = \sum_{i=1}^{N} d_i\alpha_i. \tag{5.427}$$

Thus, finite element Eqs. (5.424) are equivalent to the form (5.426) where ψ_N is an arbitrary linear combination (5.427) of finite element functions α_i.

The form (5.426) has the same mathematical structure as the Galerkin form (5.407). For this reason, by literally repeating the same line of reasoning as in the proof of the uniqueness theorem, we can establish that the function φ_N that satisfies the form (5.426) is unique. This is tantamount to the uniqueness of the solution of the finite element Eqs. (5.424). It is remarkable that the uniqueness holds for any finite element mesh (and for any N).

Now, we shall discuss the convergence of the finite element solution φ_N to the exact solution. To this end, consider the Hilbert space $\mathcal{H}^1$ with the inner product

$$< \varphi, \psi >= \int_V \nabla\psi \cdot \nabla\varphi dv + h\oint_S \nabla_S\psi \cdot \nabla_S\varphi dS, \tag{5.428}$$

and the norm

$$||\varphi|| = \left(\int_V |\nabla\varphi|^2 dv + h\oint_S |\nabla_S\varphi)^2 dS\right)^{\frac{1}{2}}. \tag{5.429}$$

Consider also subspace $\mathcal{H}^1_N$ of $\mathcal{H}^1$, which contains all linear combinations of the form (5.427). We proceed to prove the following statement.

Theorem (convergence of the finite element technique). The following inequality is valid:

$$||\varphi - \varphi_N|| \leq \frac{C}{\mu_0}||\varphi - \psi^{\varphi}_N||, \tag{5.430}$$

where ψ^{φ}_N is the best approximation to φ by functions from $\mathcal{H}^1_N$:

$$||\varphi - \psi^{\varphi}_N|| = \min_{\psi_N} ||\varphi - \psi_N||. \tag{5.431}$$

Proof. First, a few remarks concerning the meaning of this theorem. It is clear that the inequality (5.430) implies the convergence of the finite element technique. This is because the error of the best approximation of any function φ from $\mathcal{H}^1$ by functions from $\mathcal{H}^1_N$ goes to zero as the density of the finite element mesh is infinitely increased ($N \to \infty$). This inequality also characterizes the rate of convergence. Namely, this inequality means that φ_N converges to φ as fast as ψ^{φ}_N converges to φ. Furthermore, this inequality suggests the importance of the proper design of the finite element mesh. The better the design of the finite element mesh, the less the quantity $||\varphi - \psi^{\varphi}_N||$ and the better the quality of the finite element solution. This is the trite maxim that the finite element solution is as good as its finite element mesh.

The proof of inequality (5.430) is based on upper and lower estimates of the following expression:

$$I = \int_V |\varphi - \varphi_N|^2 dv - \frac{h}{\mu_0} \oint_S \nabla_S(\varphi - \varphi_N) \cdot [\mathbf{B}(-\nabla_S\varphi) - \mathbf{B}(-\nabla_S\varphi_N)]dS. \tag{5.432}$$

First, we find the lower estimate. By using the right inequality (5.408), we have

$$-\nabla_S(\varphi - \varphi_N) \cdot [\mathbf{B}(-\nabla_S\varphi) - \mathbf{B}(-\nabla_S\varphi_N)] \geq c|\nabla_S(\varphi - \varphi_N)|^2. \tag{5.433}$$

By using the last inequality in the expression (5.432) and taking into account the formula (5.429) and the fact that $c \geq \mu_0$, we derive

$$I \geq ||\varphi - \varphi_N||^2. \tag{5.434}$$

To find the upper estimate for I, we shall rewrite formula (5.432) in the following equivalent form:

$$\begin{aligned} I = & \int_V \nabla(\psi_N^\varphi - \varphi_N) \cdot \nabla(\varphi - \varphi_N) dv \\ & - \frac{h}{\mu_0} \oint_S \nabla_S(\psi_N^\varphi - \varphi_N) \cdot [\mathbf{B}(-\nabla_S \varphi) - \mathbf{B}(-\nabla_S \varphi_N)] dS \\ & + \int_V \nabla(\varphi - \psi_N^\varphi) \cdot \nabla(\varphi - \varphi_N) dv \\ & - \frac{h}{\mu_0} \oint_S \nabla_S(\varphi - \psi_N^\varphi) \cdot [\mathbf{B}(-\nabla_S \varphi) - \mathbf{B}(-\nabla_S \varphi_N)] dS. \end{aligned} \tag{5.435}$$

It turns out that

$$\begin{aligned} & \int_V \nabla(\psi_N^\varphi - \varphi_N) \cdot \nabla(\varphi - \varphi_N) dv \\ & - \frac{h}{\mu_0} \oint_S \nabla_S(\psi_N^\varphi - \varphi) \cdot [\mathbf{B}(-\nabla_S \varphi) - \mathbf{B}(-\nabla_S \varphi_N)] dS = 0. \end{aligned} \tag{5.436}$$

To prove this, we replace ψ in the Galerkin form (5.407) by an arbitrary function ψ_N from $\mathcal{H}_N^1$ and then subtract the formula (5.426) from that expression. As a result we find

$$\int_V \nabla \psi_N \cdot (\varphi - \varphi_N) dv - \frac{h}{\mu_0} \oint_S \nabla_S \psi_N \cdot [\mathbf{B}(-\nabla_S \varphi) - \mathbf{B}(-\nabla_S \varphi_N)] dS = 0. \tag{5.437}$$

Because $\psi_N^\varphi - \varphi_N$ belongs to $\mathcal{H}_N^1$ and because the identity (5.437) is valid for any ψ_N from $\mathcal{H}_N^1$, it will also be valid for $\psi_N = \psi_N^\varphi - \varphi_N$. This proves the formula (5.436).

From formulas (5.435) and (5.436), we conclude:

$$\begin{aligned} I = & \int_V \nabla(\varphi - \psi_N^\varphi) \cdot \nabla(\varphi - \varphi_N) dv \\ & - \frac{h}{\mu_0} \oint_S \nabla_S(\varphi - \psi_N^\varphi) \cdot [\mathbf{B}(-\nabla_S \varphi) - \mathbf{B}(-\nabla_S \varphi_N) dS. \end{aligned} \tag{5.438}$$

By using the Cauchy inequality, we derive

$$\begin{aligned} I \leq & \left(\int_V |\nabla(\varphi - \psi_N^\varphi)|^2 dv \right)^{\frac{1}{2}} \cdot \left(\int_V |\nabla(\varphi - \varphi_N)|^2 dv \right)^{\frac{1}{2}} \\ & + \frac{1}{\mu_0} \left(h \oint_S |\nabla_S(\varphi - \psi_N^\varphi)|^2 d \right)^{\frac{1}{2}} \cdot \left(h \oint_S |\mathbf{B}(-\nabla \varphi) - \mathbf{B}(-\nabla \varphi_N)|^2 dS \right)^{\frac{1}{2}}. \end{aligned} \tag{5.439}$$

By recalling formulas (2.295) and (2.296), we have

$$\mathbf{B}(-\nabla\varphi) - \mathbf{B}(-\nabla_S\varphi_N) = \hat{A}[-\nabla_S(\varphi - \varphi_N)], \tag{5.440}$$

where the matrix $\hat{A}$ is given by

$$\hat{A} = \int_0^1 \hat{J}(\mathbf{H}_\nu)d\nu, \quad \mathbf{H}_\nu = -\nabla_S\varphi_N + \nu\nabla_S(\varphi_N - \varphi). \tag{5.441}$$

Strictly speaking, matrix $\hat{A}$ depends on φ and φ_N. However, for all possible φ and φ_N, this matrix is symmetric and satisfies inequalities:

$$c|\vec{\xi}|^2 \leq \vec{\xi}\cdot\hat{A}\vec{\xi} \leq C|\vec{\xi}|^2. \tag{5.442}$$

These inequalities directly follow from formula (5.409). From the above inequalities and the symmetry of $\hat{A}$, we find that for any φ and φ_N we have

$$||\hat{A}|| \leq C.$$

Consequently,

$$|\mathbf{B}(-\nabla_S\varphi) - \mathbf{B}(-\nabla_S\varphi_N)| \leq C|\nabla_S(\varphi - \varphi_N)|^2. \tag{5.443}$$

By substituting the last result into formula (5.439) and taking into account that $C \geq \mu_0$, we obtain

$$\begin{aligned} I \leq \frac{C}{\mu_0}\Bigg\{&\left(\int_V |\nabla(\varphi - \psi_N^\varphi)|^2 dv\right)^{\frac{1}{2}} \cdot \left(\int_V |\nabla(\varphi - \varphi_N)|^2 dv\right)^{\frac{1}{2}} \\ &+ \left(h\oint_S |\nabla_S(\varphi - \psi_N^\varphi)|^2 dS\right)^{\frac{1}{2}} \cdot \left(h\oint_S |\nabla_S(\varphi - \varphi_N(|^2 dS\right)^{\frac{1}{2}}\Bigg\} \end{aligned} \tag{5.444}$$

By introducting vectors

$$\mathbf{a} = (a_1, a_2), \qquad \mathbf{b} = (b_1, b_2), \tag{5.445}$$

$$a_1 = \left(\int_V |\nabla(\varphi - \psi_N^\varphi)|^2 dv\right)^{\frac{1}{2}}, \quad a_2 = \left(h\oint_S |\nabla_S(\varphi - \psi_N^\varphi)^2 dS\right)^{\frac{1}{2}}, \tag{5.446}$$

$$b_1 = \left(\int_V |\nabla(\varphi - \varphi_N)|^2 dv\right)^{\frac{1}{2}}, \quad b_s = \left(h\oint_S |\nabla_S(\varphi - \varphi_N)|^2 dS\right)^{\frac{1}{2}}, \tag{5.447}$$

we can rewrite the inequality (5.444) as follows:

$$I \leq \frac{C}{\mu_0}\mathbf{a}\cdot\mathbf{b} \leq \frac{C}{\mu_0}|\mathbf{a}||\mathbf{b}|. \tag{5.448}$$

It is clear from formulas (5.429) and (5.445)–(5.447) that

$$|\mathbf{a}| = ||\varphi - \psi_N^{\varphi}||, \qquad |\mathbf{b}| = ||\varphi - \varphi_N||. \tag{5.449}$$

Thus, we have

$$I \leq \frac{C}{\mu_0}||\varphi - \psi_N^{\varphi}|| \cdot \varphi - \varphi_N||. \tag{5.450}$$

From inequalities (5.434) and (5.450), we have

$$||\varphi - \varphi_N||^2 \leq \frac{C}{\mu_0}||\varphi - \psi_N^{\varphi}|| \cdot ||\varphi - \varphi_N||, \tag{5.451}$$

which is tantamount to

$$||\varphi - \varphi_N|| \leq \frac{C}{\mu_0}||\varphi - \psi_N^{\varphi}||. \tag{5.452}$$

■

Now, we consider the **globally** convergent iterative technique for the solution of nonlinear finite element Eqs. (5.425). This technique can be mathematically formulated as follows:

$$\int_V \nabla\alpha_i \cdot \nabla\varphi_N^{(k+1)} dv + \frac{h\mu_e}{\mu_0}\oint_S \nabla_S\alpha_i \cdot \nabla_S\varphi_N^{(k+1)} dS =$$
$$\frac{h\mu_e}{\mu_0}\oint_S \nabla_S\alpha_i \cdot \left[\nabla_S\varphi_N^{(k)} + \frac{1}{\mu_e}\mathbf{B}\left(-\nabla_S\varphi_N^{(k)}\right)\right]$$
$$+\int_{S_{\text{top}}} \alpha_i T dS - \int_{S_{\text{bottom}}} \alpha_i T dS, \quad (i = 1, 2, \ldots N), \tag{5.453}$$

where

$$\mu_e = \frac{C + c}{2}. \tag{5.454}$$

Thus, on each step of iterations we have to solve linear simultaneous Eqs. (5.453) whose right-hand sides depend on the previous iteration as well as on excitation conditions. The matrix of these linear equations remains the same for all iterations. This can be clearly seen if the iterative technique (5.453) is written in the form:

$$\sum_{n=1}^{N} \gamma_{in} \varphi_N^{(k+1)} = f_i^{(k+1)}, \quad (i = 1, 2, \dots N), \tag{5.455}$$

$$\gamma_{in} = \int_V \nabla \alpha_i \cdot \nabla \alpha_n dv + \frac{h\mu_e}{\mu_0} \oint_S \nabla_S \alpha_i \cdot \nabla_S \alpha_n dS, \tag{5.456}$$

$$\begin{aligned} f_i^{(k+1)} = & \frac{h\mu_e}{\mu_0} \oint_S \nabla_S \alpha_i \cdot \left[\nabla_S \varphi_N^{(k)} + \frac{1}{\mu_e} \mathbf{B}(-\nabla_S \varphi_N^{(k)}) \right] dS \\ & + \int_{S_{\text{top}}} \alpha_i T dS - \int_{S_{\text{bottom}}} \alpha_i T dS. \end{aligned} \tag{5.457}$$

To prove the global convergence of the above technique we introduce space $\mathcal{H}^1_{N,\mu_e}$ of linear combinations (5.427) with the following inner product:

$$< \psi_N, \varphi_N >_{\mu_e} = \int_V \nabla \psi_N \cdot \nabla \varphi_N dv + \frac{h\mu_e}{\mu_0} \oint_S \nabla_S \psi_N \cdot \nabla_S \varphi_N dS, \tag{5.458}$$

and the norm

$$||\psi_N||_{\mu_e} = \int_V |\nabla \psi_N|^2 dv + \frac{h\mu_e}{\mu_0} \oint_S |\nabla_S \psi_N|^2 dS. \tag{5.459}$$

Next, we shall prove the following result.

Theorem (global convergence of iterative technique). Iterations (5.453) globally converge to the solution of nonlinear finite element Eqs. (5.425) at least as fast as a geometric series with the ratio:

$$q = \frac{C - c}{C + c}. \tag{5.460}$$

Proof. By using simple transformations, from formulas (5.453) we derive

$$\begin{aligned} & \int_V \nabla \psi_N \cdot \nabla (\varphi_N^{(k+1)} - \varphi_N^{(k)}) dv + \frac{h\mu_e}{\mu_0} \oint_S \nabla_S \psi_N \cdot \nabla_S (\varphi_N^{(k+1)} - \varphi_N^{(k)}) dS = \\ & \frac{h\mu_e}{\mu_0} \oint_S \nabla_S \psi_N \cdot \Big[\nabla_S (\varphi_N^{(k)} - \varphi_N^{(k+1)}) \\ & + \frac{1}{\mu_e} (\mathbf{B}(-\nabla_S \varphi_N^{(k)}) - \mathbf{B}(-\nabla_S \varphi_N^{(k+1)})) \Big] dS, \end{aligned} \tag{5.461}$$

where ψ_N is an arbitrary function from $\mathcal{H}^1_{N,\mu_e}$. By taking

$$\psi_N = \varphi_N^{(k+1)} - \varphi_N^{(k)}, \tag{5.462}$$

from formulas (5.459) and (5.461) we find

$$\begin{aligned} &||\varphi_N^{(k+1)} - \varphi_N^{(k)}||^2_{\mu_e} = \\ &\frac{h\mu_e}{\mu_0} \oint_S \nabla_S(\varphi_N^{(k+1)} - \varphi_N^{(k)}) \cdot \Big[\nabla_S(\varphi_N^{(k)} - \varphi_N^{(k-1)}) \\ &+ \frac{1}{\mu_e}(\mathbf{B}(-\nabla_S\varphi_N^{(k)}) - \mathbf{B}(-\nabla_S\varphi_N^{(k+1)}))\Big] dS. \end{aligned} \tag{5.463}$$

By recalling formulas (2.295) and (2.296), we have

$$\mathbf{B}(-\nabla_S\varphi_N^{(k)}) - \mathbf{B}(-\nabla_S\varphi_N^{(k+1)}) = -\hat{A}_{k,k-1}\nabla_S(\varphi_N^{(k)} - \varphi_N^{(k-1)}), \tag{5.464}$$

where

$$\hat{A}_{k,k-1} = \int_0^1 \hat{J}(\mathbf{H}_\nu)d\nu, \quad \mathbf{H}_\nu = -\nabla_S\varphi_S^{(k)} + \nu\nabla_S(\varphi_N^{(k-1)} - \varphi_N^{(k)}), \tag{5.465}$$

and subscript "$k, k-1$" indicates that the symmetric matrix $\hat{A}_{k,k-1}$ depends on $\varphi_N^{(k)}$ and $\varphi_N^{(k-1)}$. By using the expression (5.464) in the formula (5.463), we obtain

$$\begin{aligned} &||\varphi_N^{(k+1)} - \varphi_N^{(k)}||^2_{\mu_e} = \\ &\quad \frac{h\mu_e}{\mu_0} \oint_S \nabla_S(\varphi_N^{(k+1)} - \varphi_N^{(k)}) \cdot (\hat{I} - \frac{1}{\mu_e}\hat{A}_{k,k-1})\nabla_S(\varphi_N^{(k)} - \varphi_N^{(k-1)})dS, \end{aligned} \tag{5.466}$$

where $\hat{I}$ is the identity matrix.

Next, we shall prove that

$$||\hat{I} - \frac{1}{\mu_e}\hat{A}_{k,k-1}|| \leq \frac{C-c}{C+c}. \tag{5.467}$$

Indeed, according to inequalities (5.409) and formula (5.465), we have

$$c|\vec{\xi}|^2 \leq \vec{\xi} \cdot \hat{A}_{k,k-1}\vec{\xi} \leq C|\vec{\xi}|^2. \tag{5.468}$$

By using this fact and formula (5.454) for μ_e, we find

$$\frac{c-C}{C+c}|\vec{\xi}|^2 \leq \vec{\xi} \cdot \left(\hat{I} - \frac{1}{\mu_e}\hat{A}_{k,k-1}\right)\vec{\xi} \leq \frac{C-c}{C+c}|\vec{\xi}|^2. \tag{5.469}$$

Because matrix $\hat{I} - \frac{1}{\mu_e}\hat{A}_{k,k-1}$ is symmetric, the last inequalities imply the validity of inequality (5.467).

By using inequality (5.467), we derive

$$|(\hat{I} - \frac{1}{\mu_e}\hat{A}_{k,k-1})\nabla_S(\varphi_N^{(k)} - \varphi_N^{(k-1)})| \leq \frac{C-c}{C+c}|\nabla_S(\varphi_N^{(k)} - \varphi_N^{(k-1)})|, \quad (5.470)$$

Now, we turn back to formula (5.466) and, by using (5.470) and the Cauchy inequality, we obtain:

$$\begin{aligned}
&||\varphi_N^{(k+1)} - \varphi_N^{(k)}||^2_{\mu_e} \leq \\
&\frac{C-c}{C+c}\cdot\frac{h\mu_e}{\mu_0}\oint_S|\nabla_S(\varphi_N^{(k+1)} - \varphi_N^{(k)})|\cdot|\nabla_S(\varphi_N^{(k)} - \varphi_N^{(k-1)})|dS \\
&= \frac{C-c}{C+c}\left(\frac{h\mu_e}{\mu_0}\oint_S|\nabla_S(\varphi_N^{(k+1)} - \varphi_N^{(k)})|^2 dS\right)^{\frac{1}{2}} \\
&\times\left(\frac{h\mu_e}{\mu_0}\oint_S|\nabla_S(\varphi_N^{(k)} - \varphi_N^{(k-1)})|^2 dS\right)^{\frac{1}{2}} \\
&\leq \frac{C-c}{C+c}||\varphi_N^{(k+1)} - \varphi_N^{(k)}||_{\mu_e}\cdot||\varphi_N^{(k)} - \varphi_N^{(k-1)}||_{\mu_e},
\end{aligned} \quad (5.471)$$

which is tantamount to

$$||\varphi_N^{(k+1)} - \varphi_N^{(k)}||_{\mu_e} \leq \frac{C-c}{C+c}||\varphi_N^{(k)} - \varphi_N^{(k-1)}||_{\mu_e}. \quad (5.472)$$

That means that iterations $\varphi_N^{(k)}$ form a contracting sequence. This sequence converges at least as fast as a geometric series with the ratio given by the formula (5.460). ■

It is evident from the proof of the theorem that the actual rate of convergence may be substantially higher than the one specified by inequality (5.472). This is because we have used rough estimates to guarantee the validity of inequality (5.472) under all possible circumstances. It is also important to keep in mind that we have proved the global convergence of the iterative technique (5.455)–(5.457) in the norm (5.459), which is stronger than the norm (5.429). This means that the iterative technique (5.455)-(5.457) will converge faster in the norm (5.429) than is predicted by the inequality (5.472).

The proven theorem also implies the **existence** of the solution of nonlinear finite element Eqs. (5.424) for any finite element mesh (for any N).

Furthermore, using the iterations

$$\begin{aligned}\int_V \nabla\psi \cdot \nabla\varphi^{(k+1)} dv + \frac{h\mu_e}{\mu_0} \oint_S \nabla_S\psi \cdot \nabla_S\varphi^{(k+1)} dS = \\ \frac{h\mu_e}{\mu_0} \oint_S \nabla_S\psi \cdot [\nabla_S\varphi^{(k)} + \frac{1}{\mu_e}\mathbf{B}(-\nabla_S\varphi^{(k)})] dS \\ + \int_{S_{\text{top}}} \psi T dS - \int_{S_{\text{bottom}}} \psi T dS,\end{aligned} \tag{5.473}$$

and by literally repeating the same line of reasoning as in the proof of the last theorem, we can establish that iterations (5.473) globally converge to the exact solution φ specified by the weak Galerkin form (5.407). This implies the **existence** of the weak solution of the nonlinear boundary value problem (5.397)–(5.403).

It is instructive to compare the iterative technique (5.455)–(5.457) with the Newton method. The above iterative technique converges globally (i.e., for any choice of initial guess), whereas the Newton method converges locally (i.e., for initial guesses that are sufficiently close to the actual solution). The above iterative technique requires the solution of linear equations with the same stiffness matrix $\{\gamma_{ni}\}$ at every step of iterations, whereas in the case of the Newton method the matrix of the simultaneous equations to be solved at each iteration is a Jacobian of the nonlinear finite element equations and this Jacobian must be reevaluated by using the previous iteration. Thus, this Jacobian changes from one iteration to another. This makes the computation of every new iteration more costly. However, in a sufficiently close vicinity of the actual solution, this additional conputational cost is paid for by the **quadratic** rate of convergence of the Newton method, whereas the iterative technique (5.455)–(5.457) has only the **linear** rate of convergence. This suggests that the combination of the above iterative technique and the Newton method may be desirable.

Finally, we remark that the results presented in this section can also be obtained for static fields in the presence of bulky (voluminous) magnetic objects. This is actually accomplished in the author's book [7] albeit published in Russian.

REFERENCES

[1] W.E. Brittin, W.K. Smythe, and W. Wyss, American Journal of Physics, **50**, No. 8, 693, (1982).

[2] A. Friedman, "Foundation of Modern Analysis," Dover, New York, (1982).

[3] S.R.H. Hoole, "Computer Aided Analysis and Design of Electromagnetic Devices," Elsevier, New York, (1989).
[4] A. Ishimaru, "Electromagnetic Wave Propagation, Radiation and Scatterings," Prentice-Hall, New Jersey, (1991).
[5] J.D. Jackson, "Classical Electrodynamics," John Wiley and Sons, New York, (1975).
[6] P.R. Kotiuga, IEEE Transactions on Magnetics, **25**, 2925, (1989).
[7] I.D. Mayergoyz, "Iterative Techniques for the Analysis of Static Fields in Inhomogeneous Anisotropic and Nonlinear Media," Naukova Dumka (Publisher), Kiev, (1978).
[8] I.D. Mayergoyz, IEEE Transactions on Magnetics, **18**, 536, (1982).
[9] I.D. Mayergoyz, in the book "Computational Electromagnetics" (editor Z. Cendes), North-Holland Publishers, 163–171, (1986).
[10] I.D. Mayergoyz and G. Bedrosian, IEEE Transactions on Magnetics, **29**, 2335, (1993).
[11] I.D. Mayergoyz and G. Bedrosian, Journal of Applied Physics, **75**, No. 10, 6027, (1994).
[12] I.D. Mayergoyz and G. Bedrosian, IEEE Transaction on Magnetics, **31**, 1319, (1995).
[13] S.G. Mihlin, "Linear Partial Differential Equations," Vyshya Shcola (Publisher), Moscow, (1997).
[14] J.C. Nedelec, SIAM Journal of Numerical Analysis, **15**, 580 (1978).
[15] S.M. Rytov, Journal of Technical Physics, No. 10, 935 (1940).
[16] B.K. Skagerstam, American Journal of Physics, **51**, No. 12, 1148, (1983).
[17] O.V. Tozoni and I.D. Mayergoyz, "Analysis of Three Dimensional Electromagnetic Fields," Technika (Publisher), Kiev, 1974.

Appendix A

The Preisach Model of Hysteresis

Suppose that we want to model a hysteretic rate-independent relation between some physical quantity $u(t)$, which will be called the input, and another physical quantity $f(t)$, which will be called the output. It is assumed that the $f(t)$ vs. $u(t)$ relation exhibits history-dependent branching. This means that the realization of a particular branch of this hysteretic relation is determined by the past extremum values of the input $u(t)$. To model such a hysteretic relation, we need a mathematical tool that itself (due to its intrinsic structure) will be able to detect and accumulate input extrema and then to choose appropriate branches of the hysteretic relation according to the accumulated histories. One such tool is the Preisach model, which we will proceed to define next.

Consider an infinite (continuous) set of the simplest hysteretic nonlinearities (operators) $\hat{\gamma}_{\alpha\beta}$. Each of these operators can be represented by a rectangular loop on the input-output diagram (see Fig. A.1). Numbers α and β correspond to "up" and "down" switching values of input, respectively. It will be assumed in the sequel that $\alpha \geq \beta$. Outputs of these elementary hysteresis operators may assume only two values: $+1$ and -1. As the input $u(t)$ is monotonically increased above α, the ascending branch *abcde* is traced. When the input is monotonically decreased below β, the descending branch *edfba* is followed. The horizontal parts of these rectangular loops are fully reversible. Thus, irreversibility occurs as a result of switchings. Along with the infinite set of operators $\hat{\gamma}_{\alpha\beta}$ consider an arbitrary weight function $\mu(\alpha, \beta)$. Then the Preisach model can be written as

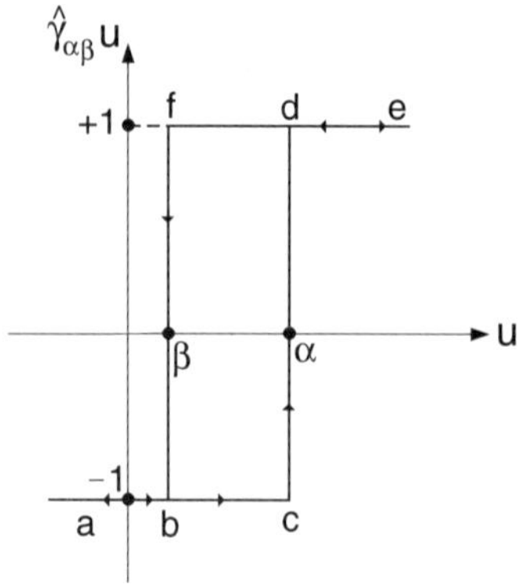

Fig. A.1

$$f(t) = \iint_{\alpha \geq \beta} \mu(\alpha, \beta)\hat{\gamma}_{\alpha\beta}u(t)d\alpha d\beta. \tag{A.1}$$

Thus, the Preisach model is constructed as a superposition of the simplest hysteretic operators $\hat{\gamma}_{\alpha\beta}$. These operators can be viewed as main building blocks of the model. In this sense, the Preisach model can be viewed as a spectral decomposition of complicated hysteretic operators into rectangular loop operators $\hat{\gamma}_{\alpha\beta}$. There is also an interesting parallel between the Preisach model and wavelet transforms, which are currently very popular in the area of signal processing. Indeed, all rectangular loop operators $\hat{\gamma}_{\alpha\beta}$ can be obtained by translating and dilating the rectangular loop operator $\hat{\gamma}_{1,-1}$, which can be regarded as the "mother loop operator." Thus, the Preisach model can be viewed as a "wavelet operator transform."

It is important to note that rectangular loop operators $\hat{\gamma}_{\alpha\beta}$ are hysteretic operators with local memories. This means that for these operators the values of output at some instant of time t_0 and the values of input $u(t)$ at all subsequent instants of time $t \geq t_0$ uniquely predetermine the value of output for all $t > t_0$. In other words, for hysteretic operators with local memories the past exerts its influence upon the future through the current values of output. Although the Preisach model is constructed as a superposition of rectangular loop operators with local memories, it usually has a nonlocal memory. In other words, for the Preisach model the current value of output $f(t_0)$ is not sufficient in order to predict the future values of $f(t)$ for $t > t_0$ given the values of $u(t)$ for $t \geq t_0$. It turns out that the past extremum values of $u(t)$ that occurred for $t < t_0$ affect the future values of $f(t)$ for $t > t_0$, and this is the essence of nonlocal memory. It is remarkable that a new qualitative property of nonlocal memory emerges as a collective property of a system having a huge (infinite) number of simple and qualitatively similar components $\hat{\gamma}_{\alpha\beta}$ with local memories.

To understand the formation of nonlocal memory of the Preisach model as well as its other properties, the special diagram technique can be employed. This diagram technique is based on the simple fact that there is a one-to-one correspondence between operators $\hat{\gamma}_{\alpha\beta}$ and points (α, β) of the half-plane $\alpha \geq \beta$ (see Fig. A.2). Consider a right triangle T. Its hypotenuse is a part of the line $\alpha = \beta$, while the vertex of its right angle has the coordinates α_0 and β_0 with $\beta_0 = -\alpha_0$. In the sequel, this triangle will be called the limiting triangle and the case when $\mu(\alpha, \beta)$ is a finite function with a support within T will be discussed. In other words, it will be assumed that the function $\mu(\alpha, \beta)$ is equal to zero outside the triangle T. This case covers the important class of hysteresis nonlinearities with closed major loops.

To start the discussion, we first assume that the input $u(t)$ at some instant of time t_0 has a value that is less than β_0. Then, the outputs of all $\hat{\gamma}$-operators that correspond to the points of the triangle T are equal to -1. In other words, all $\hat{\gamma}$-operators are in the "down" position. This corresponds to the state of "negative saturation" of the hysteresis nonlinearity represented by the Preisach model.

Now, we assume that the input is monotonically increased until it reaches at time t_1 some maximum value u_1. As the input is being increased, all $\hat{\gamma}$-operators with "up" switching values α less than the current input value $u(t)$ are being turned into the "up" position. This means that their outputs become equal to $+1$. Geometrically, it leads to the subdivision of the triangle T into two sets: $S^+(t)$ consisting of points (α, β) for which the corresponding $\hat{\gamma}$-operators are in the "up" position, and $S^-(t)$ consisting of points (α, β) such that the corresponding $\hat{\gamma}$-operators are still in the "down" position. This subdivision is made by the line $\alpha = u(t)$ (see Fig. A.3), which moves upward as the input is being increased. This upward

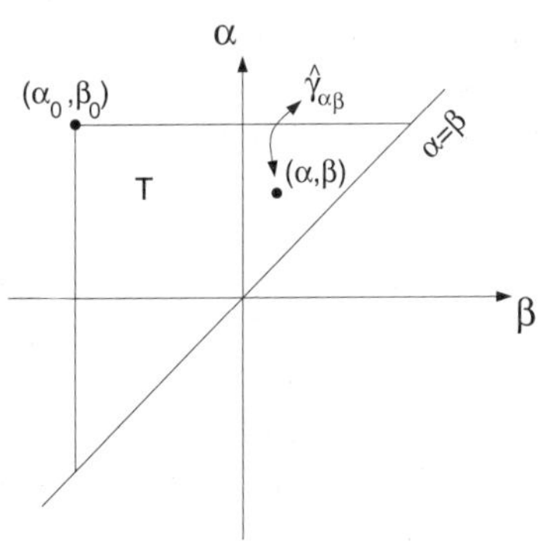

Fig. A.2

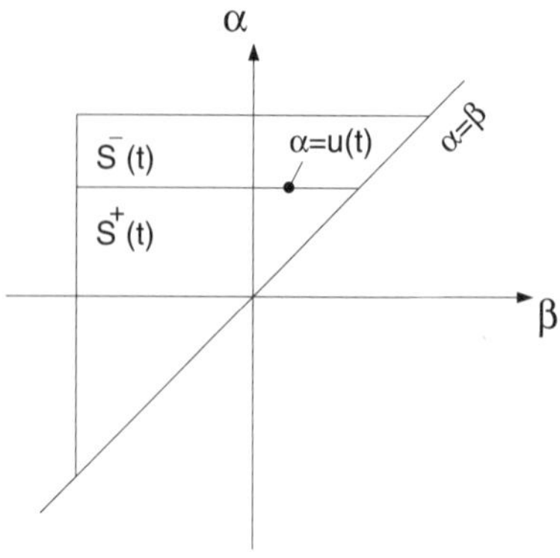

Fig. A.3

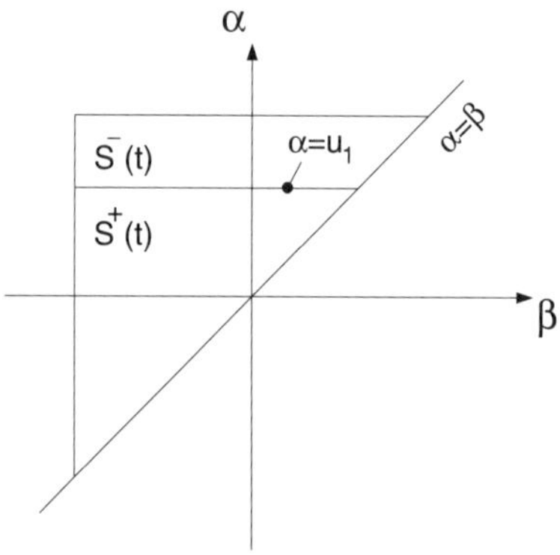

Fig. A.4

motion is terminated when the input reaches the maximum value u_1. The subdivision of the triangle T into $S^+(t)$ and $S^-(t)$ for this particular instant of time is shown in Fig. A.4.

Next, we assume that the input is monotonically decreased until it reaches at time t_2 some minimum value u_2. As the input is being decreased, all $\hat{\gamma}$-operators with "down" switching values β above the current input value, $u(t)$, are being turned back to the "down" position. This changes the previous subdivision of T into positive and negative sets. Indeed, the interface $L(t)$ between $S^+(t)$ and $S^-(t)$ now has two links, the horizontal and vertical ones. The vertical link moves from right to left and its motion is specified by the equation $\beta = u(t)$. This is illustrated by Fig. A.5. The above motion of the vertical link is terminated when the input reaches its minimum value u_2. The subdivision of the triangle T for this particular instant of time is shown in Fig. A.6. The vertex of the interface $L(t)$ at this instant of time has the coordinates $\alpha = u_1$ and $\beta = u_2$.

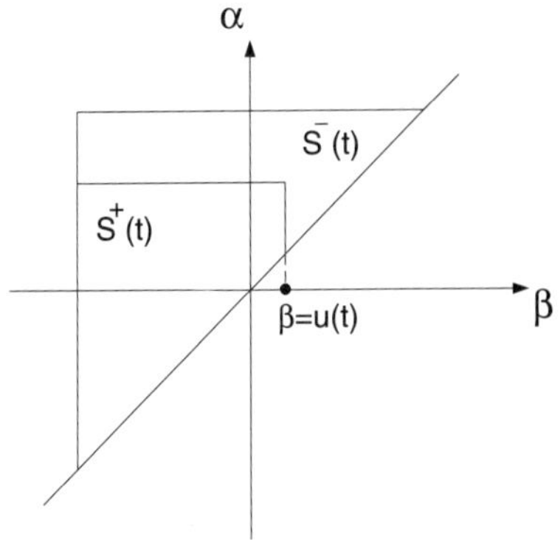

Fig. A.5

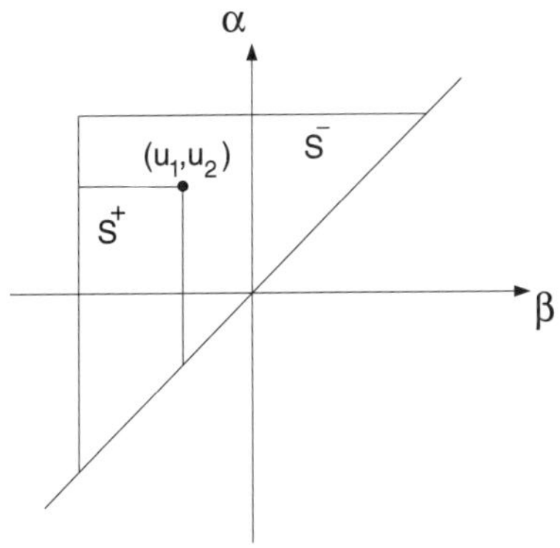

Fig. A.6

Now, we assume that the input is increased again until it reaches at time t_3 some maximum value u_3, which is less than u_1. Geometrically, this increase results in the formation of a new horizontal link of $L(t)$ that moves upwards. This upward motion is terminated when the maximum u_3 is reached. This is shown in Fig. A.7.

Next, we assume that the input is decreased again until it reaches at time t_4 some minimum value u_4, which is above u_2. Geometrically, this input variation results in the formation of a new vertical link that moves from right to left. This motion is terminated as the input reaches its minimum value u_4. As a result, a new vertex of $L(t)$ is formed that has the coordinates $\alpha = u_3$ and $\beta = u_4$. This is illustrated by Fig. A.8. By generalizing the previous analysis, the following conclusion can be reached. At any instant of time, the triangle T is subdivided into two sets: $S^+(t)$

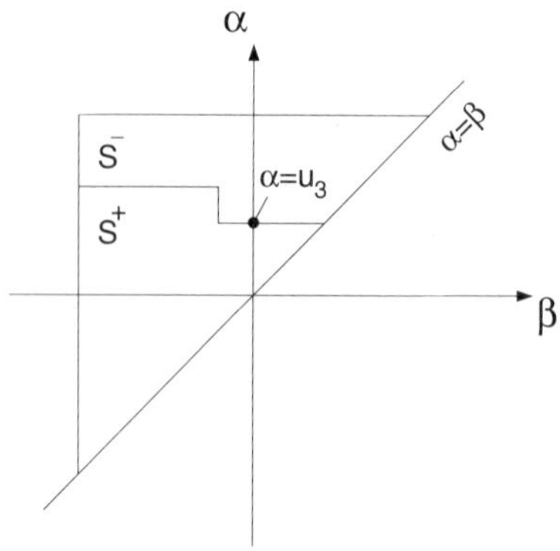

Fig. A.7

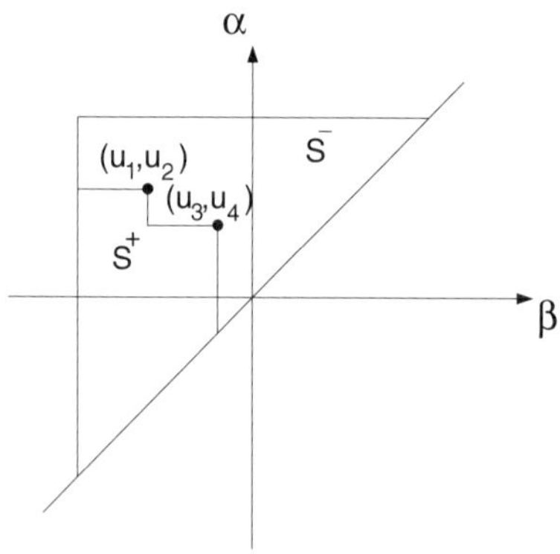

Fig. A.8

consisting of points (α, β) for which the corresponding $\hat{\gamma}$-operators are in the "up" position, and $S^-(t)$ consisting of points (α, β) for which the corresponding $\hat{\gamma}$-operators are in the "down" position. The interface $L(t)$ between $S^+(t)$ and $S^-(t)$ is a staircase line whose vertices have α and β coordinates coinciding respectively with local maxima and minima of input at previous instants of time. The final link of $L(t)$ is attached to the line $\alpha = \beta$ and it moves when the input is changed. This link is a horizontal one and it moves upwards as the input is increased (see Fig. A.9). The final link is a vertical one and it moves from right to left as the input is decreased (see Fig. A.10).

According to the above conclusion, at any instant of time the integral in (A.1) can be subdivided into two integrals, over $S^+(t)$ and $S^-(t)$, respectively:

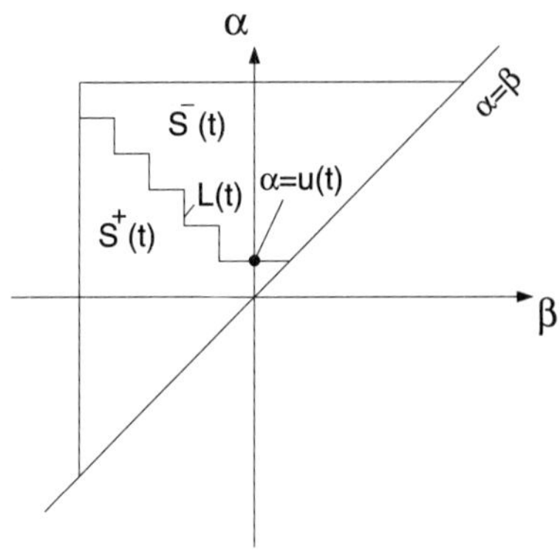

Fig. A.9

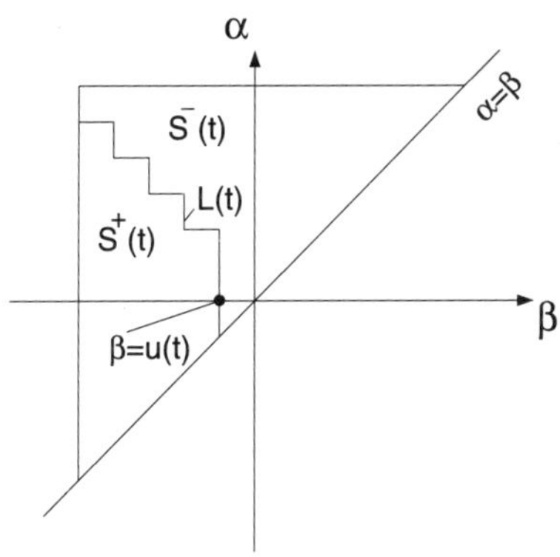

Fig. A.10

$$f(t) = \hat{\Gamma}u(t) = \iint\limits_{S^+(t)} \mu(\alpha,\beta)\hat{\gamma}_{\alpha\beta}u(t)d\alpha d\beta$$

$$+ \iint\limits_{S^-(t)} \mu(\alpha,\beta)\hat{\gamma}_{\alpha\beta}u(t)d\alpha d\beta. \tag{A.2}$$

Since

$$\hat{\gamma}_{\alpha\beta}u(t) = +1, \quad \text{if } (\alpha,\beta)\epsilon S^+(t) \tag{A.3}$$

and

$$\hat{\gamma}_{\alpha\beta}u(t) = -1, \quad \text{if } (\alpha,\beta)\epsilon S^-(t), \tag{A.4}$$

from (A.2) we find

$$f(t) = \iint\limits_{S^+(t)} \mu(\alpha,\beta)d\alpha d\beta - \iint\limits_{S^-(t)} \mu(\alpha,\beta)d\alpha d\beta. \tag{A.5}$$

From this expression, it follows that an instantaneous value of output depends on a particular subdivision of the limiting triangle, T, into positive and negative sets $S^+(t)$ and $S^-(t)$. This subdivision is determined by a particular shape of the interface $L(t)$. This shape, in turn, depends on the past extremum values of input because these extremum values are the coordinates of the vertices of $L(t)$. Consequently, the past extremum values of input shape the staircase interface, $L(t)$, and in this way they leave their mark upon the future.

To make this point perfectly clear, consider two inputs $u_1(t)$ and $u_2(t)$ with two different past histories for $t < t'$. This means that they had different local extrema for $t < t'$. It is next assumed that these inputs coincide for $t \geq t'$. Then according to (A.5), the outputs $f_1(t)$ and $f_2(t)$ corresponding to the above inputs are given by the formulae:

$$f_1(t) = \iint_{S_1^+(t)} \mu(\alpha, \beta) d\alpha d\beta - \iint_{S_1^-(t)} \mu(\alpha, \beta) d\alpha d\beta, \tag{A.6}$$

$$f_2(t) = \iint_{S_2^+(t)} \mu(\alpha, \beta) d\alpha d\beta - \iint_{S_2^-(t)} \mu(\alpha, \beta) d\alpha d\beta, \tag{A.7}$$

where $S_1^+(t)$ and $S_1^-(t), S_2^+(t)$ and $S_2^-(t)$ are positive and negative sets of two subdivisions of T associated with $u_1(t)$ and $u_2(t)$, respectively.

These two subdivisions are different because they correspond to two different input histories. Thus, from (A.6) and (A.7) we conclude that

$$f_1(t) \neq f_2(t) \qquad \text{for } t > t'. \tag{A.8}$$

It is clear that the last inequality (A.8) holds even if the outputs $f_1(t')$ and $f_2(t')$ are somehow the same at time t'. This means that the Preisach model describes, in general, hysteresis nonlinearities with nonlocal memories.

The previous discussion reveals the mechanism of memory formation in the Preisach model. The memory is formed as a result of two different rules for the modification of the interface $L(t)$. Indeed, for a monotonically increasing input, we have a horizontal final link of $L(t)$ moving upwards, whereas, for a monotonically decreasing input, we have a vertical final link of $L(t)$ moving from right to left. These two different rules result in the formation of the staircase interface, $L(t)$, whose vertices have coordinates equal to past input extrema.

We next proceed to the discussion of an interesting property that further elucidates the mechanism of memory formation in the Preisach model. It turns out that this model does not accumulate all past extremum values of input. Some of them can be wiped out by subsequent input variations.

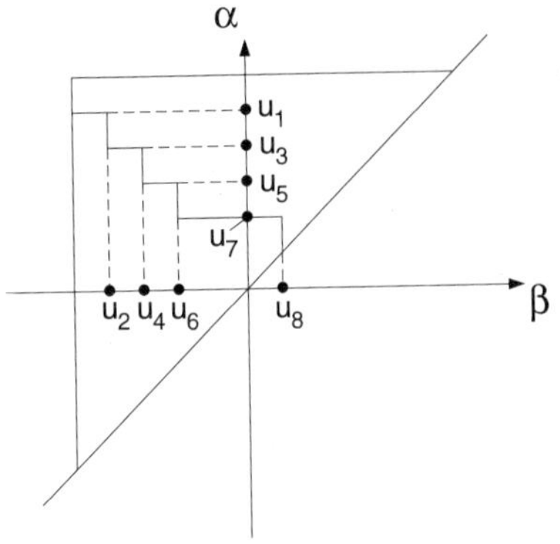

Fig. A.11

To make this property clear, consider a particular past history that is characterized by a finite decreasing sequence $\{u_1, u_3, u_5, u_7\}$ of local input maxima and an increasing sequence $\{u_2, u_4, u_6, u_8\}$ of local input minima. A typical $\alpha - \beta$ diagram for this kind of history is shown in Fig. A.11. Now, we assume that the input $u(t)$ is monotonically increased until it reaches some maximum value u_9, which is above u_3. This monotonic increase of input $u(t)$ results in the formation of a horizontal final link of $L(t)$, which moves upwards until the maximum value u_9 is reached. This results in a modified $\alpha - \beta$ diagram shown in Fig. A.12. It is evident that all vertices whose α-coordinates were below u_9 have been wiped out. It is also clear that the wiping out of vertices is equivalent to the erasing of the history associated with these vertices. Namely, the past input maxima and minima that were respectively equal to α and β-coordinates of the erased vertices have been wiped out. We have illustrated how the wiping out of vertices occurs for monotonically increasing inputs. However, it is obvious

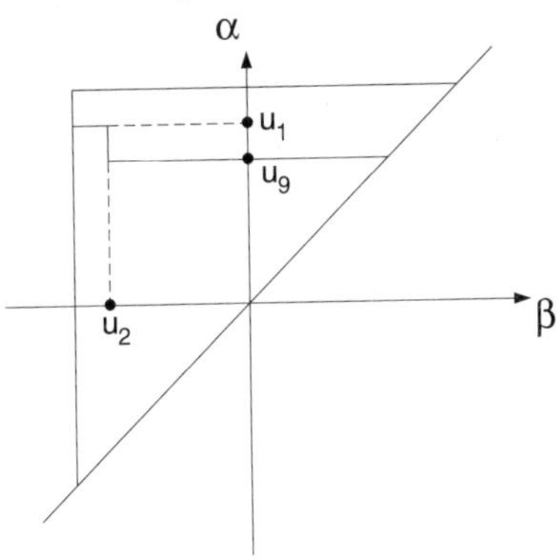

Fig. A.12

that the wiping out of vertices occurs in a similar manner for monotonically decreasing inputs as well. Thus, we can formulate the following property of the Preisach model, which is called the **wiping-out property**.

Each local input maximum wipes out the vertices of $L(t)$ whose α-coordinates are below this maximum, and each local minimum wipes out the vertices whose β-coordinate are above this minimum.

The wiping-out property is asserted here in purely geometric terms. This property can also be described in analytical terms.

Consider a particular input variation shown in Fig. A.13 for the time interval $t_0 \leq t \leq t'$. We assume that at the initial instant of time t_0 the input value $u(t_0)$ was below β_0. This means that the initial state is the state of negative saturation. Consequently, the whole history has been written by the input variation after time t_0. We would like to specify explicitly which local input extremum values will be stored by the Preisach model at time t'. Consider the global maximum of the input at the time interval $[t_0, t']$. We will use the notation M_1 for this maximum and t_1^+ for the instant of time the maximum was reached:

$$M_1 = \max_{[t_0, t']} u(t), \qquad u(t_1^+) = M_1. \tag{A.9}$$

It is clear that all previous input extrema were wiped out by this maximum. Now, consider the global minimum of the input at the interval $[t_1^+, t']$. We will use the notation m_1 for this minimum and t_1^- for the time it was reached:

$$m_1 = \min_{[t_1^+, t']} u(t), \qquad u(t_1^-) = m_1. \tag{A.10}$$

It is apparent that all intermediate input extrema that occurred between t_1^+ and t_1^- were erased by the minimum m_1. Next, consider the global maximum of the input at the interval $[t_1^-, t']$. The notations M_2 and t_2^+ are appropriate for this maximum and the time it occurred:

$$M_2 = \max_{[t_1^-, t']} u(t), \qquad u(t_2^+) = M_2. \tag{A.11}$$

It is obvious that this maximum wiped out all intermediate input extrema that occurred between t_1^- and t_2^+. As before, consider the global minimum of input at the time interval $[t_2^+, t']$ and the notations

$$m_2 = \min_{[t_2^+, t']} u(t), \qquad u(t_2^-) = m_2. \tag{A.12}$$

It is clear that this minimum wiped out all intermediate input extrema.

Continuing this line of reasoning, we can inductively introduce the global maxima M_k and global minima m_k:

$$M_k = \max_{[t_{k-1}^-, t']} u(t), \qquad u(t_k^+) = M_k. \tag{A.13}$$

$$m_k = \min_{[t_k^+, t']} u(t), \qquad u(t_k^-) = m_k. \tag{A.14}$$

Only these input extrema are accumulated by the Preisach model, while all intermediate input extrema are erased. It is natural to say that M_k and m_k $(k = 1, 2, ...)$ form an alternating series of dominant maxima and minima.

It is evident from this analysis that α- and β- coordinates of vertices of the interface $L(t')$ are equal to M_k and m_k, respectively. It is also clear that the alternating series of dominant extrema is modified as time goes by. This means that new dominant extrema can be introduced by the time-varying input, while the previous ones can be erased. In other words, M_k and m_k are functions of t' as it is clearly suggested by their definitions.

Now, the wiping-out property can be stated in the following form. *Only the alternating series of dominant input extrema are stored by the Preisach model. All other input extrema are wiped out.*

It is worth noting that the wiping-out property is somewhat natural and consistent with experimental facts. Indeed, experiments in the area of magnetics show the existence of major hysteresis loops whose shapes do not depend on how these loops are arrived at. In other words, the major hysteresis loops are well defined. This means that past history is wiped out by input oscillations of sufficiently large magnitude. This is in complete agreement with the wiping-out property.

Consider another characteristic property of the Preisach model that is valid for periodic input variations. Let $u_1(t)$ and $u_2(t)$ be two inputs that may have different past histories (different alternating series of dominant extrema). However, starting from some instant of time t_0, these inputs vary back and forth between the same two consecutive extremum values, u_+ and u_-. It can be shown that these periodic input variations result in minor hysteresis loops. Let Figs. A.14 and A.15 represent $\alpha - \beta$ diagrams for the inputs $u_1(t)$ and $u_2(t)$, respectively. As the inputs vary back and forth between u_+ and u_-, the final links of staircase interfaces $L_1(t)$ and $L_2(t)$ move within the identical triangles T_1 and T_2. This results in periodic shape variations for $L_1(t)$ and $L_2(t)$, which in turn produce periodic variations of the outputs, $f_1(t)$ and $f_2(t)$. This means that some minor hysteresis loops are traversed in the $f - u$ diagram for both inputs (see Fig. A.16). The positions of these two loops with respect to the f-axis are different. This is because these two inputs have different past histories, which lead to different shapes for staircase interfaces, $L_1(t)$ and $L_2(t)$. As a result, the values of outputs for the same values of inputs are different. This is easily seen from the formula (A.5). However, it can be proven that these two hysteresis loops are congruent. This means that the coincidence of these loops can be achieved by the appropriate translation of these loops along the f-axis.

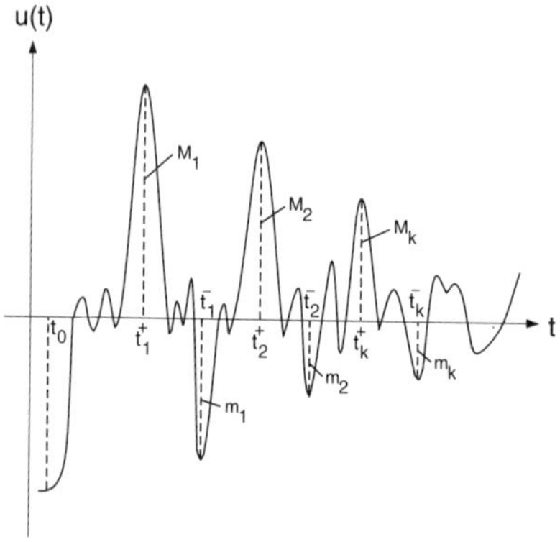

Fig. A.13

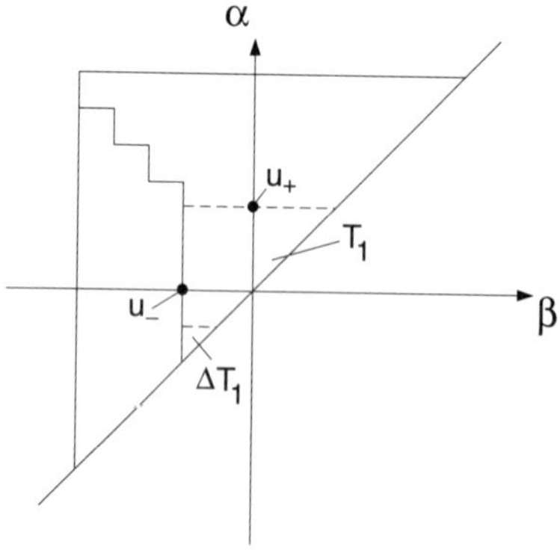

Fig. A.14

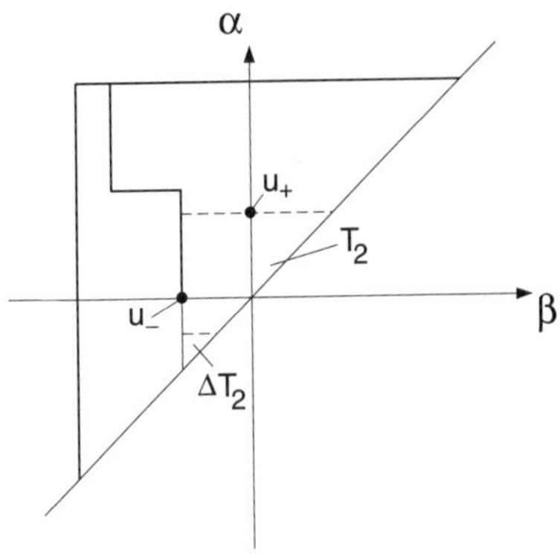

Fig. A.15

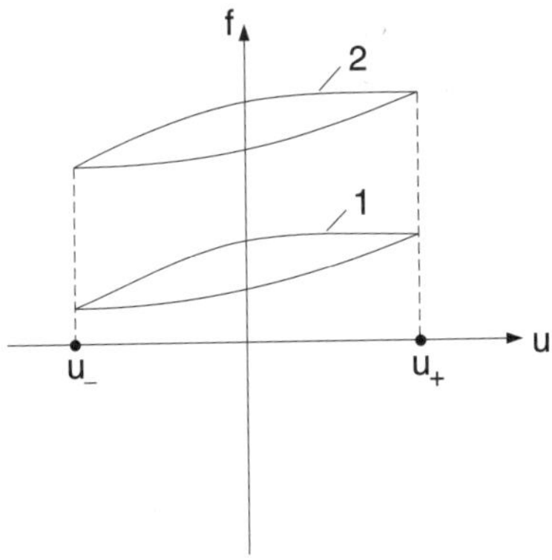

Fig. A.16

The proof of the congruency of these loops is equivalent to showing that any equal increments of inputs $u_1(t)$ and $u_2(t)$ result in equal increments of outputs $f_1(t)$ and $f_2(t)$. To this end, let us assume that both inputs after achieving the same minimum value u_- are increased by the same amount: $\Delta u_1 = \Delta u_2 = \Delta u$. As a result of these increases, the identical triangles ΔT_1 and ΔT_2 are added to the positive sets $S_1^+(t)$ and $S_2^+(t)$ and subtracted from the negative sets $S_1^-(t)$ and $S_2^-(t)$ (see Figs. A.14 and A.15). Now, using the formula (A.5), we find that the corresponding output increments are given by

$$\Delta f_1 = 2 \iint_{\Delta T_1} \mu(\alpha, \beta) d\alpha d\beta, \tag{A.15}$$

$$\Delta f_2 = 2 \iint_{\Delta T_2} \mu(\alpha, \beta) d\alpha d\beta. \tag{A.16}$$

Since $\Delta T_1 = \Delta T_2$, we conclude that

$$\Delta f_1 = \Delta f_2. \tag{A.17}$$

The equality (A.17) has been proven for the case when inputs $u_1(t)$ and $u_2(t)$ are monotonically increased by the same amount after achieving the same minimum value u_-. Thus, this equality means congruency for the ascending branches of these minor loops. By literally repeating the previous reasoning, we can prove that the same equality (A.17) holds when the inputs $u_1(t)$ and $u_2(t)$ are monotonically decreased by the same amount Δu after achieving the maximum value u_+. This means that the descending branches of these minor loops are congruent as well. Thus, we have established the following property of the Preisach model, which is called the **congruency property**.

All minor hysteresis loops corresponding to back and forth variations of inputs between the same two consecutive extremum values are congruent.

Next, we proceed to the discussion of the identification problem for the Preisach model. The essence of this problem is in the determination of weight function $\mu(\alpha, \beta)$. The set of first-order transition curves will be used for this purpose. These curves can be defined as follows. First, the input $u(t)$ should be decreased to the value that is less than β_0. This brings a hysteresis nonlinearly to the state of negative saturation. Next, the input is monotonically increased until it reaches some value α'. The corresponding $\alpha - \beta$ diagram is shown in Fig. A.17. As the input is increased, an ascending branch of a major loop is followed (see Fig. A.18). This branch will also be called the limiting ascending branch because usually there is no

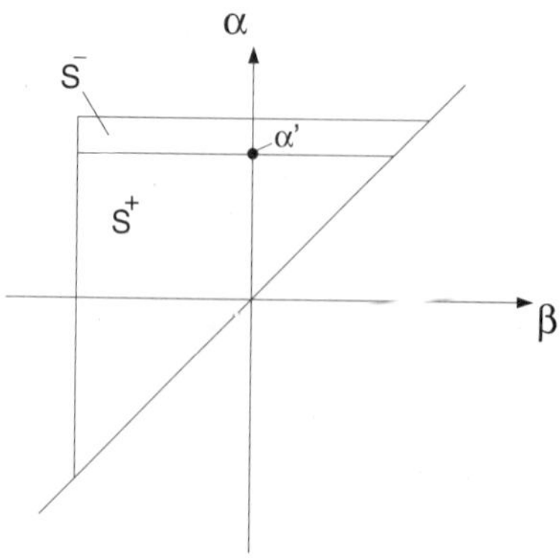

Fig. A.17

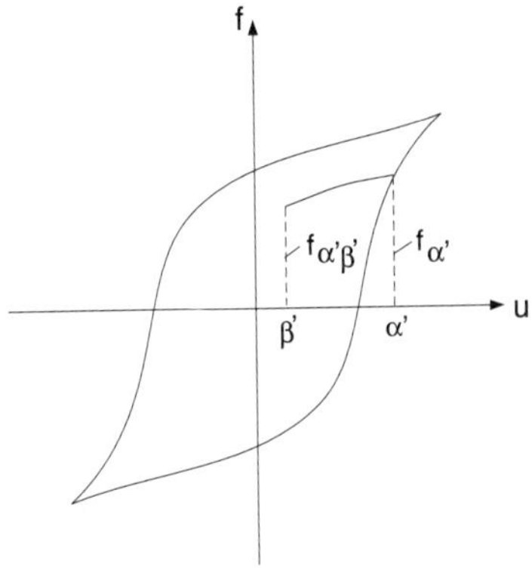

Fig. A.18

branch below it. The notation $f_{\alpha'}$ will be used for the output value on this branch, which corresponds to the input value $u = \alpha'$. The first-order transition (reversal) curves are attached to the limiting ascending branch. Each of these curves is formed as the above monotonic increase of the input is followed by a subsequent monotonic decrease. The term "first-order" is used to emphasize the fact that each of these curves is formed after the first reversal of input. The notation $f_{\alpha'\beta'}$ will be used for the output value on the transition curve attached to the limiting ascending branch at the point $f_{\alpha'}$. This output value corresponds to the input value $u = \beta'$ (see Fig. A.18). The above monotonic decrease of input modifies the previous $\alpha - \beta$ diagram shown in Fig. A.17. A new $\alpha - \beta$ diagram for the instant of time when the input reaches the value β' is illustrated by Fig. A.19.

Now, we define the function:

$$F(\alpha', \beta') = \frac{1}{2}(f_{\alpha'} - f_{\alpha'\beta'}). \tag{A.18}$$

This function is equal to one half of the output increments along the first-order transition curves. The next step is to express this function in terms of the weight function $\mu(\alpha, \beta)$. To this end, we compare the $\alpha - \beta$ diagrams shown in Figs. A.17 and A.19. It is clear from these diagrams that the triangle $T(\alpha', \beta')$ is added to the negative set S^- and subtracted from the positive set S^+ as a result of the monotonic input decrease from the value $u = \alpha'$ to the value $u = \beta'$. Using this fact and the formula (A.5), we find that the Preisach model will match the output increments along the first-order transition curves if the function $\mu(\alpha, \beta)$ satisfies the equation:

$$f_{\alpha'\beta'} - f_{\alpha'} = -2 \iint\limits_{T(\alpha',\beta')} \mu(\alpha, \beta) d\alpha d\beta. \tag{A.19}$$

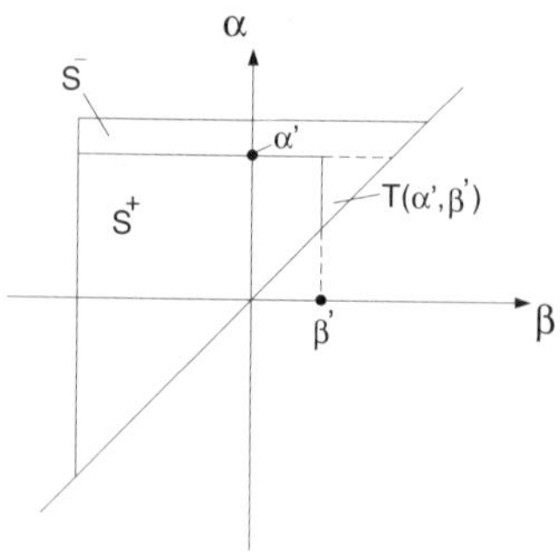

Fig. A.19

By comparing the formulas (A.18) and (A.19), we obtain

$$F(\alpha',\beta') = \iint\limits_{T(\alpha',\beta')} \mu(\alpha,\beta)d\alpha d\beta. \tag{A.20}$$

The integral over the triangle $T(\alpha',\beta')$ can be written as the following double integral:

$$F(\alpha',\beta') = \int_{\beta'}^{\alpha'} \left(\int_{\beta}^{\alpha'} \mu(\alpha,\beta)d\alpha \right) d\beta. \tag{A.21}$$

By differentiating the last expression two times (first with respect to β' and then with respect to α'), we find:

$$\boxed{\mu(\alpha',\beta') = -\frac{\partial^2 F(\alpha',\beta')}{\partial\alpha'\partial\beta'}.} \tag{A.22}$$

Invoking (A.18), the expression (A.22) can be written in another equivalent form:

$$\boxed{\mu(\alpha',\beta') = \frac{1}{2}\frac{\partial^2 f_{\alpha'\beta'}}{\partial\alpha'\partial\beta'}.} \tag{A.23}$$

Thus, the complete identification of the Preisach model can be performed by using the first-order transition curves shown in Fig. A.20.

The Preisach model can be numerically implemented by using the formula (A.5) for the computation of the output, $f(t)$, and the formula (A.22) for the determination of the weight function, $\mu(\alpha,\beta)$. Although this approach is straightforward, it encounters two main difficulties. First,

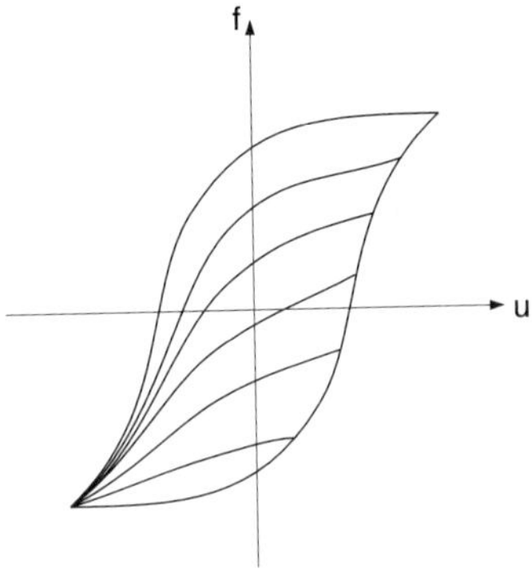

Fig. A.20

it requires the numerical evaluation of double integrals in (A.5). This is a time-consuming procedure, which may impede the use of the Preisach model in practical applications. Second, the determination of the weight function $\mu(\alpha, \beta)$ by employing the formula (A.22) requires differentiations of experimentally (or numerically) obtained data. These differentiations may strongly amplify errors (noise) inherently present in any experimental data. It turns out that another approach can be developed for the numerical implementation of the Preisach model. This approach completely circumvents these difficulties. It is based on the explicit formula for the integrals in (A.5). This formula directly involves (without any differentiation) the data used for the identification of $\mu(\alpha, \beta)$.

The starting point for the derivation of the explicit formula for $f(t)$ is the expression (A.5).

By adding and subtracting the integral of $\mu(\alpha, \beta)$ over $S^+(t)$, the expression (A.5) can be represented in the form

$$f(t) = -\iint_{T} \mu(\alpha, \beta) d\alpha d\beta + 2 \iint_{S^+(t)} \mu(\alpha, \beta) d\alpha d\beta, \qquad (A.24)$$

where, as before, T is the limiting triangle.

According to (A.20), we find

$$\iint_{T} \mu(\alpha, \beta) d\alpha d\beta = F(\alpha_o, \beta_o). \qquad (A.25)$$

The positive set, $S^+(t)$, can be subdivided into n trapezoids Q_k (see Fig. A.21). As a result, we have

$$\iint_{S^+(t)} \mu(\alpha, \beta) d\alpha d\beta = \sum_{k=1}^{n(t)} \iint_{Q_k(t)} \mu(\alpha, \beta) d\alpha d\beta. \qquad (A.26)$$

It is clear that the number, n, of these trapezoids and their shapes may change with time. For this reason, n and Q_k are shown in (A.26) as functions of time.

Each trapezoid Q_k can be represented as a difference of two triangles, $T(M_k, m_{k-1})$ and $T(M_k, m_k)$. Thus, we obtain

$$\iint_{Q_k(t)} \mu(\alpha,\beta)d\alpha d\beta = \iint_{T(M_k,m_{k-1})} \mu(\alpha,\beta)d\alpha d\beta - \iint_{T(M_k,m_k)} \mu(\alpha,\beta)d\alpha d\beta, \tag{A.27}$$

where, for the case of $k = 1$, m_o in (A.27) is naturally equal to β_o.

According to (A.20), we find

$$\iint_{T(M_k,m_{k-1})} \mu(\alpha,\beta)d\alpha d\beta = F(M_k, m_{k-1}), \tag{A.28}$$

and

$$\iint_{T(M_k,m_k)} \mu(\alpha,\beta)d\alpha d\beta = F(M_k, m_k). \tag{A.29}$$

From (A.27), (A.28), and (A.29), we derive

$$\iint_{Q_k(t)} \mu(\alpha,\beta)d\alpha d\beta = F(M_k, m_{k-1}) - F(M_k, m_k). \tag{A.30}$$

From (A.24), (A.25), (A.26), and (A.30) we obtain

$$f(t) = -F(\alpha_o, \beta_o) + 2\sum_{k=1}^{n(t)} [F(M_k, m_{k-1}) - F(M_k, m_k)]. \tag{A.31}$$

It is clear from Fig. A.21 that m_n is equal to the current value of input:

$$m_n = u(t). \tag{A.32}$$

Consequently, the expression (A.31) can be written as

$$f(t) = -F(\alpha_o, \beta_o) + 2\sum_{k=1}^{n(t)-1} [F(M_k, m_{k-1}) - F(M_k, m_k)] + 2[F(M_n, m_{n-1}) - F(M_n, u(t))]. \tag{A.33}$$

The last expression has been derived for monotonically decreasing input, that is, when the final link of interface $L(t)$ is a vertical one. If the input $u(t)$ is being monotonically increased, then the final link of $L(t)$ is a horizontal one and the $\alpha - \beta$ diagram shown in Fig. A.21 should be slightly modified. The appropriate diagram is shown in Fig. A.22. This diagram can be considered as a particular case of the previous one. This case is realized when

$$m_n(t) = M_n(t) = u(t). \tag{A.34}$$

According to the definition (A.20) of $F(\alpha, \beta)$, we find

$$F(M_n, m_n) = F(u(t), u(t)) = 0. \tag{A.35}$$

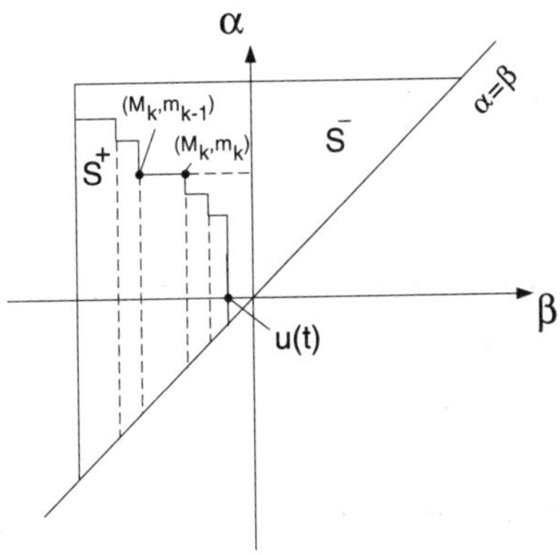

Fig. A.21

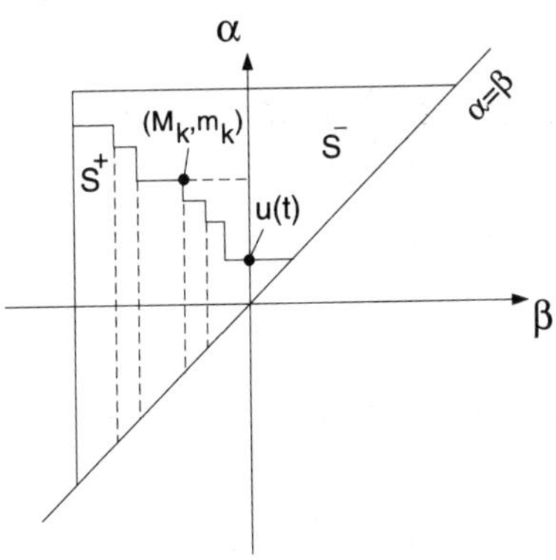

Fig. A.22

From formulas (A.31), (A.34), and (A.35), we derive the following expression for $f(t)$ in the case of monotonically increasing input:

$$\begin{aligned} f(t) = &- F(\alpha_o, \beta_o) + 2 \sum_{k=1}^{n(t)-1} [F(M_k, m_{k-1}) - F(M_k, m_k)] \\ &+ 2F(u(t), m_{n-1}). \end{aligned} \tag{A.36}$$

Thus, we have derived the explicit formulas (A.33) and (A.36) for output $f(t)$ in terms of the data used for the identification of the Preisach model. These formulas constitute the basis for the numerical implementation of the Preisach model. These formulas can also be useful for the evaluation of the time derivative of $f(t)$:

$$\frac{df(t)}{dt} = \begin{cases} -2\frac{\partial F(M_n, u(t))}{\partial u} \cdot \frac{du(t)}{dt}, & \text{if } \frac{du(t)}{dt} < 0, \\ 2\frac{\partial F(u(t), M_{n-1})}{\partial u} \cdot \frac{du(t)}{dt}, & \text{if } \frac{du(t)}{dt} > 0. \end{cases} \tag{A.37}$$

We next proceed to the formulation and the proof of the fundamental theorem, which gives the necessary and sufficient conditions for the representation of actual hysteresis nonlinearities by the Preisach model.

Representation Theorem. The wiping-out property and the congruency property constitute the necessary and sufficient conditions for a hysteresis nonlinearity to be represented by the Preisach model on the set of piecewise monotonic inputs.

Proof

Necessity. Let a hysteresis nonlinearity be representable by the Preisach model. Then, this nonlinearity should have the same properties as the model. In particular, it should have the wiping-out and congruency properties.

Sufficiency. Consider a hysteresis nonlinearity that has both the wiping-out property and the congruency property. We intend to prove that the hysteresis nonlinearity can be represented by the Preisach model.

The proof of the last statement is constructive. First, it is assumed that the weight function, $\mu(\alpha, \beta)$, is found for the given hysteresis nonlinearity by matching its first-order transition curves. This can be accomplished by using the formula (A.19). This formula is equivalent to (A.22), which means that the integrals of $\mu(\alpha, \beta)$ over triangles $T(\alpha', \beta')$ are equal to one half of output increments, $\frac{1}{2}\Delta f = \frac{1}{2}(f_{\alpha'} - f_{\alpha'\beta'})$, along the first-order transition curves. Next, it will be proven that if this weight function is substituted in (A.1), then the Preisach model and the given nonlinearity will have the same input-output relationships. This statement is true for the first-order transition curves due to the very way by which the weight function, $\mu(\alpha, \beta)$, is determined. The induction argument will be used to prove that the same statement holds for higher-order transition curves as well. Let us assume that this statement is true for transition curves with number $1, 2, \dots k$. Then, for the induction inference, we need to prove that this statement holds for a transition curve number $k + 1$.

Let a be a point at which the transition curve number $k + 1$ starts (see Fig. A.23). The point, a, corresponds to some input value $u = \alpha'$. According to the induction assumption, the output values of the hysteretic nonlinearity and the Preisach model coincide at this point. Thus, it remains to be proven that the output increments along the transition curve number $k + 1$ are the same for the actual nonlinearity and the Preisach model. Consider an arbitrary input value $u = \beta' < \alpha'$. The output increment for the nonlinearity will be equal to the increment of f along some curve ab (see Fig. A.23). Let Figs. A.24 a and A.24 b represent $\alpha - \beta$ diagrams for the Preisach model at the time instants when $u = \alpha'$ and $u = \beta'$, respectively. From these diagrams we find that the input decrease from α' to β' results in adding the triangle $T(\alpha', \beta')$ to the negative set, S^-, and subtracting the same triangle from the positive set, S^+. Using this fact and the formula (A.5), we find that for the Preisach model the output increment along the transition curve number $k + 1$ is given by:

$$\Delta f = 2 \iint\limits_{T(\alpha', \beta')} \mu(\alpha, \beta) d\alpha d\beta. \tag{A.38}$$

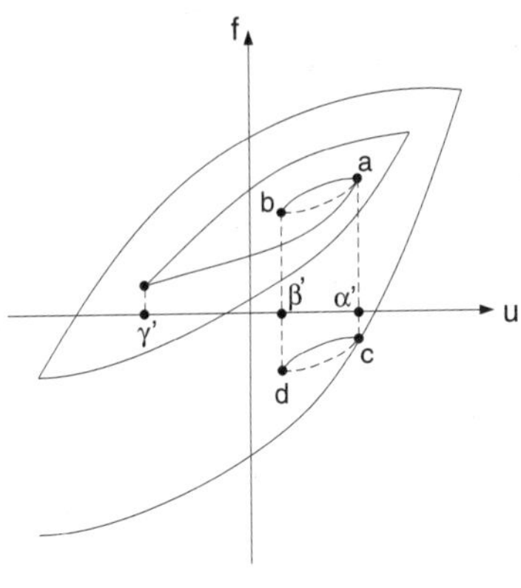

Fig. A.23

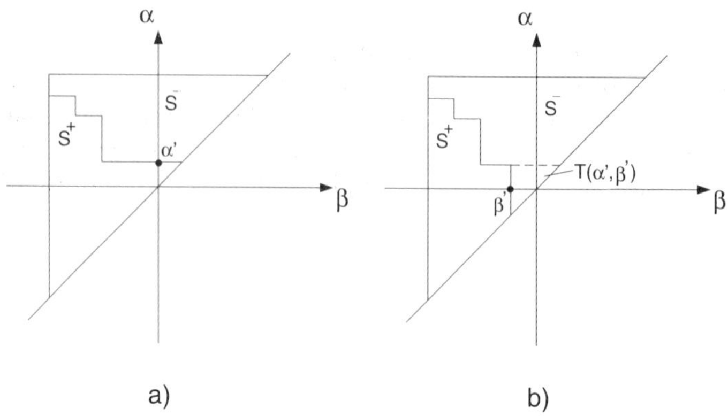

Fig. A.24

However, according to the way the function, $\mu(\alpha, \beta)$, is defined, the right-hand side of (A.38) is equal to the increment of the output along the first-order transition curve cd (see Fig. A.23). Thus, it remains to be shown that the output increments along the curves ab and cd are the same. It is here that the wiping-out and congruency properties will be used. The proof proceeds as follows. If starting from the point b we monotonically increased the input value from β' back to α', then, according to the wiping-out property, we would arrive at the same point a by moving along some curve ba (see Fig. A.23). Indeed, the wiping-out property implies that as soon as the input exceeded the value α', the history associated with back-and-forth input variations between α' and β' should be erased and the subsequent output variation should follow the transition curve number k. However, this would be possible if only for $u = \alpha'$ we arrived back at the point a. Similarly, if starting from the point d we monotonically increased the input value from β' to α', then, according to the same wiping-out property, we would arrive at the point c moving along some curve dc. Now, invoking the congruency property, we conclude that the hysteresis loops bab and dcd are congruent. This is true because both loops result from back and forth input variations between the same two consecutive input extrema, α' and β'. From the congruency of these loops, we find that the output increments along the curves ab and cd are the same. Consequently, the output values for the hysteretic nonlinearity and for the Preisach model coincide along the transition curve number $k+1$. The last fact has been proven for any $\beta' \geq \gamma'$. However, according to the wiping-out property, the transition curve number $k+1$ should coincide with the transition curve number $k-1$ for $\beta' < \gamma'$. Thus, the case $\beta' < \gamma'$ falls in the domain of the induction assumption. ∎

It is easy to see that the essence of the given proof is in the reduction of higher-order transition curves to the first-order transition curves. This reduction rests on both the wiping-out property and congruency property.

The proven theorem is very important because it clearly establishes the limits of applicability of the Preisach model. These limits are formulated in purely phenomenological terms, without any reference to the actual physical nature of hysteresis. This reveals the physical universality of the Preisach model.

INDEX

E

F

G

H

I